Application of Remote Sensing and GIS for Promoting Sustainable Geoenvironment

Application of Remote Sensing and GIS for Promoting Sustainable Geoenvironment

Guest Editors

Hariklia D. Skilodimou
George D. Bathrellos
Konstantinos G. Nikolakopoulos

MDPI

Basel • Beijing • Wuhan • Barcelona • Belgrade • Novi Sad • Cluj • Manchester

Guest Editors
Hariklia D. Skilodimou
Department of Geology
University of Patras
Patras
Greece

George D. Bathrellos
Department of Geology
University of Patras
Patras
Greece

Konstantinos G. Nikolakopoulos
Department of Geology
University of Patras
Patras
Greece

Editorial Office
MDPI AG
Grosspeteranlage 5
4052 Basel, Switzerland

This is a reprint of the Special Issue, published open access by the journal *Sustainability* (ISSN 2071-1050), freely accessible at: https://www.mdpi.com/journal/sustainability/special_issues/80O8M5V4O5.

For citation purposes, cite each article independently as indicated on the article page online and as indicated below:

Lastname, A.A.; Lastname, B.B. Article Title. *Journal Name* **Year**, *Volume Number*, Page Range.

ISBN 978-3-7258-6121-7 (Hbk)
ISBN 978-3-7258-6122-4 (PDF)
https://doi.org/10.3390/books978-3-7258-6122-4

Contents

About the Editors

Hariklia D. Skilodimou

Hariklia D. Skilodimou is a Ph.D. Researcher and Teaching Staff Member at the University of Patras, Department of Geology. Her research interests include the following: geomorphology (pure and applied); natural hazards; modeling; and GIS. Dr. Skilodimou has taught Geomorphology, Physical Geography, GIS, Remote Sensing, Urban Planning, Environmental Management and Monitoring, and Hydrology in graduate and post-graduate courses in three Greek universities. She has contributed to over 200 publications in international conferences and journals. She has participated in several national and international research projects.

George D. Bathrellos

George D. Bathrellos is a full-time Professor at the University of Patras, Department of Geology. His research interests include the following: geomorphology (pure and applied); natural hazards; urban planning; modeling; and GIS. Dr. George Bathrellos has taught Geomorphology, Physical Geography, GIS, Urban Planning, Environmental Management and Monitoring, and Hydrology in graduate and post-graduate courses in four Greek universities. He has published over 200 scientific papers in international and Greek journals and conference proceedings, achieving significant international recognition. Dr. Bathrellos serves as Associate Editor for the journal *Natural Hazards* (Springer) and as Editor-in-Chief for *Edelweiss Applied Science and Technology*. He has served as Guest Editor for numerous Special Issues in various scientific journals. In 2025, he was awarded the "Editor Distinction Award" by Springer Nature. He is also an Editorial Board member and reviewer for multiple scientific journals. He has participated in several national and international research projects.

Konstantinos G. Nikolakopoulos

Konstantinos Nikolakopoulos is a full-time Professor at the Department of Geology, University of Patras. His areas of specialty are remote sensing, photogrammetry and geographic information system (GIS) for geological applications. He has more than 250 publications (200 in Scopus) in peer reviewed journals and conference proceedings with an h-index of 27 and more than 2500 references (according to Scopus). He has teaching experience spanning more than 13 years at the University of Patras. He is currently a reviewer for more than 40 international journals. He was elected Chairman of the Special Interest Group Geological Applications of the European Association of Remote Sensing Laboratories. He has been co-chairing the SPIE conference "Earth Resources and Environmental Remote Sensing - GIS Applications for the last ten years. He has served as Associate Editor of the *European Journal of Remote Sensing* since 2016. Professor Nikolakopoulos is on the list of 2% of the top cited scientists, according to the "Updated science-wide author databases of standardized citation indicators" from the Meta Research Innovation Center of Stanford University for the last six years.

Prof. Nikolakopoulos is the founder and head of the GIS and Remote Sensing Laboratory, established in 2019. This laboratory is dedicated to advancing the understanding and monitoring of Earth surface processes through cutting-edge technologies in Geographic Information Systems (GISs) and Remote Sensing.

Preface

This reprint collects nine scientific papers focusing on the applications of Remote Sensing (RS) and Geographical Information Systems (GIS) for the study of the geoenvironment. The published contributions cover a diverse range of topics, including vegetation assessment, land use and land-cover change detection, ecological quality evaluation, and the assessment and prevention of natural hazards. Together, these papers highlight the importance of integrating geospatial technologies to better understand environmental processes and to support sustainable development strategies.

All articles included in this reprint were originally published in the Special Issue of the journal *Sustainability*, entitled "Application of Remote Sensing and GIS for Promoting Sustainable Geoenvironment," published between 2023 and 2025. The aim of this Special Issue was to bring together recent advances and innovative applications of RS and GIS that contribute to environmental analysis, spatial planning, and the implementation of sustainable geoenvironmental practices. This collection is intended for scientists, researchers, and practitioners who are engaged in the use of geospatial methods and technologies for addressing geoenvironmental and sustainability challenges.

The Guest Editors would like to express their deepest gratitude to all authors for their valuable contributions and to the reviewers for their insightful comments and rigorous evaluations, which ensured the scientific quality and integrity of this volume. Special thanks are also extended to the editorial team of *Sustainability* for their continuous assistance, professionalism, and dedication throughout the publication process.

Looking forward, the research presented in this reprint provides a solid foundation for future investigations into the sustainable management of natural and human-modified environments. The integration of RS and GIS continues to offer powerful tools for monitoring environmental dynamics, mitigating natural hazards, and supporting evidence-based policy and planning. It is our hope that this collection will serve as both a reference and an inspiration for further scientific progress, fostering collaboration and innovation in the field of geospatial science and sustainable geoenvironmental development.

Hariklia D. Skilodimou, George D. Bathrellos, and Konstantinos G. Nikolakopoulos
Guest Editors

sustainability

MDPI

Editorial

Application of Remote Sensing and GIS for Promoting Sustainable Geoenvironment

Hariklia D. Skilodimou *, George D. Bathrellos and Konstantinos G. Nikolakopoulos

Department of Geology, University of Patras, 26504 Patras, Greece; gbathrellos@upatras.gr (G.D.B.); knikolakop@upatras.gr (K.G.N.)
* Correspondence: hskilodimou@upatras.gr

The continuous growth of Earth's population poses increasingly complex challenges to the physical environment. Over recent decades, urban areas have undergone accelerated growth, leading to significant landscape alterations, heightened demand for natural resources, and increased stress on ecological systems [1]. According to projections by the United Nations, the proportion of the world's population residing in urban areas, which currently exceeds 50%, is expected to rise to nearly 70% by 2050, with the majority of this growth occurring in developing regions of Asia and Africa [2]. These demographic and spatial dynamics are expected to promote continued urban sprawl.

Urban Expansion and Geoenvironmental Consequences

Urban expansion has profoundly transformed the geoenvironment and has been consistently linked to a wide range of detrimental consequences. These include the intensification of energy demand, increased levels of environmental pollution, air, water, and soil degradation, traffic congestion, natural resource depletion, water and soil salinization, biodiversity loss, waste, desertification fires, deforestation, and climate change [3,4]. Over the past decades, the rapid expansion of urban areas has led to new environmental challenges including the development of urban heat islands, increased greenhouse gas emissions from transportation and buildings, and greater pressure on water supply, sewage, and waste management systems [5].

The expansion growth of urban areas has triggered significant modifications in land use practices and coastal morphology [6]. Land use changes can exert significant influences on Earth. Urban sprawl not only has substantial impacts on the climate, but also contributes to ecosystem disruption and environmental degradation [7]. The conversion of naturally vegetated land, such as forests, grasslands, or shrublands, into impervious surfaces intensifies surface runoff and increases the likelihood of flooding [8]. Moreover, coastal regions are projected to experience a 160% increase in urban extent between 2000 and 2030 [9]. In Mediterranean coastal areas, such as Italy, each year, more than 10 km^2 of natural and agricultural land types are converted into anthropogenic surfaces [10]. In Greece, the coastal areas adjacent to the Athens Metropolitan Area demonstrate that over the past 76 years, artificial land cover has increased by 40% [11]. The loss of natural land due to urbanization is a global phenomenon, potentially exposing many cities to various natural hazards.

Furthermore, it should be noted that urban sprawl has been associated with the amplification of natural hazard occurrences and the exacerbation of disaster impacts on a global scale [12]. The unregulated expansion of built-up areas into flood prone zones has heightened both human and infrastructural vulnerability [13,14]. A global assessment revealed that urban exposure to flooding increased more than fourfold between 1985 and 2018, with a significant proportion of new urban development taking place within floodplain areas [15]. Urban expansion also alters topography and increases susceptibility to landslides. In hilly or sloped areas, construction activities frequently destabilize slopes, thereby elevating the likelihood of landslide events [16–19].

Overall, while urbanization continues to serve as a catalyst for economic development and social progress, its uncontrolled expansion presents significant challenges to environmental sustainability and disaster resilience. Consequently, the integration of spatial planning, ecosystem-based management, and climate adaptation policies is imperative to mitigate the adverse geoenvironmental impacts associated with urban growth.

Environmental System Pressures and the Necessity of Continuous Monitoring

Pressures on environmental systems are continuously intensifying as population growth drives extensive land-use and land-cover transformations. These transformations frequently result in irreversible consequences. As human activities reshape the Earth's surface at unprecedented rates, understanding these processes necessitates comprehensive baseline data, continuous environmental monitoring, and the systematic evaluation of spatial and temporal dynamics [20].

Regional-scale assessments are particularly vital, as they enable the identification of localized environmental stresses that may remain undetected at the global scale. Through systematic observation and the use of long-term datasets, researchers can quantify the cumulative impacts of anthropogenic activities on geomorphological processes, hydrological dynamics, and climatic variables [21].

Furthermore, the continuous monitoring of environmental parameters provides critical feedback to policymakers and stakeholders, facilitating the development of adaptive, evidence-based management strategies. Such approaches form the cornerstone of sustainable geoenvironmental management, ensuring that development policies respect ecological thresholds and promote long-term environmental resilience [22]. Ultimately, the systematic observation and analysis of environmental data are essential for maintaining a balance between human development and the sustainability of the geoenvironment.

Remote Sensing and GIS in Geoenvironmental Research

Over the past several decades, the proliferation of advanced remote sensing (RS) technologies has facilitated the acquisition of high-resolution, multi-sensor, and multi-temporal geospatial datasets. Simultaneously, geographical information systems (GIS) have evolved into sophisticated computational platforms for the integration, management, and quantitative analysis of heterogeneous geospatial data. The synergistic fusion of RS and GIS enables the generation of multi-dimensional, dynamic datasets that are critical for systematic environmental monitoring, change detection, and predictive modeling [14–16].

The advancement of modern RS technologies has allowed researchers to obtain highly detailed data regarding land cover characteristics, vegetation dynamics, hydrological patterns, urban expansion, and environmental degradation at unprecedented spatial and temporal scales [23–26]. The use of high-resolution satellite imagery, synthetic aperture radar (SAR), and LiDAR technologies now permits the identification of subtle environmental changes that were previously undetectable through conventional field-based techniques.

Simultaneously, geographical information systems (GIS) have evolved into sophisticated computational platforms capable of integrating, managing, and quantitatively analyzing heterogeneous geospatial datasets. Modern GIS supports advanced modeling, statistical analysis, and scenario-based simulations of environmental processes, enhancing visualization and spatial querying capabilities [27,28].

The integration of remote sensing (RS) and geographic information systems (GIS) significantly improves the capacity to analyze spatial and temporal variations in environmental systems. This combined approach offers essential insights for policymakers, resource managers, and stakeholders, facilitating data-driven decision-making in sustainable land-use planning, environmental protection, and geoenvironmental management.

The application of RS and GIS spans a wide spectrum of geoenvironmental studies. These include but are not limited to:

- Precise geological and geomorphological mapping for understanding terrain evolution [29,30];
- Quantitative assessment of natural resources such as soil, water, vegetation, and mineral deposits [31–33];
- Evaluation of contaminant dispersal and environmental hazard assessment [34,35];
- Ecosystem and ecological vulnerability assessment [36,37];
- Modeling of natural hazard susceptibility and identification of vulnerable communities [13–19];
- Urbanization and climate changes [38,39];
- Land-use planning and resource management for sustainable development [40–42].

RS and GIS allow for the integration of remotely acquired data with in situ measurements, supporting predictive modeling and decision-making in both applied and theoretical geosciences. These technologies enhance the precision and efficiency of environmental assessments and provide the foundation for adaptive, data-driven strategies in sustainable geoenvironmental management. Consequently, such capabilities ensure that geoenvironmental management remains responsive and evidence-based, effectively addressing the complex challenges arising from rapid environmental change and urbanization.

Consequently, RS and GIS constitute indispensable methodological tools in contemporary geoenvironmental research. They underpin the quantitative analysis of human–environment interactions and support the sustainable management of geological, ecological, and socio-environmental systems [8,27,43–47]. The continuous advancement of sensor technologies, data analytics, and spatial modeling algorithms further underscores their critical role in addressing complex environmental challenges in an era of rapid global change [9,22,48–50]. By integrating high-resolution geospatial data, real-time monitoring, and predictive modeling, RS and GIS provide essential tools for mitigating environmental degradation, reducing disaster risk, and promoting resilient and sustainable geoenvironmental management.

The main aim of this Special Issue is to emphasize the wide-ranging applications of RS and GIS in advancing sustainable geoenvironmental research. It encompasses systematic monitoring, evaluation, and the spatiotemporal analysis of environmental conditions. The Issue also addresses strategies for enhancing the long-term sustainability of geoenvironmental systems, reducing anthropogenic impacts, and supporting the preservation of high-quality human life. Additionally, natural hazard assessment and prevention are presented.

Conflicts of Interest: The authors declare no conflict of interest.

References

1. Grimm, N.B.; Faeth, S.H.; Golubiewski, N.E.; Redman, C.L.; Wu, J.; Bai, X.; Briggs, J.M. Global change and the ecology of cities. *Science* **2008**, *319*, 756–760. [CrossRef] [PubMed]
2. DESA, U. *World Urbanization Prospects: The 2018 Revision*; United Nations: New York, NY, USA, 2020; Available online: https://population.un.org/wup/ (accessed on 22 September 2025).
3. Bathrellos, G.D.; Skilodimou, H.D. Estimation of sand and gravel extraction sites. *Z. Geomorphol.* **2022**, *63*, 313–328. [CrossRef]
4. Semenov, S.M. Intergovernmental Panel on Climate Change: Results, Problems, and Prospects. *Izv. Atmos. Ocean. Phys.* **2024**, *60* (Suppl. 3), S323–S330. [CrossRef]
5. Li, X.; Zhou, Y.; Ouyang, Z. Ecological and environmental impacts of urban expansion: A review. *Land Use Policy* **2020**, *91*, 104380. [CrossRef]
6. Bathrellos, G.D.; Skilodimou, H.D. Land use planning for natural hazards. *Land* **2019**, *8*, 128. [CrossRef]

7. Hong, C.; Burney, J.A.; Pongratz, J.; Nabel, J.E.; Mueller, N.D.; Jackson, R.B.; Davis, S.J. Global and regional drivers of land-use emissions in 1961–2017. *Nature* **2021**, *589*, 554–561. [CrossRef]
8. Skilodimou, H.D.; Bathrellos, G.D.; Alexakis, D.E. Flood Hazard Assessment Mapping in Burned and Urban Areas. *Sustainability* **2021**, *13*, 4455. [CrossRef]
9. Seto, K.; Fragkias, M.; Güneralp, B.; Reilly, M. A Meta-Analysis of Global Urban Land Expansion. *PLoS ONE* **2011**, *6*, e23777. [CrossRef]
10. Smiraglia, D.; Cavalli, A.; Giuliani, C.; Assennato, F. The increasing coastal urbanization in the Mediterranean environment: The state of the art in Italy. *Land* **2023**, *12*, 1017. [CrossRef]
11. Skilodimou, H.D.; Antoniou, V.; Bathrellos, G.D.; Tsami, E. Mapping of coastline changes in Athens Riviera over the past 76 year's measurements. *Water* **2021**, *13*, 2135. [CrossRef]
12. Huppert, H.E.; Sparks, R.S.J. Extreme natural hazards: Population growth, globalization and environmental change. *Phil. Trans. R. Soc. A* **2006**, *364*, 1875–1888. [CrossRef]
13. Hemmati, M.; Ellingwood, B.R.; Mahmoud, H.N. The role of urban growth in resilience of communities under flood risk. *Earth's Future* **2020**, *8*, e2019EF001382. [CrossRef]
14. Maranzoni, A.; D'Oria, M.; Rizzo, C. Quantitative flood hazard assessment methods: A review. *J. Flood Risk Manag.* **2023**, *16*, e12855. [CrossRef]
15. Tellman, B.; Sullivan, J.A.; Kuhn, C.; Kettner, A.J.; Doyle, C.S.; Brakenridge, G.R.; Erickson, T.A.; Slayback, D.A. Satellite imaging reveals increased proportion of population exposed to floods. *Nature* **2021**, *596*, 80–86. [CrossRef] [PubMed]
16. Alcántara-Ayala, I. Landslides in a changing world. *Landslides* **2025**, *22*, 2851–2865. [CrossRef]
17. Bathrellos, G.D.; Koukouvelas, I.K.; Skilodimou, H.D.; Nikolakopoulos, K.G.; Vgenopoulos, A.-L. Landslide causative factors evaluation using GIS in the tectonically active Glafkos River area, northwestern Peloponnese, Greece. *Geomorphology* **2024**, *461*, 109285. [CrossRef]
18. Bathrellos, G.D.; Kalivas, D.P.; Skilodimou, H.D. Landslide Susceptibility Assessment Mapping: A Case Study in Central Greece. In *Remote Sensing of Hydrometeorological Hazards*; Petropoulos, G.P., Islam, T., Eds.; CRC Press, Taylor & Francis Group: London, UK, 2017; pp. 493–512. ISBN -13: 978-1498777582. [CrossRef]
19. Kechebour, B.E. Relation between stability of slope and the urban density: Case study. *Procedia Eng.* **2015**, *114*, 824–831. [CrossRef]
20. Turner, B.L.; Lambin, E.F.; Reenberg, A. The emergence of land change science for global environmental change and sustainability. *Proc. Natl. Acad. Sci. USA* **2007**, *104*, 20666–20671. [CrossRef]
21. Pettorelli, N.; Laurance, W.F.; O'Brien, T.G.; Wegmann, M.; Nagendra, H.; Turner, W. Satellite remote sensing for applied ecologists: Opportunities and challenges. *J. Appl. Ecol.* **2014**, *51*, 839–848. [CrossRef]
22. Rockström, J.; Steffen, W.; Noone, K.; Persson, Å.; Chapin, F.S., III; Lambin, E.F.; Lenton, T.M.; Scheffer, M.; Folke, C.; Schellnhuber, H.J.; et al. A safe operating space for humanity. *Nature* **2009**, *461*, 472–475. [CrossRef]
23. Manapragad, N.V.S.K.; Mandelmilch, M.; Roitberg, E.; Kizel, F.; Natanian, J. Remote Sensing for Environmentally Responsive Urban Built Environment: A Review of Tools, Methods and Gaps. *Remote Sens. Appl.* **2025**, *38*, 101529. [CrossRef]
24. Ahmad, M.N.; Almutlaq, F.; Huq, M.E.; Islam, F.; Javed, A.; Skilodimou, H.D.; Bathrellos, G.D. Enhanced urban impervious surface land use mapping using a novel multi-sensor feature fusion method and remote sensing data. *Environ. Earth Sci.* **2025**, *84*, 228. [CrossRef]
25. Foody, G.M. Remote sensing in landscape ecology. *Landsc. Ecol.* **2023**, *38*, 2711–2716. [CrossRef]
26. Ahmad, M.N.; Shao, Z.; Javed, A.; Israr, A.; Islam, F.; Skilodimou, H.D.; Bathrellos, G.D. Optical-SAR data fusion based on Simple Layer Stacking and XGBoost Algorithm to extract urban impervious surface in diverse Urban Environments. *Remote Sens.* **2024**, *16*, 873. [CrossRef]
27. Goodchild, M.F. Scale in GIS: An overview. *Geomorphology* **2011**, *130*, 5–9. [CrossRef]
28. Nezhad, M.M.; Moradian, S.; Guezgouz, M.; Shi, X.; Avelin, A.; Wallin, F. A GIS-portal platform from the data perspective to energy hub digitalization solutions-A review and a case study. *Renew. Sustain. Energ. Rev.* **2025**, *223*, 116019. [CrossRef]
29. Skilodimou, H.D.; Bathrellos, G.D.; Maroukian, H.; Gaki-Papanastassiou, K. Late Quaternary evolution of the lower reaches of Ziliana stream in south Mt. Olympus, Greece. *Geogr. Fis. Din. Quat.* **2014**, *37*, 43–50. [CrossRef]
30. Kokinou, E.; Skilodimou, H.D.; Bathrellos, G.D.; Antonarakou, A.; Kamberis, E. Morphotectonic analysis, structural evolution/pattern of a contractional ridge: Giouchtas Mt., Central Crete, Greece. *J. Earth Syst. Sci.* **2015**, *124*, 587–602. [CrossRef]
31. Huang, X.; Li, Z.; Liu, X. Urban expansion and its impacts on surface temperature in metropolitan regions: A multi-sensor analysis. *Remote Sens. Environ.* **2019**, *233*, 111374. [CrossRef]
32. Islam, F.; Ahmad, M.N.; Janjuhah, H.T.; Ullah, M.; Islam, I.U.; Kontakiotis, G.; Skilodimou, H.D.; Bathrellos, G.D. Modelling and Mapping of Soil Erosion Susceptibility of Murree, Sub-Himalayas Using GIS and RS Based Model. *Appl. Sci.* **2022**, *12*, 12211. [CrossRef]
33. Gandhi, G.M.; Parthiban, S.; Thummalu, N.; Christy, A. Ndvi: Vegetation change detection using remote sensing and gis—A case study of Vellore District. *Procedia Comput. Sci.* **2015**, *57*, 1199–1210. [CrossRef]

34. Makri, P.; Stathopoulou, E.; Hermides, D.; Kontakiotis, G.; Zarkogiannis, S.D.; Skilodimou, H.D.; Bathrellos, G.D.; Antonarakou, A.; Scoullos, M. The Environmental Impact of a Complex Hydrogeological System on Hydrocarbon-Pollutants' Natural Attenuation: The Case of the Coastal Aquifers in Eleusis, West Attica, Greece. *J. Mar. Sci. Eng.* **2020**, *8*, 1018. [CrossRef]
35. Bathrellos, G.D.; Skilodimou, H.D.; Gamvroula, D.E.; Alexakis, D.E. Evaluate the spatial distribution of trace elements in soil of a karst terrain. *Carbonate Evaporite* **2024**, *39*, 41. [CrossRef]
36. Ding, Q.; Wang, L.; Fu, M.; Huang, N. An integrated system for rapid assessment of ecological quality based on remote sensing data. *Environ. Sci. Pollut. Res.* **2020**, *27*, 32779–32795. [CrossRef]
37. Kamran, M.; Yamamoto, K. Evolution and use of remote sensing in ecological vulnerability assessment: A review. *Ecol. Indic.* **2023**, *148*, 110099. [CrossRef]
38. Lu, Y.; Wang, J.; Xie, Y. Urbanization impacts on coastal morphology and sustainability: A global perspective. *Sustainability* **2021**, *13*, 6071. [CrossRef]
39. Humbal, A.; Chaudhary, N.; Pathak, B. Urbanization Trends, Climate Change, and Environmental Sustainability. In *Climate Change and Urban Environment Sustainability*; Pathak, B., Dubey, R.S., Eds.; Disaster Resilience and Green Growth; Springer: Singapore, 2017; pp. 151–166. [CrossRef]
40. Karpouza, M.; Skilodimou, H.D.; Kaviris, G.; Zymvragakis, A.; Antonarakou, A.; Bathrellos, G.D. Escape routes and safe points in natural hazards. A case study for soil. *Eng. Geol.* **2024**, *340*, 107683. [CrossRef]
41. Chen, S.; Guo, Q.; Li, L. Sustainable land use dynamic planning based on GIS and symmetric algorithm. *Adv. Civ. Eng.* **2022**, *2022*, 4087230. [CrossRef]
42. Kingra, P.K.; Majumder, D.; Singh, S.P. Application of Remote Sensing and Gis in Agriculture and Natural Resource Management Under Changing Climatic Conditions. *Agric. Res. J.* **2016**, *53*, 295–302. [CrossRef]
43. Bareth, G. GIS-and RS-based spatial decision support: Structure of a spatial environmental information system (SEIS). *Int. J. Digit. Earth* **2009**, *2*, 134–154. [CrossRef]
44. Paul, P.; Aithal, P.S.; Bhimali, A.; Kalishankar, T.; Saavedra, R.; Aremu, P.S.B. Geo information systems and remote sensing: Applications in environmental systems and management. *IJMTS* **2020**, *5*, 11–18. [CrossRef]
45. Ahmadi, H.; Pekkan, E. Fault-based geological lineaments extraction using remote sensing and GIS—A review. *Geosciences* **2021**, *11*, 183. [CrossRef]
46. Chaminé, H.I.; Pereira, A.J.; Teodoro, A.C.; Teixeira, J. Remote sensing and GIS applications in earth and environmental systems sciences. *SN Appl. Sci.* **2021**, *3*, 870. [CrossRef]
47. Rosa, A.G.F.; Silva, W.D.O.; Fontana, M.E.; Levino, N.; Guarnieri, P. A GIS-based multi-criteria approach for identifying areas vulnerable to subsidence in the world's largest ongoing urban socio-environmental mining disaster. *Extr. Ind. Soc.* **2024**, *19*, 101500. [CrossRef]
48. Goodchild, M.F.; Haining, R.P. GIS and spatial data analysis: Converging perspectives. *Pap. Reg. Sci.* **2004**, *83*, 363–385. [CrossRef]
49. Kumar, S.S.; Arivazhagan, S.; Rangarajan, N. Remote sensing and GIS applications in Environmental Sciences—A review. *J. Environ. Nanotechnol.* **2013**, *2*, 92–101. [CrossRef]
50. Hogland, J.; Anderson, N. Function modeling improves the efficiency of spatial modeling using big data from remote sensing. *Big Data Cogn. Comput.* **2017**, *1*, 3. [CrossRef]

sustainability

MDPI

Article

Remote Sensing-Based Assessment of Eco-Environmental Quality Dynamics and Driving Forces in the Anhui Section of the Yangtze-to-Huaihe Water Diversion Project (2015–2024)

Xiaoming Qi [1,2,*], Qian Li [1], Qiang Han [1], Bowen Li [1], Le Liu [3], Zhikong Shi [3], Yuanchao Ou [1] and Dejian Wang [1]

1 College of Civil and Hydraulic Engineering, Bengbu University, Bengbu 233000, China; hanq@hhu.edu.cn (Q.H.)
2 Anhui Rural Ecological Environment Protection and Restoration Research Center, Bengbu University, Bengbu 233030, China
3 Anhui Provincial Institute of Ecology and Environmental Sciences, Hefei 230000, China
* Correspondence: xmqi_hydro@163.com

Abstract: The water source protection areas of the Yangtze-to-Huaihe Water Diversion Project (YHWDP) in Anhui Province serve as crucial ecological barriers to water quality protection. Quantifying their eco-environmental quality (EEQ) dynamics and driving mechanisms is critical for sustainable management. This paper calculated the Remote Sensing Ecological Index (RSEI) for the study area using Landsat satellite data (2015–2024). Temporal and spatial variation characteristics were analyzed using the Theil–Sen estimator, Mann–Kendall test, and coefficient of variation. Future trends were predicted using the Hurst exponent. Finally, the Geodetector model was applied to assess the impact of driving factors. EEQ exhibited a declining trend ($p < 0.05$), with significant intra-regional heterogeneity. Mean RSEI values ranked as follows: (1) Yangtze River–Huaihe River Connection < Yangtze River Water Northward Conveyance < Yangtze River–Chaohu Lake Water Diversion. (2) From 2015 to 2024, eco-environmental quality improved significantly, showing a spatial pattern of "south > north, east > west." (3) Overall EEQ changes were characterized by slight to moderate fluctuations. Stability rankings: Yangtze River–Huaihe River Connection > Yangtze River–Chaohu Lake Water Diversion > Yangtze River Water Northward Conveyance. (4) Geodetector analysis identified precipitation, impervious area, and vegetation coverage as the primary factors influencing EEQ in the YHWDP's water source protection areas. This study reveals ecological changes in the YHWDP region and validates the effectiveness of the comprehensive evaluation method. The findings provide actionable insights for ecological protection in large-scale water diversion projects.

Keywords: eco-environmental quality; Remote Sensing Ecological Index (RSEI); Geodetector; Yangtze-to-Huaihe Water Diversion Project

1. Introduction

A favorable ecological environment provides advantageous natural conditions for human survival and development [1]. The rapid economic development has led to increasing conflicts between humans and nature, gradually disrupting the ecological balance. In recent years, a series of ecological issues, such as air pollution and land degradation, have garnered widespread attention among scholars. A comprehensive assessment of regional ecological conditions, coupled with analysis of ecological quality trends and their driving mechanisms, provides a crucial foundation for regional ecological conservation and sustainable socio-economic development [2–4].

The Normalized Difference Vegetation Index (NDVI) has become an indispensable analytical tool in the comprehensive assessment of spatiotemporal variations in vegetation and analysis of environmental changes [5,6]. Methods such as the Pressure–State–Response (PSR) [7] and the Ecological Index (EI) [8] have also been applied to evaluate the evolutionary status of regional natural conditions. However, these research methods are relatively limited and have not been able to integrate natural and social factors in a single evaluation. The Remote Sensing Ecological Index (RSEI) [9,10] offers distinct advantages over conventional single-indicator methods and statistical evaluation approaches. By systematically integrating multiple ecological indicators and automatically weighting them according to their principal component contributions, the RSEI provides a more robust and objective framework for assessing spatiotemporal variations in ecological quality. Thus, the RSEI has become the dominant assessment model in recent ecological studies [11,12]. Chinese scholars have adopted the RSEI to assess ecological changes in various climatically representative regions, including the Yellow River Basin in the north [13], the Dianchi Lake Basin in the south [14], and the Lanxi urban agglomeration in the east [15], all yielding highly accurate results. At the same time, there are also good application examples in urban environmental assessment in Japan [16] and the eastern region of India [17]. This method has been widely applied around the world.

Previous studies have primarily focused on ecological assessments within individual river basins, while comprehensive evaluations of ecological impacts in inter-basin water diversion areas have been scarce. The Yangtze-to-Huaihe Water Diversion Project (YHWDP) region serves as a critical corridor connecting eastern and central China. Local water scarcity, coupled with high population density, has intensified the conflict between human development and the natural environment [18,19]. The purpose of this study is to evaluate the spatiotemporal changes in the ecological environment of the inter-basin area using the RSEI, explore the driving factors behind them, and predict future trends. This provides a scientific basis for the local government in the YHWDP region to implement targeted policies for protecting vegetation and the ecological environment

2. Study Area and Datasets

2.1. Study Area

The YHWDP is a major inter-catchment hydraulic engineering initiative that establishes hydrological connectivity between the Yangtze and Huaihe River basins [20]. It provides water supplies to 55 counties and districts across 15 cities in Anhui and Henan provinces. Construction of the YHWDP commenced in December 2016, with the trial operation of Phase I in the Anhui section initiated in December 2023. This project optimizes regional water resource allocation, alleviates water scarcity in parts of the Yangtze River Delta in Anhui, and enhances regional soil–water ecosystems and ecological functions [21].

The Anhui section of the Yangtze-to-Huaihe River Water Diversion Project is systematically divided into three hydraulically and geographically distinct segments (Figure 1), reflecting spatial heterogeneity in water resource allocation and engineering design principles. The Yangtze River–Chaohu Lake Water Diversion spans 7676.16 km^2, encompassing critical nodes such as Tongcheng City (Anqing), Zongyang County (Tongling), and Lujiang County (Hefei). Adjacent to this, the Yangtze River–Huaihe River Connection covers 12,394.50 km^2, linking Feixi County and Chaohu City (Hefei) with Huainan's Tianjia'an District, Xiejiaji District, and Shou County. The Yangtze River Water Northward Conveyance extends across 12,437.29 km^2, integrating water delivery infrastructure across Bozhou's Qiaocheng District, Guoyang/Lixin/Mengcheng Counties, Fuyang's Taihe County/Jieshou City/Linquan County/Yingquan–Yingdong Districts, and Huainan's Fengtai County/Maoji Experimental Zone/Panji District. This tripartite division optimizes hydraulic efficiency through terrain-adaptive conduit geometries and flow parameter calibration, balancing large-area coverage with localized hydraulic stability while maintaining water transfer capacity across varying topographic gradients.

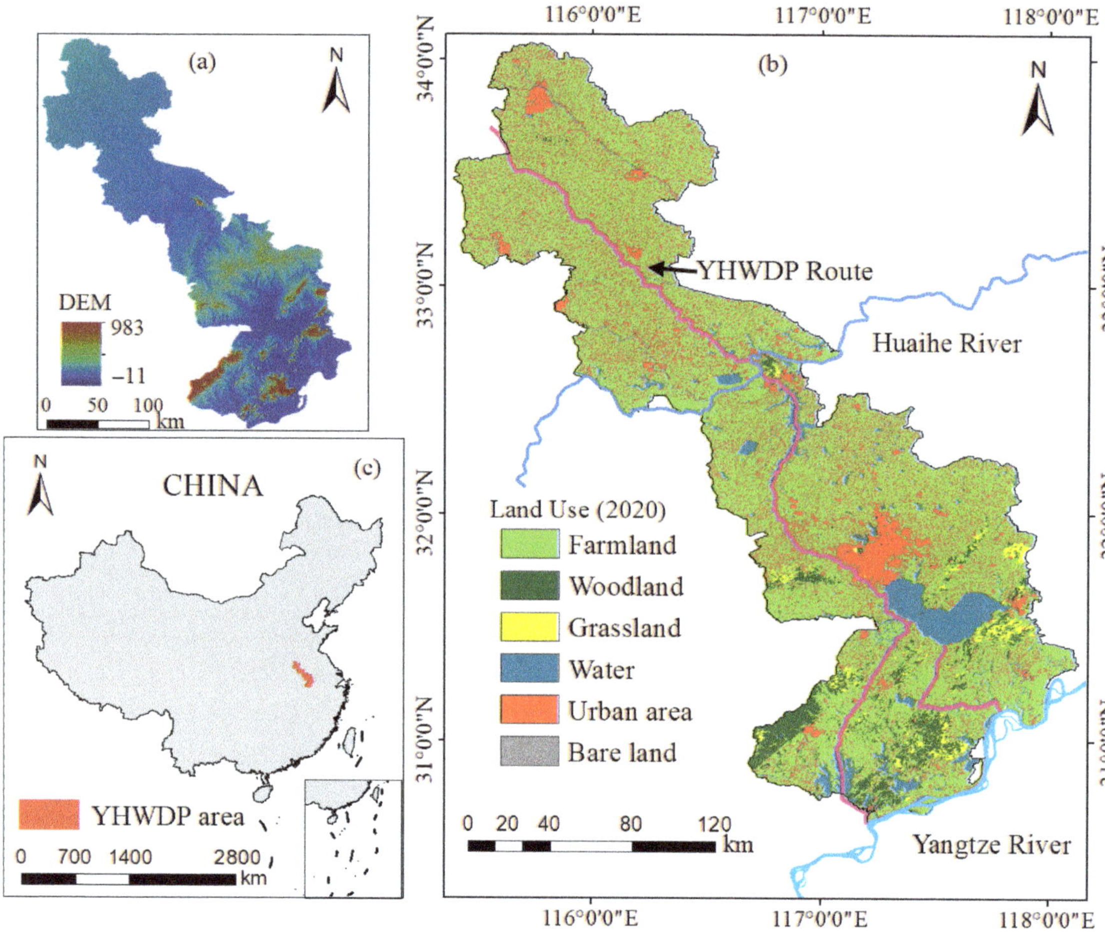

Figure 1. Location of the Anhui section of the Yangtze-to-Huaihe Water Diversion Project: DEM map of the study area (**a**); route map of the Yangtze-to-Huaihe Water Diversion Project (**b**); and location of the study area (**c**).

2.2. Datasets

The remote sensing data utilized in this study comprises Landsat 8 OLI satellite imagery procured from the Geospatial Data Cloud (https://www.gscloud.cn) spanning the vegetation growing season (April to July) from 2015 to 2024. Elevation data were obtained from the ASTER GDEM, which has a 30 m resolution. Using ENVI (5.6 version) software, preprocessing steps such as cloud removal, radiometric correction, atmospheric correction, and cropping were conducted. Additionally, water bodies were masked using the Modified Normalized Difference Water Index (MNDWI). All data were georeferenced to WGS 1984 UTM Zone 49 N with a spatial resolution of 30 m $\times$ 30 m to ensure consistency in spatial resolution and projection. Socioeconomic data, collected at the county/district level, were derived from statistical yearbooks and water resources bulletins of Anhui Province and its cities.

3. Methods

This study employs the Sen slope, Mann–Kendall trend analysis, and Hurst exponent to assess spatiotemporal variations in RSEI from 2015 to 2024 and predict future ecological trends. The geographical detector method was applied to quantitatively identify the spatial heterogeneity and underlying driving mechanisms of RSEI variations (Figure 2).

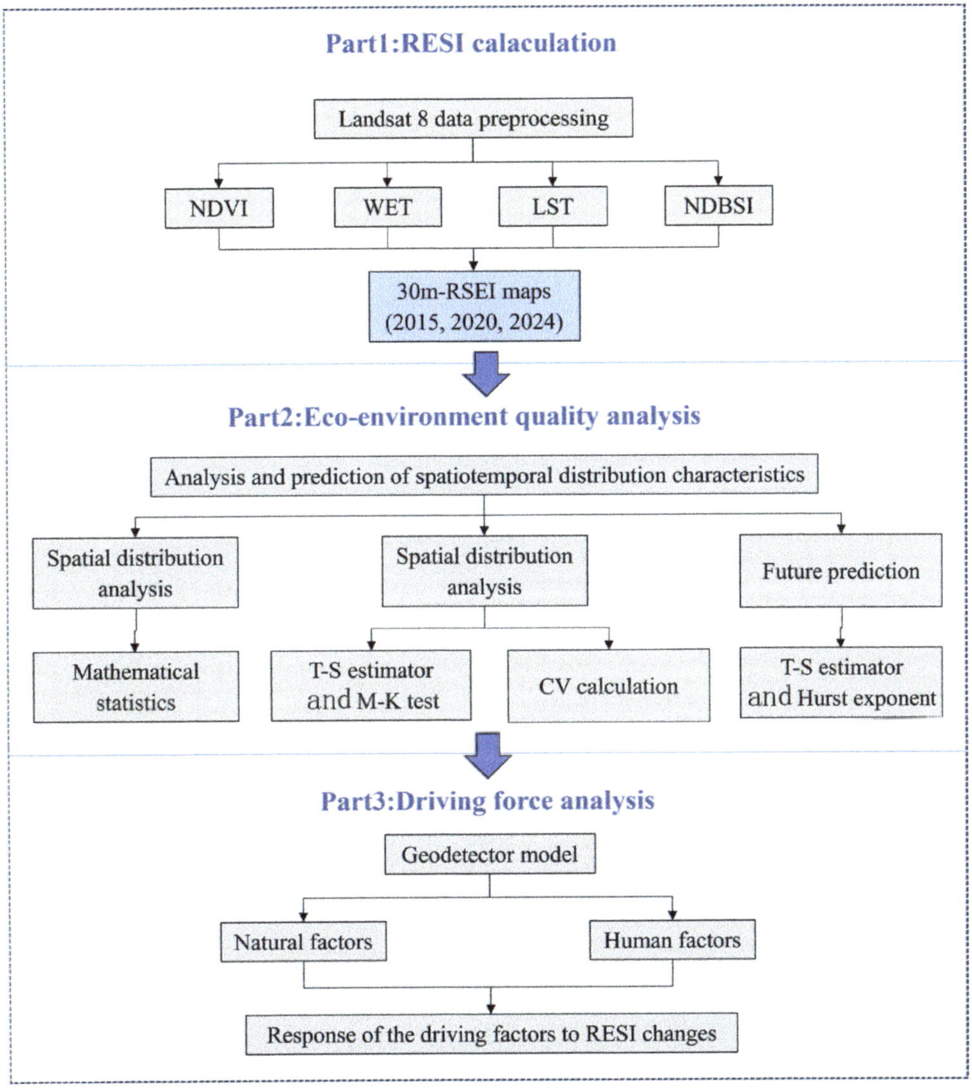

Figure 2. Technology flowchart.

3.1. RSEI Model

As a comprehensive ecosystem assessment tool, the RSEI algorithm combines four fundamental environmental components: greenness, wetness, dryness, and heat, all of which are directly measurable through spectral analysis of satellite imagery. NDVI serves as an indicator of vegetation greenness and is utilized to demonstrate the state of the environment. WET stands for wetness. The Normalized Difference Built-up and Soil Index (NDBSI) characterizes land surface dryness. Land Surface Temperature (LST) reflects the local climate variations induced by environmental changes.

(1) Retrieval of vegetation

NDVI serves as a widely adopted remote sensing metric for quantifying and monitoring vegetation growth and coverage status, mathematically defined as follows:

$$NDVI = \frac{\rho_{NIR} - \rho_R}{\rho_{NIR} + \rho_R} \tag{1}$$

where ρ_{NIR} and ρ_R represent the reflectance values in the near-infrared and red spectral bands, respectively.

(2) Retrieval of land surface moisture

The wetness component serves as a spectral indicator of hydrological conditions, quantitatively characterizing moisture levels in three key surface features: soil matrix water retention, liquid water bodies, and vegetation water content. For Landsat OLI sensors, this component is derived through the following tasseled cap transformation formula:

$$WET_{OLI} = 0.151\rho_B + 0.1973\rho_G + 0.3283\rho_R + 0.3407\rho_{NIR} - 0.7117\rho_{SWIR1} - 0.4559\rho_{SWIR2} \tag{2}$$

(3) Retrieval of land surface moisture

The accelerated urbanization process and intensive anthropogenic disturbances have triggered land surface modifications. Impervious surfaces and exposed soils progressively supplant natural vegetation cover, which induces ecological degradation. Hu and Xu developed NDBSI through spectral feature fusion, integrating the Index-based Built-up Index (IBI) and Soil Index (SI) as follows:

$$SI = \frac{(\rho_{SWIR1} + \rho_R) - (\rho_{NIR} + \rho_B)}{(\rho_{SWIR1} + \rho_R) + (\rho_{NIR} + \rho_B)} \tag{3}$$

$$IBI = \frac{2 \times \rho_{SWIR1}/(\rho_{SWIR1} + \rho_{NIR}) - [\rho_{NIR}/(\rho_{NIR} + \rho_R) + \rho_G/(\rho_G + \rho_{SWIR1})]}{2 \times \rho_{SWIR1}/(\rho_{SWIR1} + \rho_{NIR}) + (\rho_{NIR}/(\rho_{NIR} + \rho_R) + \rho_G/(\rho_G + \rho_{SWIR1}))} \tag{4}$$

$$NDBSI = \frac{SI - IBI}{2} \tag{5}$$

(4) Retrieval of land surface temperature

LST was evaluated as follows:

$$L = gain \times DN + bias \tag{6}$$

$$T_b = \frac{K_2}{ln\left(\frac{K_1}{L} - 1\right)} \tag{7}$$

$$LST = \frac{T}{1 + \frac{\lambda T}{\rho} ln\varepsilon} \tag{8}$$

λ denotes the spectral wavelength of terrestrial emitted radiance, ρ represents the first radiation constant (1.438×10^{-2} m·K), ε characterizes the dimensionless land surface emissivity ranging from 0 to 1, and T indicates the at-satellite brightness temperature (K) measured through atmospheric window channels.

(5) Acquisition of RSEI

Principal component analysis (PCA) was performed on the four components of the RSEI, namely the NDVI, WET, LST, and NDBSI, and the first principal component was defined as the $RSEI_0$. The $RSEI_0$ was then standardized to obtain the RSEI to facilitate the measurement as follows:

$$RSEI_0 = 1 - \{PC_1[f(NDVI, WET, NDBSI, LST)]\} \quad (9)$$

After normalization, RSEI is obtained, which is a unitless value between 0 and 1. The larger the value, the better the ecological quality of the region. The formula is as follows:

$$RSEI = \frac{RSEI_0 - RSEI_{0,min}}{RSEI_{0,max} - RSEI_{0,min}} \quad (10)$$

where $RSEI_0$ is the first principal component of the four indicators; $RSEI_{0,min}$ is the minimum value of the $RSEI_0$; and $RSEI_{0,max}$ is the maximum value of the $RSEI_0$.

RSEI adopts a normalized scale ranging from 0 (degraded) to 1(optimal). The numerical magnitude exhibits a positive correlation with ecosystem health status. With an interval of 0.2, the levels of RSEI were classified into five groups: poor (0–0.2), fair (0.2–0.4), moderate (0.4–0.6), good (0.6–0.8), and excellent (0.8–1.0).

3.2. Combined T–S Estimator with M–K Test

The combined application of the Theil–Sen median trend analysis and the Mann–Kendall test provides a robust method for assessing trend changes in time series data. This method exhibits stronger resistance to errors and outliers in the data, thereby enhancing analytical accuracy. It evaluates trends by calculating the median of slopes between all pairwise data points, with the mathematical formula expressed as follows:

$$\beta = Median\left(\frac{RSEI_j - RSEI_i}{j - i}\right),\ 2015 \leq \mathrm{i} < \mathrm{j} \leq 2024 \quad (11)$$

where median denotes the median value; β is the slope of RSEI change. When the coefficient β is greater than zero, it indicates an upward (improving) trend in RSEI; conversely, a negative β value ($\beta < 0$) suggests a downward trend.

The Mann–Kendall test is a non-parametric statistical method that can be applied to determine the significance of trends in time series data. It imposes no distributional assumptions on the data and is robust against outliers. The test statistic is calculated as follows:

$$Z = \begin{cases} (S-1)/\sqrt{var(S)}, S > 0 \\ (S+1)/\sqrt{var(S)}, S < 0 \end{cases} \quad (12)$$

$$S = \sum_{i=1}^{n-1} \sum_{j=i+1}^{n} sgn(RSEI_j - RSEI_i) \quad (13)$$

$$\mathrm{sgn}(RSEI_j - RSEI_i) = \begin{cases} 1,\ RSEI_j - RSEI_i > 0 \\ 0, RSEI_j - RSEI_i = 0 \\ -1, RSEI_j - RSEI_i < 0 \end{cases} \quad (14)$$

$$var(S) = n(n-1)(2n+5)/18 \quad (15)$$

where n is the length of the time series, sgn () denotes the sign function; Z is the significance test statistic, whose value ranges from $(-\infty, +\infty)$.

The trend categories for the Theil–Sen (T–S) estimator and Mann–Kendall (M–K) test are summarized in Table 1.

Table 1. Trend categories of the Theil–Sen estimator and Mann–Kendall method.

β	Z	Trend Type	Trend Features
<0	≤−1.96	SD	Significant decrease
<0	−1.96 to 1.96	NSD	No significant decrease
0	−1.96 to 1.96	NC	No change
>0	−1.96 to 1.96	NSI	No significant increase
>0	≥1.96	SI	significant increase

NOTE: β is the T–S slope coefficient, which quantifies the magnitude of the temporal trend in RSEI values; Z is the M–K standardized test statistic, which represents the significance level of detected trends following M–K hypothesis testing.

3.3. *Coefficient of Variation (CV)*

The coefficient of variation (CV) reflects the relative variation in geographic data and serves as a critical indicator for measuring the stability of a time series. By calculating the ratio of the standard deviation to the mean, it quantifies the relative variability of a variable, making it widely applicable for assessing the degree of variation in time series data. Therefore, the CV is employed to analyze the stability of the RSEI. The calculation formula is as follows:

$$\mathrm{CV} = \frac{\sigma}{\mu} \tag{16}$$

CV denotes the coefficient of variation, σ represents the standard deviation of the RSEI time series, and μ represents the average value of the RSEI time series. A lower CV value indicates more stable interannual variations in the RSEI, while a higher CV value suggests greater volatility in its year-to-year fluctuations.

3.4. *Hurst Exponent*

The Hurst exponent (H), obtained through rescaled range analysis, serves as a metric for characterizing the long-term dependence and persistence of time series data. The Hurst exponent (H, 0 < H < 1) is generally divided into three cases: 0.5 < H < 1 indicates long-term memory effects, meaning the future trend will align with the past; H = 0.5 indicates that the time series is an independent random sequence; and 0 < H < 0.5 exhibits anti-persistent dynamics, suggesting that the future trend will likely reverse past patterns (Table 2).

Table 2. Classification of RSEI change trends.

Hurst	Sen's Slope	Future Trends
0.5 < H < 1	$\beta > 0$ $\beta < 0$	Improvement Degradation
H = 0.5	-	Uncertain
0 < H < 0.5	$\beta > 0$ $\beta < 0$	Improvement Degradation

3.5. *Quantitative Analysis Methods for Geospatial Differentiation*

The Geographical Detector (GeoDetector) represents a spatially explicit statistical framework designed to quantify spatial heterogeneity and identify its underlying driving factors. The formula for spatial differentiation and driving force detection is expressed mathematically as follows:

$$q = 1 - \frac{\sum_{h=1}^{L} N_h \sigma_h^2}{N\sigma^2} \tag{17}$$

L denotes the stratification (or classification/partitioning) of either variable Y or explanatory factor X, N_h represents the sample size within stratum h and N corresponds to the total population size, respectively. The terms σh^2 and σ^2 indicate the variances of Y-values within stratum h and across the entire study region, respectively. The q-statistic spans a range from 0 to 1, where a higher value indicates a greater degree of spatial heterogeneity in Y. If the stratification is based on the independent variable X, elevated q-values further imply greater explanatory capacity of X regarding Y's spatial distribution. Detailed interaction effects are presented in Table 3.

Table 3. Interaction relation.

Criteria	Interaction
q(X1 ∩ X2) < Min (q(X1), q(X2))	Nonlinear Weakening
Min(q(X1), q(X2)) < q(X1 ∩ X2) < Max(q(X1), q(X2))	Single-Factor Nonlinear Weakening
q(X1 ∩ X2) > Max(q(X1), q(X2))	Two-Factor Enhancement
q(X1 ∩ X2) = q(X1) + q(X2)	Mutual Independence
q(X1 ∩ X2) > q(X1) + q(X2)	Nonlinear Enhancement

4. Results

4.1. RSEI Model Applicability Validation

PCA was employed to reduce the dimensionality of the dataset while retaining the most significant variance among the ecological indicators, facilitating a comprehensive assessment of the RSEI framework. The first principal component (PC_1) accounts for 78.0% to 84% of the total variance, as shown in Table 4, indicating that it effectively captures the dominant features of the four indicators.

Table 4. Mean values of each indicator and the contribution of PC_1 from 2015 to 2024.

Year	RSEI Value	NDVI Value	WET Value	NDBSI Value	LST Value	PC_1 Value	PC_1 Contribution Rate
2015	0.6232	0.4412	0.5248	0.6984	0.6095	0.2863	79.40%
2020	0.5374	0.5625	0.4815	0.6563	0.5333	0.2605	83.74%
2024	0.5247	0.3879	0.4481	0.6952	0.4757	0.2369	78.17%

4.2. Spatial Distribution of EEQ

The RSEI values were categorized into five distinct ecological quality tiers using the Jenks optimization method, with the following categorical designations: Poor (0–0.2), Fair (0.2–0.4), Moderate (0.4–0.6), Good (0.6–0.8), and Excellent (0.8–1.0). The RSEI in the water source protection areas along the Anhui section of the YHWDP exhibits a spatial distribution pattern characterized by lower values in the western and northern regions and higher values in the eastern and southern regions (Figure 3).

Significant differences and dynamic trends are observed among the three segments: Yangtze River–Chaohu Lake Water Diversion, the Yangtze River–Huaihe River Connection, and the Yangtze River Water Northward Conveyance. According to Table 5, the proportions of poor and fair RSEI zones in 2024 are 8.45% and 21.97%, respectively, while the combined area of Grade 1–3 zones (representing better ecological conditions) accounts for 65.98%. The mean RSEI values across grades initially increased and then declined over time. For example, Grade 4 (moderate ecological status) shows a decreasing trend: 0.6232 (2015), 0.5374 (2020), and 0.5247 (2024).

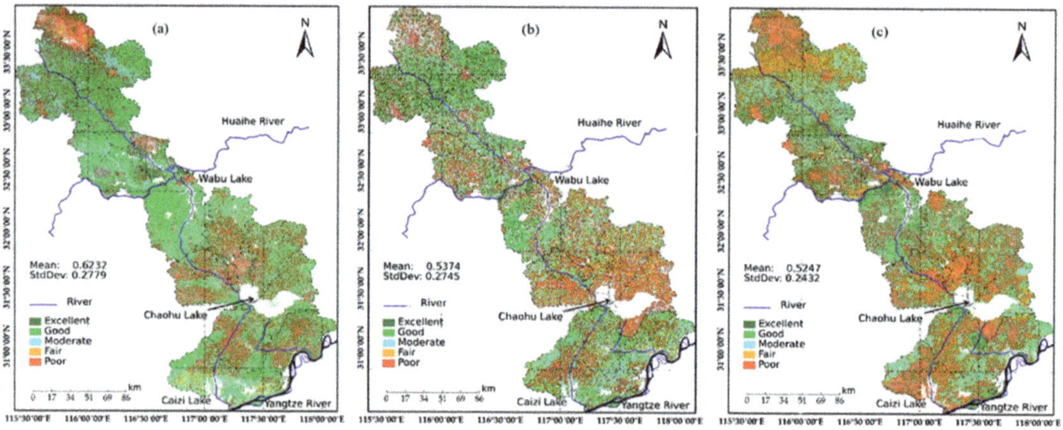

Figure 3. Spatial distribution of RSEI in the Anhui section of the Yangtze-to-Huaihe Water Diversion Project: 2015 (**a**), 2020 (**b**) and 2024 (**c**).

Table 5. RSEI classification area and proportion.

Class	Range	RSEI (2015)		RSEI (2020)		RSEI (2024)	
		Area (km^2)	Proportion (%)	Area (km^2)	Proportion (%)	Area (km^2)	Proportion (%)
Excellent	0.8–1.0	10924.20	37.44	3526.90	12.42	3637.24	12.43
Good	0.6–0.8	6474.50	22.19	7118.85	23.13	7608.52	26.34
Moderate	0.4–0.6	4637.73	15.89	8837.25	30.31	9018.88	30.82
Fair	0.2–0.4	4838.39	16.58	6766.31	23.47	6444.21	21.97
Poor	0.0–0.2	2305.91	7.9	2930.69	10.67	2471.84	8.45

The ecological quality assessed by RSEI remains at a median quality level (RSEI = 0.4–0.6), with mean values of 0.4026 in 2015, 0.6917 in 2020, and 0.6917 in 2024. However, the Yangtze River–Chaohu Lake Water Diversion shows an overall improving trend in RSEI-based ecological quality (Figure 4). Spatially, high RSEI values are concentrated along the YHWDP channels, the Yangtze River banks, the southern shore of Chaohu Lake, and surrounding water bodies.

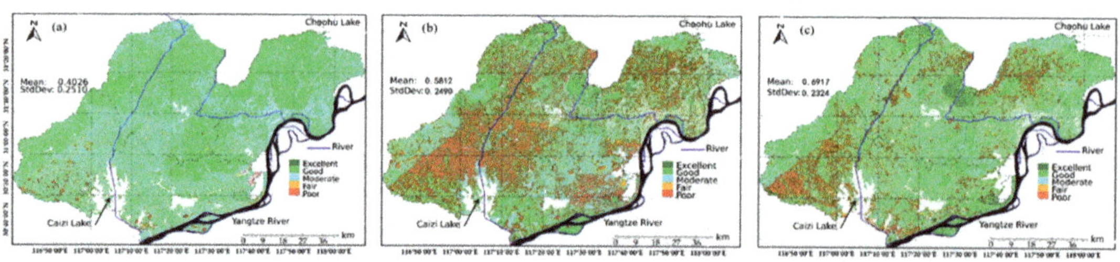

Figure 4. Spatial distribution of RSEI in the Yangtze River–Chaohu Lake Water Diversion (2015–2024): 2015 (**a**), 2020 (**b**) and 2024 (**c**).

As shown in Figure 5, the ecological quality assessed by RSEI remains at a poor level, though the area with significantly low RSEI in the Yangtze River–Huaihe River Connection shows a shrinking trend. The mean RSEI values for the Yangtze River–Huaihe River Connection are 0.4821 (2015), 0.5147 (2020), and 0.4789 (2024). Notably, Wuwei County,

Shushan District, and Xiejiaji District exhibit a yearly decline in RSEI, indicating a persistent deterioration in their ecological environments. Spatially, RSEI degradation is concentrated in the southern regions. For example, the Binhu New District shows fragmented ecological quality distribution (marked by red patches), likely linked to lake reclamation or over-tourism development. In contrast, areas with high ecological quality are predominantly clustered along the southern shore of Lake Wabu.

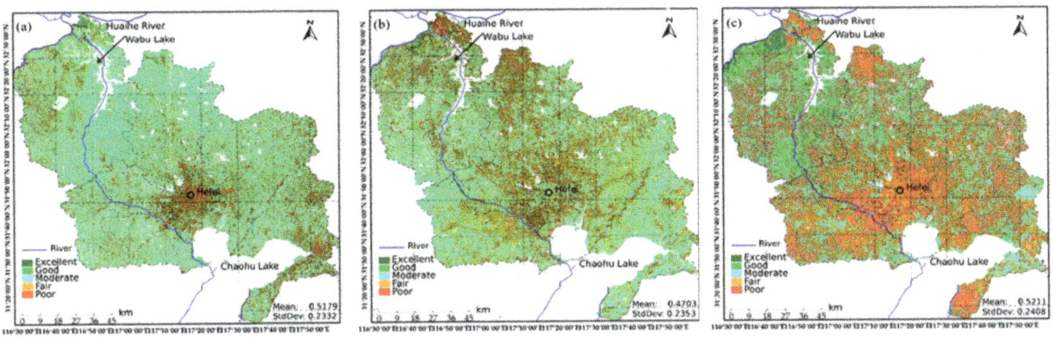

Figure 5. Spatial distribution of RSEI in the Yangtze River–Huaihe River Connection (2015–2024): 2015 (**a**), 2020 (**b**) and 2024 (**c**).

As shown in Figure 6, the RSEI index reflects an overall declining trend in regional ecological quality. The mean RSEI values are 0.6552 (2015), 0.5577 (2020), and 0.4842 (2024), indicating progressive deterioration. However, certain areas still exhibit significantly poor ecological conditions. JiaoCheng District, Wuoyang County, and Yingshang County experienced a transition from "good" to "moderate" and finally to "poor" RSEI levels, signaling sustained degradation in ecological quality. Panji District shows a year-on-year decline in RSEI, with its ecological environment worsening markedly.

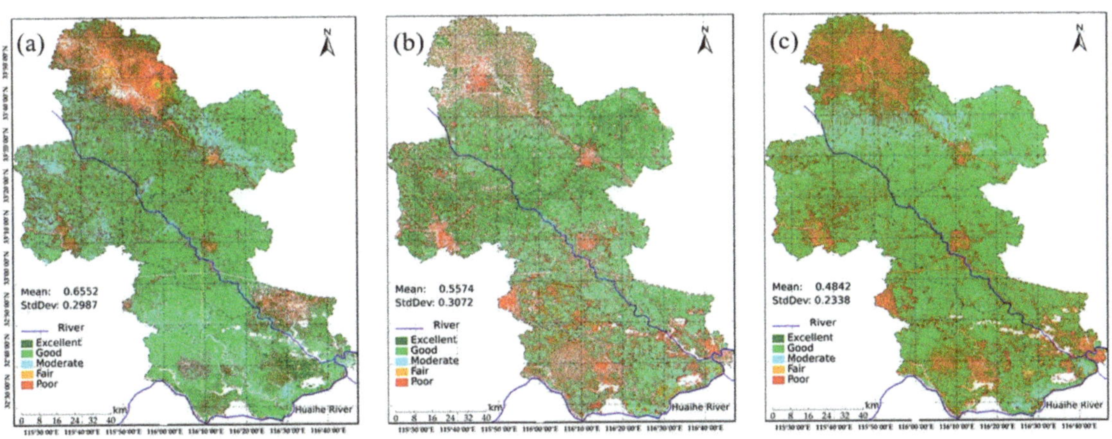

Figure 6. Spatial distribution of RSEI in the Yangtze River Water Northward Conveyance (2015–2024): 2015 (**a**), 2020 (**b**) and 2024 (**c**).

4.3. Spatial Change Characteristics of EEQ

A combined non-parametric approach integrating the T–S estimator and M–K test was systematically applied to investigate spatiotemporal variations in Ecological Environment

Quality (EEQ) along the Anhui segment of the YHWDP during the period 2015–2024 (Figure 7, Table 6).

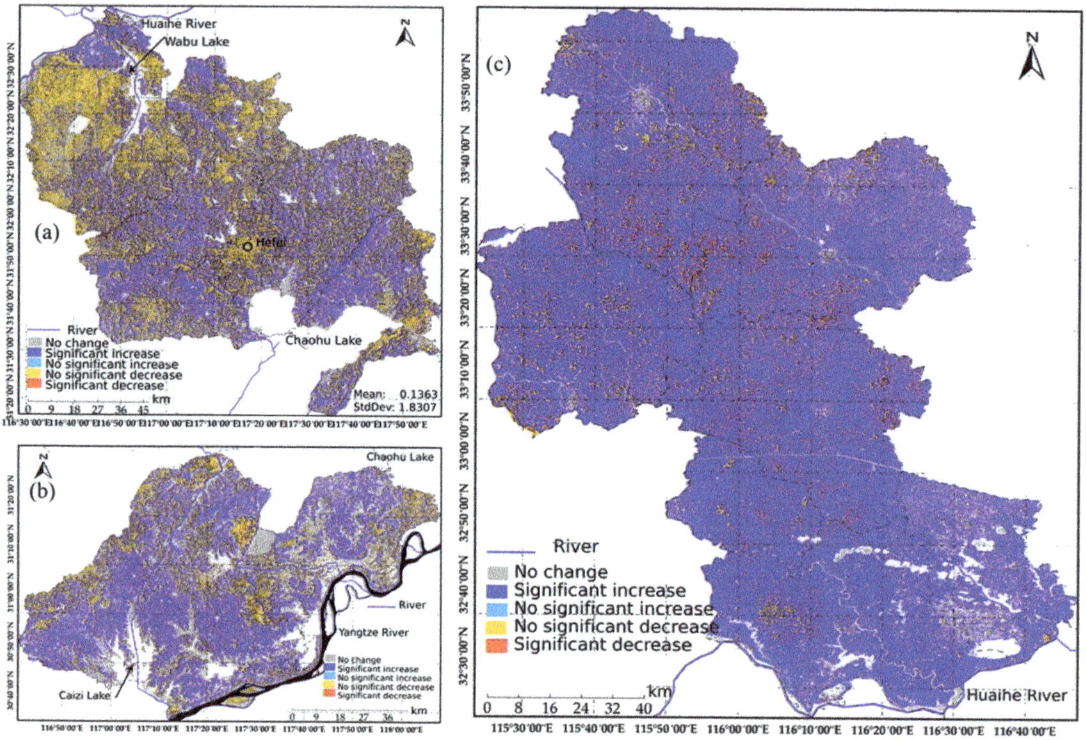

Figure 7. Spatial distribution of RSEI change characteristics in Yangtze River–Huaihe River Connection (**a**), Yangtze River–Chaohu Lake Water Diversion (**b**) and Yangtze River Water Northward Conveyance (**c**) during the period 2015–2024.

Table 6. Grading table of RSEI change amplitude in the Yangtze-to-Huaihe Water Diversion Project.

Segment	Trend Type	2015–2024	
		Area/km^2	Proportion/%
Yangtze River–Chaohu Lake Water Diversion	No change	1968.36	25.64
	No significant increase	2936.97	38.26
	Significant increase	57.33	0.75
	No significant decrease	2057.56	26.80
	Significant decrease	26.93	0.35
Yangtze River–Huaihe River Connection	No change	1896.21	15.30
	No significant increase	4633.03	37.38
	Significant increase	34.06	0.27
	No significant decrease	5146.40	41.52
	Significant decrease	41.88	0.34
Yangtze River Water Northward Conveyance	No change	895.43	7.20
	No significant increase	8366.28	67.27
	Significant increase	202.44	1.63
	No significant decrease	2056.50	16.53
	Significant decrease	91.76	0.74

4.4. Spatial Stability of the RSEI

The EEQ fluctuations in the water source protection areas along the Anhui segment of the YHWDP main water conveyance route were analyzed using the coefficient of variation (CV) of the RSEI. The variability was categorized into five levels via the natural breaks method (Jenks): Stability (CV ≤ 0.30); Slight changes (0.30 < CV ≤ 0.60); Moderate variation (0.60 < CV ≤ 0.90); Significant changes (0.90 < CV ≤ 1.20); and Drastic changes (1.20 < CV ≤ 1.50). This classification procedure yielded a comprehensive spatial-temporal visualization of RSEI CV patterns throughout the study region during the period 2015–2024 (Figure 8).

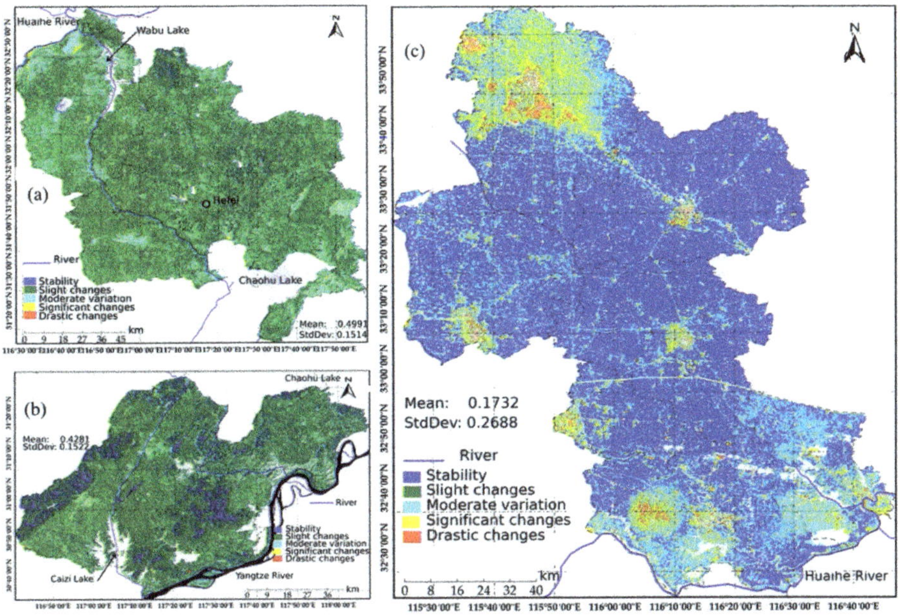

Figure 8. Spatial distribution of the coefficient of variation (CV) for RSEI in Yangtze River–Huaihe River Connection (**a**), Yangtze River–Chaohu Lake Water Diversion (**b**) and Yangtze River Water Northward Conveyance (**c**) (2015–2024).

The average coefficient of variation (CV) for RSEI changes was 0.4281, indicating that slight fluctuation (66.57%) and moderate fluctuation (9.24%) dominated the overall variability, collectively accounting for 75.81% of the study area (Table 7). Extreme fluctuation zones (CV > 1.2) represented 0.2% of the total area and were concentrated near Caizi Lake, water conveyance channels, and the north bank of the Yangtze River, particularly in the paddy fields highlighted in Figure 8a.

Table 7. Statistics of RSEI coefficient of variation (CV) area and proportion in the study area from 2015 to 2024.

CV	Yangtze River–Chaohu Lake Water Diversion		Yangtze River–Huaihe River Connection		Yangtze River Water Northward Conveyance	
	Area (km^2)	Proportion (%)	Area (km^2)	Proportion (%)	Area (km^2)	Proportion (%)
Stability	1141.11	14.87	862.64	6.96	5928.66	47.67
Slight changes	5109.67	66.57	8215.37	66.28	116.11	0.93
Moderate variation	709.46	9.24	2507.26	20.23	3705.16	29.79
Significant changes	68.21	0.89	151.61	1.22	1486.79	11.95
Drastic changes	15.58	0.20	5.51	0.04	282.68	2.27

As shown in Figure 8b, the mean coefficient of variation (CV) for RSEI changes was 0.4991, indicating slight fluctuation (66.28%) as the dominant variability category, followed by moderate fluctuation (20.23%), collectively accounting for 86.51% of the study area. Pronounced RSEI variability was primarily distributed in lake-, river-, and pond-dense regions, particularly around Hefei City. These patterns likely stem from the interplay of topographic complexity, climatic variability, and intense human activities (e.g., urban sprawl).

As shown in Figure 8c, the average coefficient of variation (CV) for RSEI changes was 0.1732, with slight fluctuation, moderate fluctuation, and higher fluctuation collectively accounting for 41.74% of the study region. Pronounced RSEI variability was predominantly found in the northern Huaihe River Basin, where intensive agricultural activities have led to significant anthropogenic disturbances.

The ecological environment quality across the study region predominantly exhibited slight to moderate fluctuations, with coefficient of variation (CV) values of 0.4281 for the Yangtze River–Chaohu Lake Water Diversion, 0.4991 for the Yangtze River–Huaihe River Connection, and 0.1732 for the Yangtze River Water Northward Conveyance. By aggregating stable and slightly fluctuating zones, the stability ranking was: Yangtze River–Huaihe River Connection (86.51%) > Yangtze River–Chaohu Lake Water Diversion (78.81%) > Yangtze River Water Northward Conveyance (29.79%).

4.5. Prediction of Future Trends in Ecological Environment Quality

Yangtze River–Chaohu Lake Water Diversion: The RSEI exhibited an average Hurst index of 0.5319, indicating a moderate persistence in ecological quality trends. It suggests that 27.78% of the study area is likely to experience enhanced ecological conditions, while 38.38% may face sustained deterioration (Figure 9).

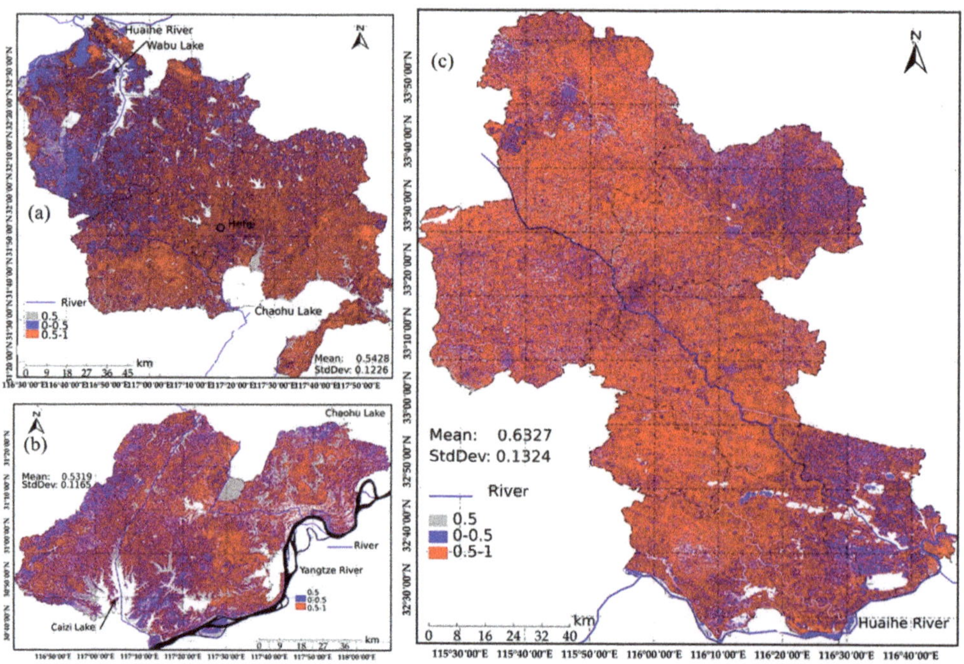

Figure 9. Spatial distribution of RSEI Hurst index in Yangtze River–Huaihe River Connection (**a**), Yangtze River–Chaohu Lake Water Diversion (**b**) and Yangtze River Water Northward Conveyance (**c**) from 2015 to 2024.

Yangtze River–Huaihe River Connection: The RSEI analysis reveals an average Hurst index value of 0.5428. Approximately 30.02% of the study area is anticipated to undergo ecological improvement, while a more substantial proportion (49.49%) is expected to experience continued degradation (Table 8).

Table 8. Statistics of Sen–Hurst coefficient area and proportion for RSEI in the study area (2015–2024).

CV	Yangtze River–Chaohu Lake Water Diversion		Yangtze River–Huaihe River Connection		Yangtze River Water Northward Conveyance	
	Area (km^2)	Proportion (%)	Area (km^2)	Proportion (%)	Area (km^2)	Proportion (%)
Uncertain	1968.36	25.64	1896.21	15.30	895.43	7.20
Improvement	2132.66	27.78	3720.89	30.02	1754.32	14.11
Degradation	2946.12	38.38	6134.14	49.49	8962.36	72.06

Yangtze River Water Northward Conveyance: The average Hurst index of RSEI is 0.6327, with only 14.11% of areas likely to improve, while 72.06% are expected to worsen persistently. Significant RSEI variations are concentrated in the northern Huaihe, where intensive farming leads to pronounced human disturbance.

Overall Trend: The future ecological environment quality of the study region is generally pessimistic (Figure 10). The average Hurst indices are 0.5319 for Yangtze River–Chaohu Lake Water Diversion, 0.5428 for Yangtze River–Huaihe River Connection, and 0.6327 for Yangtze River Water Northward Conveyance. The projected deterioration follows this order: Yangtze River Water Northward Conveyance (72.06%) > Yangtze River–Huaihe River Connection (49.49%) > Yangtze River–Chaohu Lake Water Diversion (38.38%).

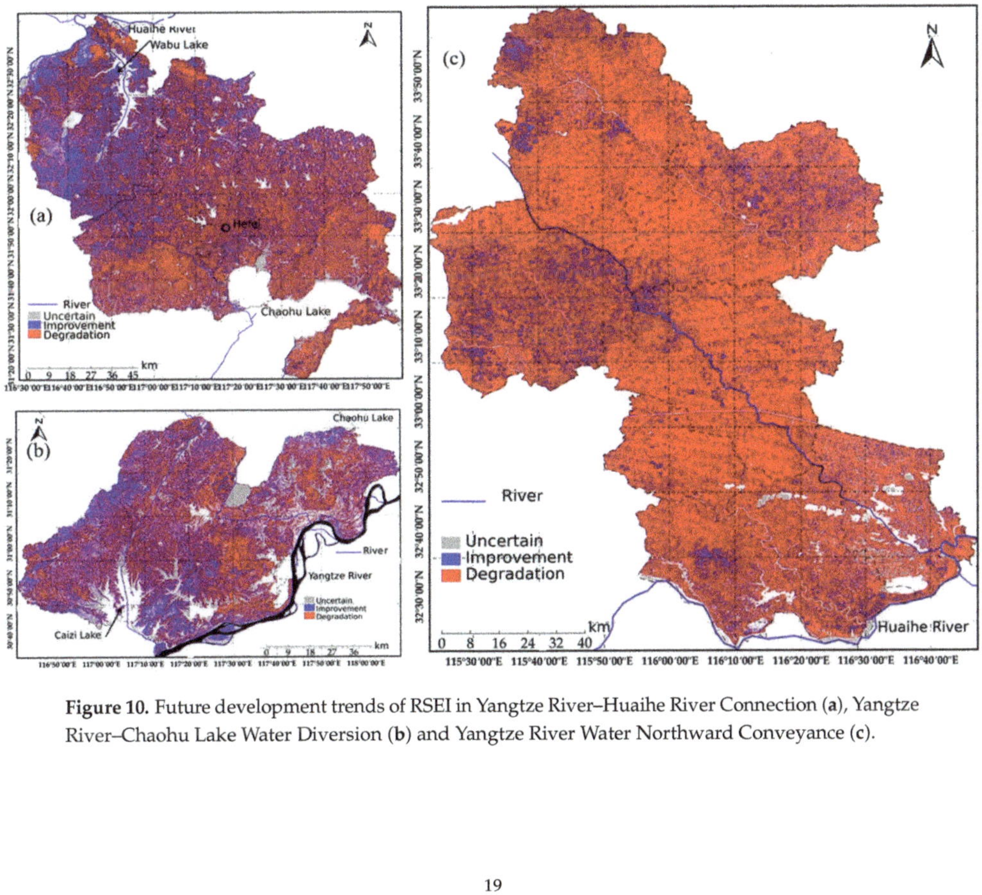

Figure 10. Future development trends of RSEI in Yangtze River–Huaihe River Connection (**a**), Yangtze River–Chaohu Lake Water Diversion (**b**) and Yangtze River Water Northward Conveyance (**c**).

4.6. Quantitative Attribution Analysis of Driving Factors for Spatial Differentiation of RSEI

Integrating the natural and socioeconomic conditions of the water source conservation areas along the Jianghuai Water Diversion Project's Anhui Section, this study selected ten variables—including population, GDP (Gross Domestic Product), built-up area, precipitation, and mean NDVI—as influencing factors for correlation analysis with the RSEI. The Geodetector method was employed to analyze the driving mechanisms behind dynamic changes in EEQ. The independent variables (X) include the following: X1: Annual mean temperature (°C); X2: Annual mean precipitation (mm); X3: Vegetation coverage (%); X4: Mean NDVI; X5: Elevation (m); X6: Slope (°); X7: Built-up area; X8: Land Surface Temperature (LST); X9: GDP per unit area (10,000 yuan/km^2); X10: Population density (persons/km^2).

The analysis of critical topographic parameters across the three major water diversion segments reveals distinct geomorphological characteristics (Table 9). The Yangtze River–Chaohu Lake Water Diversion exhibits the highest mean elevation (49.32 m) and steepest mean slope (5.16°), indicating significant terrain complexity that may influence hydraulic dynamics and sediment transport patterns. In contrast, the Yangtze River–Huaihe River Connection features a moderate elevation (38.20 m) and gentler slope (2.77°), favoring efficient water conveyance over long distances while requiring vigilance against anthropogenic disturbances in low-lying zones. The Yangtze River Water Northward Conveyance, with the lowest elevation (30.68 m) and minimal slope (2.02°), presents a flat terrain profile that heightens flood susceptibility, necessitating engineered safeguards to maintain operational reliability.

Table 9. Parameters of driving factors influencing RSEI changes.

Driving Factors		Variable	Yangtze River–Chaohu Lake Water Diversion			Yangtze River–Huaihe River Connection			Yangtze River Water Northward Conveyance		
			2015	2020	2024	2015	2020	2024	2015	2020	2024
natural factors	climatic factors	X1	18.20	17.60	17.40	17.50	16.30	17.40	16.00	16.60	14.80
		X2	1421.30	1877.90	1776.30	1234.60	1634.20	1293.10	808.60	1020.90	829.30
	vegetation factors	X3	24.86	24.86	24.86	23.43	23.43	23.43	26.91	26.91	26.91
		X4	0.50	0.56	0.21	0.35	0.34	0.22	0.26	0.56	0.29
	topographic factors	X5	49.32	49.32	49.32	38.20	38.20	38.20	30.68	30.68	30.68
		X6	5.16	5.16	5.16	2.77	2.77	2.77	2.02	2.02	2.02
	surface disturbance factors	X7	619.87	863.75	1015.04	1174.01	1443.37	2100.10	1343.59	1502.63	1623.43
		X8	0.44	0.48	0.75	0.51	0.48	0.54	0.60	0.56	0.60
socio-economic	economic factors	X9	360.00	288.00	279.00	323.00	319.00	318.00	466.00	502.00	445.00
	population size	X10	972.77	1471.97	2348.36	270.74	1824.07	2580.59	1047.92	1742.43	2495.18

The single factor detection results indicated that ecological environment quality (as dependent variable Y) was input into the optimal parameter geographical detector model. A higher q-value signifies a more robust explanatory capacity of the factor influencing ecological environment quality (Table 10). The q-values of factors ranked in descending order were: annual precipitation (0.76), land surface temperature (LST) (0.64), population density (0.50), mean annual temperature (0.37), built-up area (0.34), slope (0.30), elevation (0.17), NDVI mean (0.08), vegetation coverage (0.06), and GDP per unit area (0.00). This ranking demonstrates that annual precipitation and land surface temperature-LST serve as the primary drivers influencing the ecological environment quality in water source conservation areas along the YHWDP Route.

Table 10. Single-factor influence contribution.

	X1	X2	X3	X4	X5	X6	X7	X8	X9	X10
q statistic	0.37	0.76	0.06	0.08	0.17	0.30	0.34	0.64	0.00	0.50
p value	1.00	0.56	0.97	0.95	0.93	0.78	0.49	0.77	1.00	0.91

The interaction effects between factors demonstrated significant explanatory power for ecological environment quality. Notably, the interaction between NDVI and elevation, slope, and built-up area yielded the highest synergistic q-value (0.95), indicating their dominant combined influence on ecological conditions. Secondarily, interactions involving land surface temperature (LST) with mean annual temperature, precipitation, vegetation coverage, elevation, and slope all exhibited q-values exceeding 0.91. Other factor interactions consistently showed q-values above 0.5, representing substantial improvements over single-factor q-values. These findings emphasize that ecological restoration strategies must systematically integrate multi-factor spatial distribution patterns, prioritizing precipitation regimes, vegetation dynamics, and thermal conditions (LST) to optimize governance frameworks and enhance intervention efficacy.

As illustrated in Figure 11, ecological drivers operate through synergistic interactions rather than isolated effects. The dominant interaction patterns—nonlinear enhancement (e.g., NDVI × elevation) and bilinear enhancement (e.g., LST × slope)—reveal that spatial heterogeneity in ecosystem quality arises from compounded multi-factor processes. Particularly, vegetation-terrain interactions (NDVI × elevation/slope) demonstrated the strongest explanatory power for RSEI variations, with q-values surpassing 0.9. In contrast, anthropogenic impacts exhibited weaker direct effects, primarily manifesting indirectly through land use/cover modifications.

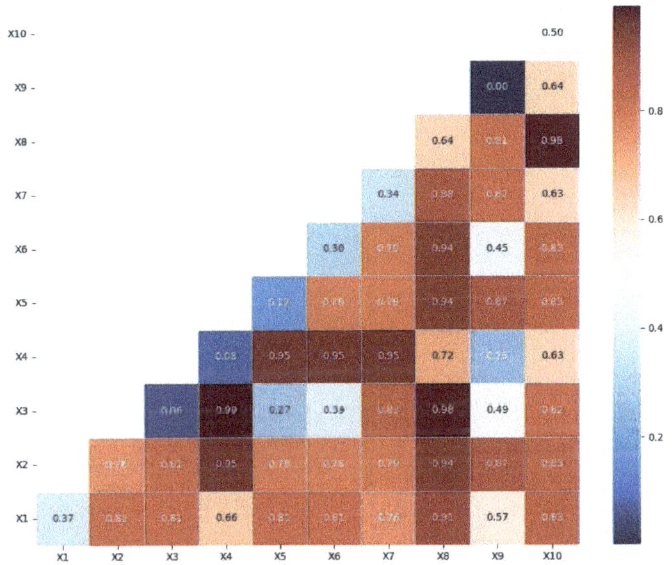

Figure 11. Interactive detection matrix.

5. Discussion

During the period 2015–2024, the ecological environment quality of water source conservation areas along the YHWDP route in Anhui Province exhibited a fluctuating yet overall deteriorating trend, necessitating heightened focus on future ecological restoration efforts. Notably, the spatial extent of ecologically degraded zones gradually decreased over this period, reflecting a progressive mitigation of environmental stress. Yangtze River–Huaihe River Connection demonstrated the most unstable Remote Sensing Ecological Index (RSEI) dynamics (mean RSEI = 0.4789), indicating heightened ecological vulnerability. Yangtze River Water Northward Conveyance ranked second in ecological fragility (RSEI = 0.4842), while the Yangtze River–Chaohu Lake Water Diversion showed relatively better conditions (RSEI = 0.6917). The spatial distribution of RSEI exhibited a pronounced latitudinal and longitudinal gradient, with elevated values distinctly clustered in southern and eastern regions, while lower magnitudes prevailed in northern and western areas. Areas with "significant degradation" dominated the ecological quality classification, particularly in urbanized regions like Hefei City, where vegetation cover was sparse. Higher-quality ecosystems clustered in densely vegetated zones, confirming vegetation coverage as a critical stabilizing factor.

This study classified the ecological environment quality of water source conservation areas along the Anhui section of the YHWDP corridor into five levels from 2015 to 2024. Statistical analysis demonstrated that both vegetation vigor (NDVI) and atmospheric moisture content (WET) significantly enhanced ecological quality indicators, while thermal stress (LST) and dryness (NDBSI) showed significant negative impacts. Single-factor analysis demonstrated the following explanatory power ranking for ecological drivers: LST > NDBSI > NDVI > WET, with LST achieving the highest q-value, highlighting its dominant role in explaining the spatial heterogeneity of the Remote Sensing Ecological Index (RSEI). Dual-factor interaction effects further indicated that LST × NDBSI had the strongest combined influence on RSEI variability, whereas NDVI × WET interactions were comparatively weaker. Overall, natural factors—including precipitation patterns, temperature gradients, and topographic elevation—were identified as primary drivers of spatial differentiation in ecological quality. In contrast, anthropogenic activities indirectly impacted the environment through land use/cover modifications, such as urbanization-driven increases in LST and expansion of built-up areas [16,17]. Meanwhile, the increase in NDVI contributed to the improvement of regional ecological environment quality [22,23]. Therefore, in the process of urban expansion and construction, attention should be paid to the reasonable protection of local natural resources. At the same time, ecological protection policies such as afforestation and water purification should be implemented, as they can help alleviate a series of environmental problems. This study is consistent with the conclusions of previous literature. These findings emphasize the need to integrate natural climatic–topographic dynamics with targeted land-use regulations to mitigate human-induced pressures and enhance ecological resilience in large-scale water diversion systems.

6. Conclusions

This research conducted a comprehensive investigation into the spatiotemporal dynamics and underlying driving mechanisms of the RSEI within water source conservation areas along the Anhui segment of the YHWDP corridor from 2015 to 2024. Key findings include the following:

(1) The RSEI in these areas exhibited a fluctuating upward trend followed by a decline, with increasing spatial heterogeneity in ecological quality. The ecological quality ranking across subregions was: Yangtze River–Huaihe River Connection (86.51%) < Yangtze River Water Northward Conveyance (78.81%) < Yangtze River–Chaohu Lake Water Diversion (29.79%). Over the nine-year period, areas classified as "Good" and "Moderate" ecological quality significantly expanded, while "Poor" and "Very Poor" categories decreased, indicating overall ecological improvement. However, spatial imbalances persist, with pronounced regional disparities in stability: Yangtze River–Huaihe River Connection (40.66% stable/slight fluctuation) > Yangtze River–Chaohu Lake Water Diversion (30.19%) > Yangtze River Water Northward Conveyance (25.02%).

(2) Yangtze River Water Northward Conveyance demonstrated relatively high ecological quality but greater RSEI volatility, attributed to vegetation dynamics in cropland protection zones. In contrast, the Yangtze River–Huaihe River Connection exhibited

lower baseline ecological quality but milder fluctuations (slight to moderate), with intense ecological changes observed near Hefei City.

(3) Annual precipitation, impervious surface area, and vegetation coverage emerged as primary drivers of ecological quality, interacting synergistically with annual average temperature and land cover to significantly enhance explanatory power over RSEI variations. Human activities, particularly land-use changes, markedly amplified RSEI dynamics, demonstrating strong collaborative effects with natural factors. All factor interactions exhibited nonlinear enhancement, emphasizing the critical role of coupled natural-anthropogenic processes in shaping ecological patterns.

This study calculated the RSEI for the YHWDP region and analyzed the temporal and spatial characteristics of changes, as well as future trends. Subsequently, the impact of different driving factors was evaluated, providing guidance for future environmental protection policy formulation. These approaches will empower watershed management departments to enhance the precision and scientific rigor of strategy formulation, effectively coordinating farmland protection, urban expansion control, and vegetation restoration. Such coordinated efforts aim to address ecological vulnerabilities in specific areas within large-scale water diversion systems. Due to space constraints, this study did not conduct a detailed analysis of the impact of different spatial scales of Landsat remote sensing data on the computational results. Future research should conduct comparative studies at different scales to better implement the comprehensive ecological environment assessment method developed in this study, and thus apply it to ecological protection in other watersheds and cross-basin water diversion work.

Author Contributions: Conceptualization, X.Q.; methodology, X.Q.; software, X.Q. and Q.H.; validation, X.Q., Q.L. and Q.H.; formal analysis, X.Q. and Q.H.; investigation, B.L., L.L. and Z.S.; resources, Y.O. and D.W.; data curation, Y.O. and D.W.; writing—original draft preparation, X.Q.; writing—review and editing, Q.L. and Q.H.; visualization, X.Q. and Q.H.; supervision, L.L. and Z.S.; project administration, X.Q.; funding acquisition, X.Q. All authors have read and agreed to the published version of the manuscript.

Funding: This research was funded by the Project on Delineation of Water Resource Projection Areas, Early-Warning System for Major Sudden Water Pollution Incidents, and Planning for the Projection of Water Sources Along Water Transmission Routes of the Diversion of Yangtze River to the Huaihe Project (Anhui Section) (No. YJJH-ZT-ZX-20230706545); Key Natural Science Research Project of Bengbu University (No. 2023ZR02zd); Bengbu University Applied Project (High-level Talent Research Fund) (No. 2024YYX66QD; No. 2024YYX65QD; No. 2025GQD039; No. 2025GQD035) and Key Projects of Natural Science Research in Universities of Anhui Province (No. 2023AH052941); and Peak Discipline of Bengbu University (No. 2025GFXK02).

Institutional Review Board Statement: Not applicable.

Informed Consent Statement: Not applicable.

Data Availability Statement: The original contributions presented in this study are included in the article. Further inquiries can be directed to the corresponding author.

Conflicts of Interest: The authors declare no conflict of interest.

References

1. Zhang, Y.; Chen, X.; Zhang, Y.; Wang, B. Quantitative contribution of climate change and vegetation restoration to ecosystem services in the Inner Mongolia under ecological restoration projects. *Ecol. Indic.* **2025**, *171*, 113240. [CrossRef]
2. Zhang, X.; Jia, W.; Lu, S.; He, J. Ecological assessment and driver analysis of high vegetation cover areas based on new remote sensing index. *Ecol. Inform.* **2024**, *82*, 102786. [CrossRef]
3. Liang, L.; Wang, Q.; Guan, Q.; Du, Q.; Sun, Y.; Ni, F.; Lv, S.; Shan, Y. Assessing vegetation restoration prospects under different environmental elements in cold and arid mountainous region of China. *Catena* **2023**, *226*, 107055. [CrossRef]

4. Han, Q.; Xue, L.; Liu, Y.; Yang, M.; Chu, X.; Liu, S. Developing a multi-objective simulation-optimization model for ecological water conveyance in arid inland river basins. *J. Hydrol.-Reg. Stud.* **2023**, *50*, 101551. [CrossRef]
5. Li, Y.; Zhang, X.; Cao, Z.; Liu, Z.; Lu, Z.; Liu, Y. Towards the progress of ecological restoration and economic development in China's Loess Plateau and strategy for more sustainable development. *Sci. Total Environ.* **2021**, *756*, 143676. [CrossRef]
6. Meng, Y.; Hou, B.; Ding, C.; Huang, L.; Guo, Y.; Tang, Z. Spatiotemporal patterns of planted forests on the Loess Plateau between 1986 and 2021 based on Landsat NDVI time-series analysis. *Gisci. Remote Sens.* **2023**, *60*, 2185980. [CrossRef]
7. Wu, Q.; Jiang, X.; Song, M.; Liu, Y.; Shi, X.; Lei, Y.; Nie, T. Study on the development trend of social-ecological systems and the drivers of sustainable development—A case study of the Loess Plateau in China. *Ecol. Indic.* **2023**, *156*, 111172. [CrossRef]
8. Xiao, Y.; Wang, R.; Wang, F.; Huang, H.; Wang, J. Investigation on spatial and temporal variation of coupling coordination between socioeconomic and ecological environment: A case study of the Loess Plateau, China. *Ecol. Indic.* **2022**, *136*, 108667. [CrossRef]
9. Long, Y.; Jiang, F.; Deng, M.; Wang, T.; Sun, H. Spatial-temporal changes and driving factors of eco-environmental quality in the Three-North region of China. *J. Arid Land* **2023**, *15*, 231–252. [CrossRef]
10. Zheng, Z.; Wu, Z.; Chen, Y.; Guo, C.; Marinello, F. Instability of remote sensing based ecological index (RSEI) and its improvement for time series analysis. *Sci. Total Environ.* **2022**, *814*, 152595. [CrossRef]
11. Kang, S.; Jia, X.; Zhao, Y.; Han, L.; Ma, C.; Bai, Y. Spatiotemporal Variation and Driving Factors of Ecological Environment Quality on the Loess Plateau in China from 2000 to 2020. *Remote Sens.* **2024**, *16*, 4778. [CrossRef]
12. Xin, J.; Yang, J.; Yu, H.; Ren, J.; Yu, W.; Cong, N.; Xiao, X.; Xia, J.; Li, X.; Qiao, Z. Towards ecological civilization: Spatiotemporal heterogeneity and drivers of ecological quality transitions in China (2001–2020). *Appl. Geogr.* **2024**, *173*, 103439. [CrossRef]
13. Zhou, M.; Li, Z.; Gao, M.; Zhu, W.; Zhang, S.; Ma, J.; Ta, L.; Yang, G. Revealing the Eco-Environmental Quality of the Yellow River Basin: Trends and Drivers. *Remote Sens.* **2024**, *16*, 2018. [CrossRef]
14. Yang, H.; Yu, J.; Xu, W.; Wu, Y.; Lei, X.; Ye, J.; Geng, J.; Ding, Z. Long-time series ecological environment quality monitoring and cause analysis in the Dianchi Lake Basin, China. *Ecol. Indic.* **2023**, *148*, 110084. [CrossRef]
15. Lv, Y.; Xiu, L.; Yao, X.; Yu, Z.; Huang, X. Spatiotemporal evolution and driving factors analysis of the eco-quality in the Lanxi urban agglomeration. *Ecol. Indic.* **2023**, *156*, 111114. [CrossRef]
16. Aurora, R.M.; Furuya, K. Spatiotemporal Analysis of Urban Sprawl and Ecological Quality Study Case: Chiba Prefecture, Japan. *Land* **2023**, *12*, 2013. [CrossRef]
17. Halder, S.; Bose, S. Comparative study on remote sensing-based indices for urban ecology assessment: A case study of 12 urban centers in the metropolitan area of eastern India. *J. Earth Syst. Sci.* **2024**, *133*, 100. [CrossRef]
18. An, L.; Liao, K.; Zhu, L.; Zhou, B. Influence of river-lake isolation on the water level variations of Caizi Lake, lower reach of the Yangtze River. *J. Geogr. Sci.* **2021**, *31*, 551–564. [CrossRef]
19. Li, C.; Li, H.; Zhang, Y.; Zha, D.; Zhao, B.; Yang, S.; Zhang, B.; de Boer, W.F. Predicting hydrological impacts of the Yangtze-to-Huaihe Water Diversion Project on habitat availability for wintering waterbirds at Caizi Lake. *J. Environ. Manag.* **2019**, *249*, 109251. [CrossRef]
20. Qi, X.; Han, Q.; Li, B.; Chen, X.; Guo, Z.; Ou, Y.; Wang, D. Developing a Multi-Objective Optimization Scheduling Method for the Yangtze to Huaihe River Water Diversion Project Considering Lake Regulation and Storage. *Water* **2025**, *17*, 1286. [CrossRef]
21. Ding, W.; Shi, G.; Zha, H.; Miao, H.; Lu, M.; Jin, J. Ecological impacts and supply demand evolution of the Yangtze to Huaihe water transfer project in Anhui section. *Sci. Rep.* **2024**, *14*, 20311. [CrossRef] [PubMed]
22. Halder, S.; Bose, S. Ecological quality assessment of five smart cities in India: A remote sensing index-based analysis. *Int. J. Environ. Sci. Technol.* **2024**, *21*, 4101–4118. [CrossRef]
23. Hasan, M.M.; Ferdous, M.T.; Talha, M.; Mojumder, P.; Roy, S.K.; Zim, M.N.F.; Akter, M.M.; Nasher, N.M.R.; Hasher, F.F.B.; Boltiziar, M.; et al. Analyzing Ecological Environmental Quality Trends in Dhaka Through Remote Sensing Based Ecological Index (RSEI). *Land* **2025**, *14*, 1258. [CrossRef]

sustainability

MDPI

Article

Drainage Network Generation for Urban Pluvial Flooding (UPF) Using Generative Adversarial Networks (GANs) and GIS Data

Muhammad Nasar Ahmad [1,2,*], Hariklia D. Skilodimou [3], Fakhrul Islam [4,5], Akib Javed [2] and George D. Bathrellos [3,*]

1 School of Information and Artificial Intelligence, Anhui Agricultural University, Hefei 230036, China
2 State Key Laboratory of Information Engineering in Surveying, Mapping and Remote Sensing, Wuhan University, Wuhan 430079, China; akibjaved@whu.edu.cn
3 Department of Geology, University of Patras, 26504 Patras, Greece; hskilodimou@upatras.gr
4 State Key Laboratory of Desert and Oasis Ecology, Xinjiang Institute of Ecology and Geography, Chinese Academy of Sciences, Urumqi 830011, China; geo.fakhar@mails.ucas.ac.cn
5 University of Chinese Academy of Sciences, Beijing 101408, China
* Correspondence: mnasarahmad@ahau.edu.cn or mnasarahmad@whu.edu.cn (M.N.A.); gbathrellos@upatras.gr (G.D.B.)

Abstract: Mapping urban pluvial flooding (UPF) in data-scarce regions poses significant challenges, particularly when drainage systems are inadequate or outdated. These limitations hinder effective flood mitigation and risk assessment. This study proposes an innovative approach to address these challenges by integrating deep learning (DL) models with traditional methods. First, deep convolutional generative adversarial networks (DCGANs) were employed to enhance drainage network data generation. Second, deep recurrent neural networks (DRNNs) and multi-criteria decision analysis (MCDA) methods were implemented to assess UPF. The study compared the performance of these approaches, highlighting the potential of DL models in providing more accurate and robust flood mapping outcomes. The methodology was applied to Lahore, Pakistan—a rapidly urbanizing and data-scarce region frequently impacted by UPF during monsoons. High-resolution ALOS PALSAR DEM data were utilized to extract natural drainage networks, while synthetic datasets generated by GANs addressed the lack of historical flood data. Results demonstrated the superiority of DL-based approaches over traditional MCDA methods, showcasing their potential for broader applicability in similar regions worldwide. This research emphasizes the role of DL models in advancing urban flood mapping, providing valuable insights for urban planners and policymakers to mitigate flooding risks and improve resilience in vulnerable regions.

Keywords: deep learning; DRNN; GANS; MCDA; remote sensing; urban pluvial flooding

1. Introduction

Floods are among the most devastating natural disasters globally, causing significant human and economic losses annually [1–7]. Urban pluvial flooding (UPF), characterized by surface water accumulation due to intense rainfall overwhelming drainage systems, has become increasingly prevalent. This escalation is attributed to climate change-induced rainfall intensification and rapid urbanization [3,8–19]. The consequences of UPF include infrastructure damage, disruption of urban life, and substantial economic losses [20]. These impacts are projected to worsen with continued urban expansion and the increasing frequency of extreme weather events [21], underscoring the urgent need for innovative solutions to mitigate UPF risks.

In many developing countries, such as Pakistan, the assessment of flood risk is hindered by the lack of comprehensive drainage network data and historical flood records. Traditional flood mapping techniques, which often rely on statistical analyses and Geographic Information Systems (GIS), struggle to capture the dynamic nature of urban flooding, particularly in areas with outdated or inadequate drainage infrastructure [22]. These methods are heavily dependent on historical data and often fail to provide accurate assessments in data-scarce environments.

Recent literature indicates a predominant focus on traditional models for urban flooding assessment, including histogram-based, threshold, statistical, and GIS models [14,16–19]. While deep learning (DL) approaches have shown promise in various flood-related applications, their implementation in UPF assessment remains limited. Ref. [23] reviewed approximately 1038 research articles related to flood disasters, finding that only about 46 employed deep learning-based modeling for flood assessment. Moreover, practical implementations of DL models in UPF contexts are scarce [24–28].

The challenges posed by data scarcity and the dynamic nature of urban systems necessitate the development of new methods to improve flood susceptibility assessment and enhance urban resilience. Conventional approaches, such as Multi-Criteria Decision Analysis (MCDA) and GIS-based models, are widely used but often lack the adaptability required for modern urban environments [29–35]. There is a pressing need for more adaptive, data-driven solutions that can provide accurate assessments even in the absence of extensive historical data.

This study aims to develop an improved flood risk assessment framework tailored for data-scarce urban regions. By leveraging alternative data sources and advanced computational approaches, this research seeks to address the limitations of traditional flood modeling methods. The findings are expected to offer valuable insights for urban planners and policymakers, facilitating the implementation of more effective flood prevention and mitigation strategies. Additionally, this research underscores the importance of integrating modern data-driven methodologies to enhance flood resilience in vulnerable urban environments.

2. Dataset

This study integrates multiple datasets to support both traditional and advanced modeling approaches to address the challenges of urban pluvial flooding (UPF) assessment in data-scarce regions.

For traditional modeling, the high-resolution ALOS PALSAR Digital Elevation Model (DEM) with a 12.5 m spatial resolution was utilized. Compared to alternatives like ASTER GDEM (30 m) and SRTM (30 m), ALOS PALSAR offers superior spatial and vertical accuracy, with a vertical accuracy of approximately ±5 m. This higher resolution ensures more precise extraction of drainage networks, which is crucial for accurate flood modeling. Additionally, ALOS PALSAR is freely available, making it a valuable resource for regions with limited access to high-quality geospatial data.

Table 1 provide the information of dataset used particularly, for the implementation of Multi-Criteria Decision Analysis (MCDA) using the Analytical Hierarchy Process (AHP), multiple datasets were integrated, including impervious surface area (ISA), land use/land cover (LULC), rainfall, slope, aspect, and other relevant factors. Each dataset contributes unique insights into the hydrological and urban characteristics of the study area, facilitating a comprehensive evaluation of flood susceptibility. Figure 1 provides an overview of the datasets used.

Table 1. Summary of datasets used.

Dataset	Source/Resolution	Purpose	Relevance to Study	URL
ALOS PALSAR DEM	12.5 m resolution (Free)	Drainage network extraction	High-resolution terrain modeling	https://www.eorc.jaxa.jp/ALOS/en/dataset/alos_open_and_free_e.htm, accessed on 7 May 2025
ASTER GDEM	30 m resolution (Comparison)	Elevation data (comparison)	Lower resolution; less accurate	https://asterweb.jpl.nasa.gov/gdem.asp, accessed on 7 May 2025
SRTM DEM	90 m resolution (Comparison)	Elevation data (comparison)	Coarser resolution; less precise	https://www.earthdata.nasa.gov/data/instruments/srtm, accessed on 7 May 2025
Sentinel-2 Imagery	10 m resolution (Free)	Land use/land cover classification	Key for understanding urbanization	https://www.esa.int/Applications/Observing_the_Earth/Copernicus/Sentinel-2, accessed on 7 May 2025
GISAI and GHSL Data	Varies	Impervious surface area mapping	Important for runoff analysis	https://data.jrc.ec.europa.eu/collection/ghsl, accessed on 7 May 2025
Precipitation Data	Government sources (Monthly)	Rainfall intensity and variability	Identifies rainfall patterns	https://nwfc.pmd.gov.pk/new/rainfall.php, accessed on 7 May 2025
Slope and Aspect	Derived from DEM	Terrain analysis	Determines water flow direction	Derived
Historical UPF Events	2–3 past events (Synthetic data included)	DRNN model training	Captures temporal flood patterns	https://www.ndma.gov.pk/urbanflooding/sdi, accessed on 7 May 2025

These datasets collectively form the foundation for both traditional and deep learning-based flood susceptibility modeling. The use of high-resolution DEM data ensures precise drainage network extraction, while artificial data generated through GANs helps mitigate the challenges posed by limited historical flood data. This integrated approach ensures that the models developed are robust, accurate, and capable of addressing the complexities of urban pluvial flooding in data-scarce regions.

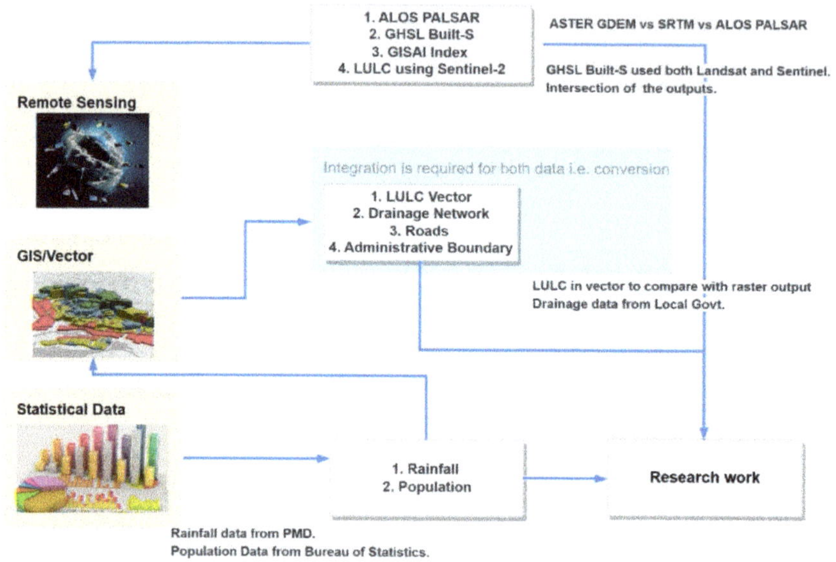

Figure 1. Overview of the dataset.

3. Methodology

This study integrates deep learning (DL) models with traditional approaches to address the challenges of urban pluvial flooding (UPF) mapping in data-scarce regions. Lahore, Pakistan, a rapidly urbanizing city frequently impacted by monsoonal UPF, was selected as the study area. The city's inadequate drainage infrastructure and extensive impervious surfaces exacerbate its vulnerability to flooding.

The research workflow for urban pluvial flooding (UPF) mapping integrates multiple datasets and methodologies to address challenges in data-scarce regions. Figure 2 provides a detailed workflow diagram, illustrating how the modules interact and the sequence of operations.

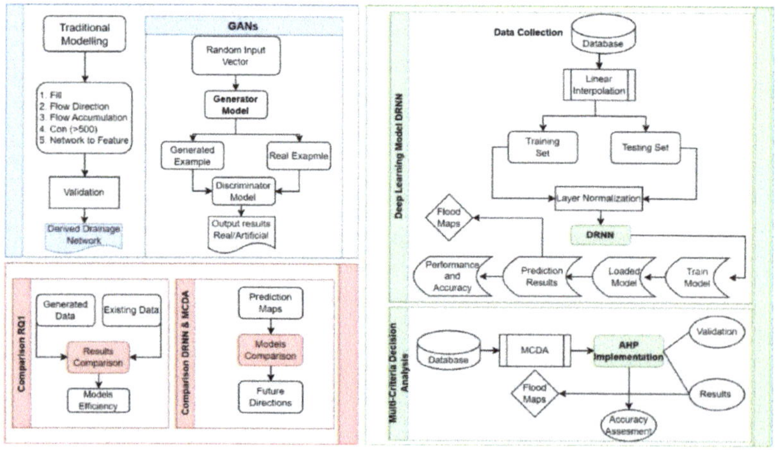

Figure 2. Workflow diagram.

3.1. Data Collection and Pre-Processing

Key datasets included the high-resolution ALOS PALSAR Digital Elevation Model (DEM) with a 12.5 m spatial resolution, utilized for natural drainage network extraction. Sentinel-2 imagery facilitated land use/land cover (LULC) classification, while GISAI and GHSL datasets provided information on impervious surface mapping. Additionally, precipitation data spanning from 2001 to 2022 were obtained from government sources for flood analysis. These datasets served as foundational inputs for both traditional and deep learning-based modeling approaches.

The ALOS PALSAR DEM data were processed using ArcGIS Pro 3.0 to extract natural drainage networks. Hydrological tools generated flow direction and accumulation maps, identifying potential flooding hotspots. These outputs also served as validation data for the deep learning models.

3.2. Synthetic Data Generation with GANs

To address data scarcity, Generative Adversarial Networks (GANs) were employed to generate synthetic drainage network data (Figure 3). GANs consist of two neural networks: a generator that creates synthetic data and a discriminator that evaluates the authenticity of the generated data. This adversarial training process enables the generation of realistic synthetic datasets. The synthetic drainage data produced by GANs supplemented the natural drainage data, ensuring robust inputs for the deep learning models.

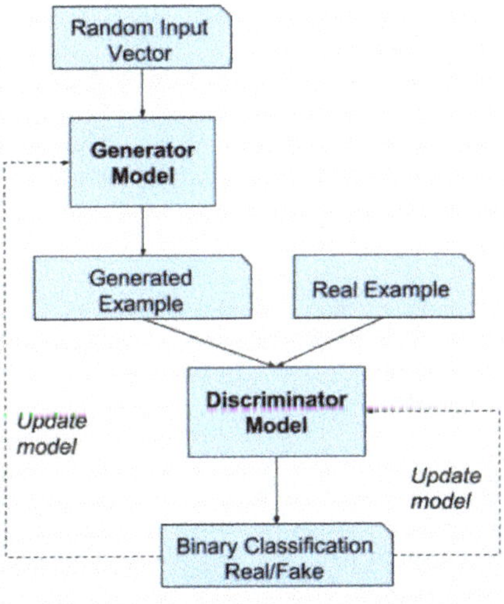

Figure 3. GANs model architecture.

3.3. Deep Learning Model (DRNN)

The Deep Recurrent Neural Network (DRNN) was utilized to model temporal dependencies using historical precipitation data and synthetic drainage data from GANs. The DRNN architecture comprised multiple stacked RNN layers and incorporated techniques such as layer normalization and Leaky Rectified Linear Unit (Leaky ReLU) activation functions to enhance prediction accuracy for flood-prone areas.

3.4. Multi-Criteria Decision Analysis (MCDA)

In parallel, the Analytical Hierarchy Process (AHP), a traditional Multi-Criteria Decision Analysis (MCDA) technique, was applied. This approach integrated multiple factors, including rainfall intensity, drainage density, and impervious surface area, to assess flood susceptibility zones. The AHP involved constructing a pairwise comparison matrix to assign weights to each criterion based on their relative importance. The MCDA results served as a benchmark for comparison with the deep learning outputs.

3.5. Hyper-Parameter Optimization

Hyper-parameter optimization was conducted for both the DRNN and GAN models to enhance predictive accuracy and model stability. For the DRNN, a grid search was performed over the following ranges: number of hidden layers (1–3), neurons per layer (50–200), learning rate (0.001–0.01), and dropout rate (0.1–0.5). The optimal configuration was determined based on validation loss and predictive performance. For the GAN, training epochs (100–500), batch size (32–128), and learning rates (0.0001–0.001) for both the generator and discriminator were varied. Final settings were selected based on the quality of generated data, assessed using Structural Similarity Index Measure (SSIM) and Root Mean Square Error (RMSE) metrics. This systematic tuning process ensured the robustness and reliability of the models.

4. Results

4.1. Impervious Surface Area (ISA) and Land Use/Land Cover (LULC) Analysis

Urbanization significantly influences urban pluvial flooding (UPF) by altering natural drainage systems and increasing impervious surface areas. Sentinel-2 imagery was used to classify land use and land cover (LULC), identifying key categories such as urban, vegetation, and water bodies. The results indicate that rapid urban expansion has led to a significant increase in impervious surfaces, reducing the natural infiltration capacity and exacerbating surface runoff. The integration of GISAI and GHSL datasets provided an accurate delineation of impervious surface areas (ISA), revealing that high-density urban zones correspond with the most flood-prone regions. Figure 4 illustrates the classified LULC map, while Figure 4 highlights impervious surface distribution, showing the temporal variation for last three decades.

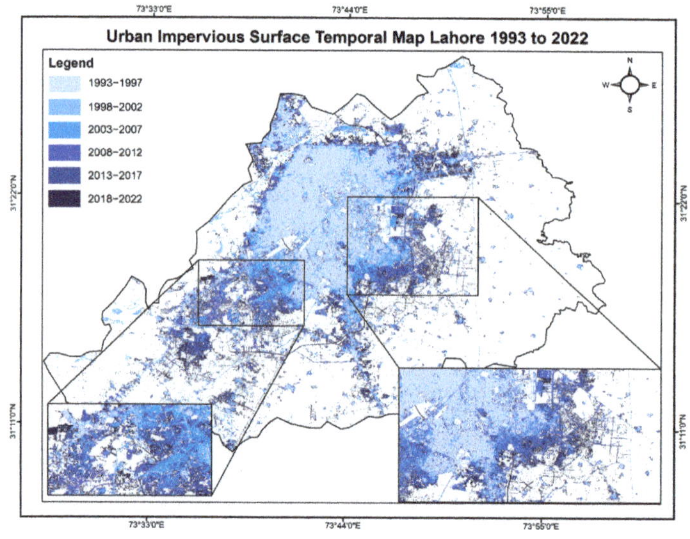

Figure 4. Impervious surface area combined through GISAI data.

The preliminary results of this study provide key insights into impervious surface area (ISA), land use/land cover (LULC), precipitation patterns, and drainage network derivation, each contributing to a comprehensive understanding of urban pluvial flooding (UPF). The impervious surface area was delineated by combining GISAI and GHSL datasets, using an intersection-based approach to yield an accurate representation of impervious surfaces in the study area. Figure 5, shows the land use map 2022 with four majorland use types.

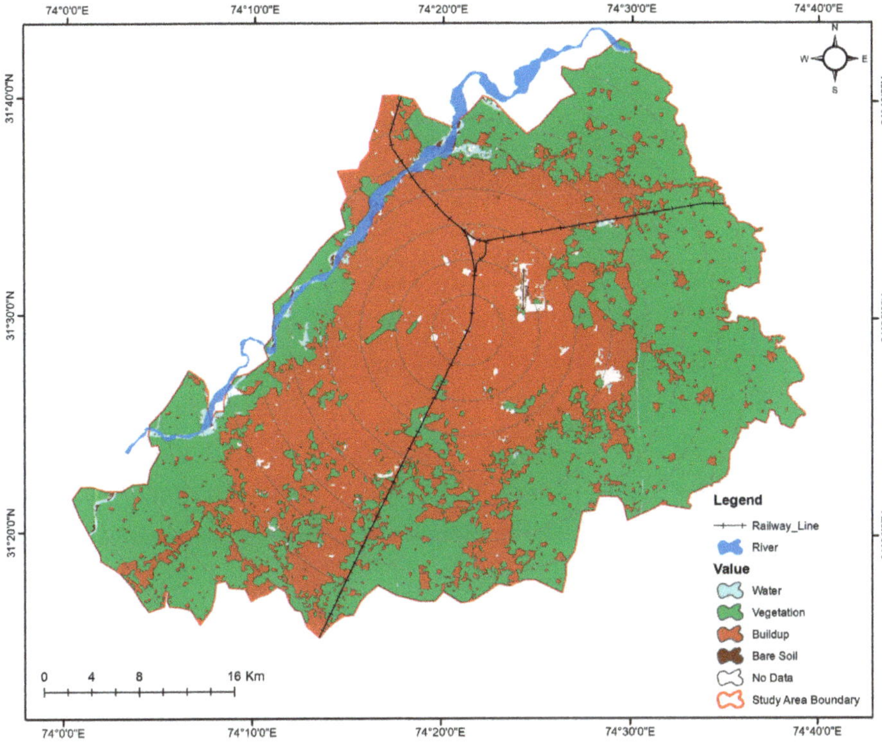

Figure 5. LULC Map Lahore 2022.

4.2. Drainage Network Extraction and Flood Hotspots Identification

The natural drainage network was extracted using ALOS PALSAR DEM data, processed through hydrological tools in ArcGIS Pro. Flow direction and accumulation analysis were performed to delineate natural water movement and identify potential drainage pathways. The results indicate that low-lying areas with poor drainage infrastructure are more prone to flooding, particularly in rapidly urbanizing districts. Figure 6 highlights flood hotspots where the drainage system is insufficient to accommodate heavy precipitation events. These findings emphasize the urgent need for improved drainage infrastructure in high-risk urban zones highlighted by red color in map.

Rainfall data spanning from 2001 to 2022 were normalized for each month and analyzed (Figure 7), revealing an increase in precipitation intensity during monsoon seasons. These findings underscore the heightened vulnerability of urban areas to pluvial flooding due to climatic changes, which align with trends observed in other developing regions.

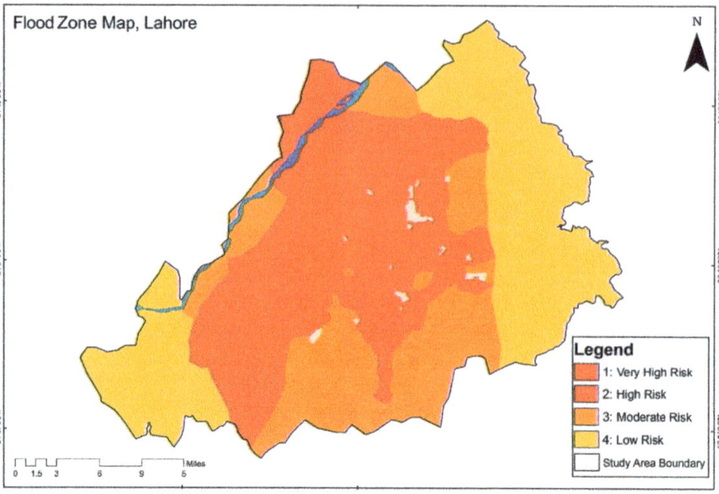

Figure 6. DRNNs based flood Zones and flood susceptibility.

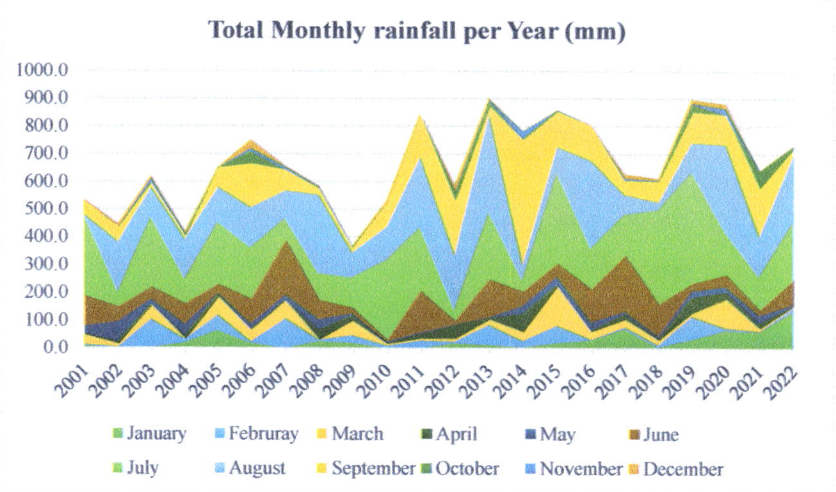

Figure 7. Precipitation Trends from 2001 to 2022.

4.3. Synthetic Data Generation with GANs and Data Quality Assessment

Due to the scarcity of high-resolution historical flood data, Generative Adversarial Networks (GANs) were implemented to generate synthetic drainage networks. The GAN model was trained using real drainage data, enabling it to produce synthetic datasets that closely mimic natural hydrological patterns.

The GAN was trained using real drainage networks extracted from ALOS PALSAR DEM data for multiple urban sectors within Lahore. The architecture followed a Deep Convolutional GAN (DCGAN) structure. The generator consisted of transposed convolutional layers with batch normalization and ReLU activations, while the discriminator was built with convolutional layers using Leaky ReLU and dropout for regularization. Binary cross-entropy was used as the loss function for both networks, optimized using the Adam optimizer (learning rate = 0.0002, $\beta_1 = 0.5$). The model was trained for 200 epochs with a batch size of 64 and image input size of 128×128 pixels. These configurations were selected based on standard best practices and preliminary tuning to ensure model stability and convergence.

To evaluate the quality of the generated data, a comparison was made between real and synthetic drainage networks using the structural similarity index measure (SSIM) and root mean square error (RMSE). The results indicate that the GAN-generated data achieved an SSIM score of 0.92 and an RMSE of 0.14, demonstrating high similarity to real-world data.

Figure 8 provides a visual comparison of real and synthetic drainage networks, while Table 2 presents validation metrics confirming the reliability of GAN-generated data. These results indicate that GANs can effectively supplement missing drainage information in data-scarce regions, enhancing the accuracy of UPF modeling.

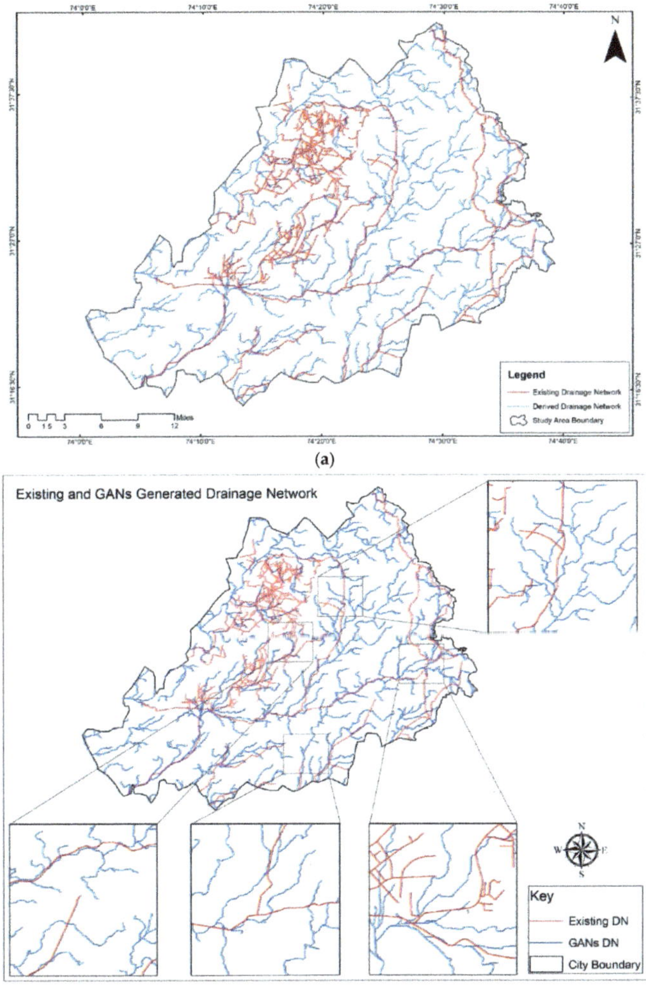

(a)

(b)

Figure 8. (**a**,**b**) Derived Natural Drainage Network.

Table 2. Validation metrics of GAN-generated data.

Metric	Value	Description
SSIM (Structural Similarity Index Measure)	0.92	Measures similarity between real and synthetic data (closer to 1 indicates high similarity).
RMSE (Root Mean Square Error)	0.14	Indicates average deviation of synthetic data from real data (lower is better).
MAE (Mean Absolute Error)	0.11	Measures average absolute difference between real and synthetic data.
Data Coverage (%)	95%	Percentage of drainage features accurately captured in synthetic data.

In addition to SSIM and RMSE, physical consistency checks were performed. The drainage network characteristics, including flow direction and accumulation patterns, were compared with those derived independently from high-resolution DEM data. These checks confirmed that the GAN-generated drainage networks closely match the physical attributes expected in natural drainage systems. Furthermore, comparative experiments were conducted by validating the synthetic drainage networks against manually extracted networks from an independent dataset, demonstrating comparable performance across key metrics.

4.4. Deep Learning Model (DRNN) Results for Flood Prediction

To predict flood-prone areas, a Deep Recurrent Neural Network (DRNN) was trained using precipitation data, LULC, and drainage network information. The model successfully identified high-risk flooding zones by capturing spatiotemporal dependencies in rainfall intensity and land characteristics. Model evaluation metrics showed an overall precision of 85%, a recall of 83%, and an F1-score of 84%, indicating a robust predictive performance. The inclusion of GAN-generated synthetic data improved the model's accuracy by 7%, compared to models trained solely on real data. Figure 8 illustrates the drainage network generated by DRNN, highlighting vulnerable urban areas, while Table 3 (a) summarizes the model's performance metrics.

Table 3. (**a**) DRNN performance metrics. (**b**) DRNN regression metrics.

(a)

Metric	Real Data Only	Real + Synthetic Data	Improvement (%)
Precision	78%	85%	+7%
Recall	76%	83%	+7%
F1-Score	75%	84%	+9%
Accuracy	77%	86%	+9%

(b)

Metric	Value
MSE	0.087
RMSE	0.295
R^2	0.81

These results confirm the effectiveness of deep learning approaches in UPF prediction, particularly when combined with synthetic data.

In addition to classification metrics, we evaluated the DRNN model using regression-based performance indicators to better reflect continuous flood susceptibility prediction. The model achieved a Mean Squared Error (MSE) of 0.087, a Root Mean Squared Error (RMSE) of 0.295, and an R^2 score of 0.81, indicating a strong fit between predicted and reference data (Table 3 (b)). These results confirm the model's capability to generalize well across different spatial conditions and flooding intensities.

4.5. Comparison of DRNN and Multi-Criteria Decision Analysis (MCDA) Results

To evaluate the effectiveness of the DRNN model, its results were compared with those from the traditional Multi-Criteria Decision Analysis (MCDA) approach. While both models successfully identified major flood-prone zones, the DRNN model exhibited higher spatial precision and adaptability to varying precipitation patterns. The comparison revealed that DRNN outperformed MCDA in capturing localized flood susceptibility, particularly in areas where historical flood data were scarce. Table 4 presents a side-by-side comparison of DRNN and MCDA and quantifies their respective performance metrics. Expert validation further confirmed that DRNN predictions aligned more closely with observed flood events. These findings highlight the advantages of deep learning in dynamic flood mapping, particularly for regions with complex urban hydrology.

Table 4. DRNN vs. MCDA performance metrics.

Metric	DRNN	MCDA	Difference
Precision	85%	78%	+7%
Recall	83%	75%	+8%
F1-Score	84%	76%	+8%
Area Under Curve (AUC)	0.88	0.81	+0.07

4.6. Uncertainties, Errors, and Model Limitations

While the results demonstrate the potential of deep learning for UPF mapping, several uncertainties and limitations must be acknowledged. The accuracy of the models is dependent on the quality and resolution of input datasets, such as DEMs, LULC classifications, and precipitation records. Additionally, while GAN-generated synthetic data improved model performance, biases may arise if the training dataset does not adequately capture the full variability of real-world drainage systems. Overfitting in DRNN models is another concern, particularly when trained on limited historical flood data. Furthermore, the computational cost of training deep learning models may be a barrier for real-time flood prediction applications in resource-constrained environments.

To mitigate these limitations, future research should focus on incorporating higher-resolution datasets, expanding GAN training with diverse urban flood scenarios, and testing model generalizability in different geographic regions. Table 5 summarizes the key sources of uncertainty in this study and their potential impact on model predictions.

A key limitation of this study is the exclusive focus on natural drainage networks derived from DEM data. Due to the unavailability of detailed maps and data on artificial drainage systems (e.g., stormwater infrastructure), these features were not incorporated into the model. However, surface permeability was partially addressed using impervious surface maps from GISAI and GHSL datasets, which capture urbanization patterns. Additionally, long-term normalized precipitation data were used to reflect seasonal and interannual rainfall variability. Future research should aim to integrate artificial drainage infrastructure and simulate dynamic hydrological processes using rainfall-runoff models to improve realism in urban flood modeling.

Table 5. Summary of model uncertainties and errors.

Source of Uncertainty/Error	Impact on Results	Mitigation Strategies
Input Data Resolution (DEM, LULC)	Reduced accuracy in small-scale flood mapping	Use higher-resolution datasets (e.g., LiDAR)
GAN Data Biases	Potential misclassification due to synthetic data inaccuracy	Expand GAN training dataset diversity
DRNN Overfitting	Reduced model generalization to new areas	Implement regularization and cross-validation
Precipitation Data Aggregation	Loss of short-term rainfall variability	Integrate real-time rainfall data
Computational Cost	Limited application in resource-constrained settings	Develop optimized, lightweight model versions

A notable limitation of this study is the reliance on remote sensing and simulation data, necessitated by the unavailability of in situ water level measurements and detailed historical flood records in the study area. The integration of such observed data is crucial for improving model calibration and validation. Future research should aim to incorporate actual hydrological measurements to enhance the model's accuracy and reliability in flood risk assessment.

These insights will be crucial in refining deep learning methodologies for improved flood risk assessment and urban resilience planning.

5. Discussion

This study demonstrates the efficacy of integrating deep learning (DL) models with traditional methods for urban pluvial flooding (UPF) assessment in data-scarce regions. The application of Deep Convolutional Generative Adversarial Networks (DCGANs) to simulate artificial drainage networks effectively addresses data scarcity challenges. When combined with Digital Elevation Model (DEM)-based natural drainage mapping, this hybrid approach offers a comprehensive framework for developing robust and adaptive flood mitigation strategies.

Comparative analyses between DL models, particularly Deep Recurrent Neural Networks (DRNNs), and traditional Multi-Criteria Decision Analysis (MCDA) techniques, such as the Analytical Hierarchy Process (AHP), indicate the superior accuracy and robustness of DRNNs. These findings align with previous research highlighting the advantages of DL methods in managing complex environmental and urban systems [23,36].

Urbanization and climatic changes are critical factors exacerbating UPF risks. The study underscores how increased impervious surfaces and outdated drainage systems heighten flooding vulnerabilities, necessitating innovative, data-driven approaches. Additionally, intensified monsoon precipitation due to climatic shifts further complicates flood management, calling for adaptive solutions.

Despite promising outcomes, data scarcity remains a significant hurdle, particularly for training DL models [24–28]. The generation of synthetic datasets using GANs and their integration into MCDA frameworks present a viable path forward. Future work will focus on validating these models with real-world flood event data and extending the methodology to other data-scarce regions globally.

Implementing DL models for UPF is challenging due to complex interactions within natural and built environments, such as drainage systems and urban infrastructure. Data limitations in the study area encompass both hydraulic datasets and historical UPF event records. This study addresses these gaps by developing synthetic data to train DL models, enhancing their predictive capabilities.

The integration of GANs and DRNNs enhances UPF prediction in data-scarce regions. GANs effectively generate synthetic drainage network data, filling critical data gaps and enabling more accurate flood susceptibility modeling. These improved predictions can inform urban planning decisions and the development of early warning systems, contributing to more effective flood mitigation strategies.

However, several uncertainties affect the reliability of model predictions. The accuracy of results depends on the quality and resolution of input datasets, such as ALOS PALSAR DEM and Sentinel-2 imagery. Limitations inherent in these datasets, like resolution constraints and potential misclassification in land use/land cover (LULC) data, can introduce errors into both GAN-generated synthetic data and DRNN flood predictions. Moreover, synthetic data may introduce biases if training data do not fully capture the variability of natural drainage patterns.

Potential sources of error include: (1) limitations in input data, such as DEM spatial resolution and LULC classification accuracy, affecting the fidelity of extracted drainage networks; (2) GAN-generated data may not perfectly replicate real-world drainage system diversity, leading to model bias; (3) limited historical flood events for training increase the risk of overfitting in the DRNN model, reducing generalizability; and (4) temporal aggregation of precipitation data into broader intervals may obscure short-term rainfall extremes critical for flash flood prediction.

To mitigate these uncertainties, future research should incorporate higher-resolution datasets, such as LiDAR-based DEMs, and expand the diversity of GAN training datasets. Integrating real-time data could enhance the dynamic performance of DRNN models, refining predictive capabilities and reducing uncertainties. These improvements are crucial for ensuring the models' applicability across diverse urban environments with varying hydrological and urbanization characteristics.

The proposed framework, integrating GANs with DRNNs, offers a promising solution for UPF prediction in data-scarce regions. By generating synthetic drainage network data, it enhances flood risk assessment, benefiting rapidly urbanizing cities with inadequate drainage infrastructure. The model's adaptability allows integration into urban planning and early warning systems.

However, the framework's accuracy depends on input data quality, such as DEMs and LULC classifications, which may introduce errors. While GAN-generated data address data scarcity, they may also introduce biases if training data lack variability [37]. Additionally, the high computational cost of DRNNs may limit applicability in resource-constrained settings, and generalization to different urban environments may require recalibration. Future research should focus on improving data resolution, expanding GAN training diversity, and optimizing model efficiency. Testing the framework in diverse urban contexts will enhance its robustness and practical applicability.

6. Conclusions

Urban pluvial flooding (UPF) remains a pressing challenge, particularly in regions where data scarcity hampers effective risk assessment and management. This study introduces an integrative framework that combines traditional methodologies with advanced deep learning (DL) techniques to address these challenges. By leveraging high-resolution ALOS PALSAR DEM data for drainage network extraction and employing Generative Adversarial Networks (GANs) for synthetic data generation, the approach effectively mitigates data limitations. Furthermore, the incorporation of Deep Recurrent Neural Networks (DRNNs) enhances flood susceptibility mapping by capturing temporal patterns in precipitation and urban features.

The findings indicate that DL models, particularly DRNNs augmented with GAN-generated datasets, surpass traditional Multi-Criteria Decision Analysis (MCDA) methods in both accuracy and robustness. This underscores the compounded impacts of urbanization and climate change on UPF risks and highlights the necessity for innovative, data-driven solutions.

Beyond technical advancements, the study offers actionable insights for urban planners. Integrating model outputs into planning strategies—such as targeted infrastructure improvements and land-use regulations—can significantly reduce UPF risks and bolster long-term resilience in vulnerable areas.

Future research should focus on validating the proposed models with real-world flood events and assessing their applicability across diverse urban environments. Expanding datasets to include real-time monitoring data and enhancing model generalizability across varying hydrological settings will further strengthen the framework's reliability. By providing a scalable, data-driven solution for proactive flood management, this framework contributes to the development of more resilient and sustainable urban landscapes.

Author Contributions: Conceptualization, F.I.; Validation, H.D.S. and F.I.; Formal analysis, M.N.A. and A.J.; Investigation, A.J.; Resources, A.J.; Writing—original draft, M.N.A.; Writing—review & editing, M.N.A. and G.D.B.; Visualization, H.D.S.; Supervision, M.N.A. and G.D.B. All authors have read and agreed to the published version of the manuscript.

Funding: This research received no funding, APC will be covered by authors.

Institutional Review Board Statement: Not applicable.

Informed Consent Statement: Not applicable.

Data Availability Statement: The data presented in this study are available on request from the corresponding author.

Conflicts of Interest: The authors declare no conflict of interest.

References

1. Luo, X.; Hu, Z.; Liu, L. Investigating the Seasonal Dynamics of Surface Water over the Qinghai–Tibet Plateau Using Sentinel-1 Imagery and a Novel Gated Multiscale ConvNet. *Int. J. Digit. Earth* **2023**, *16*, 1373–1395. [CrossRef]
2. Qiu, J.; Cao, B.; Park, E.; Yang, X.; Zhang, W.; Tarolli, P. Flood Monitoring in Rural Areas of the Pearl River Basin (China) Using Sentinel-1 SAR. *Remote Sens.* **2021**, *13*, 1384. [CrossRef]
3. Mason, D.C.; Dance, S.L.; Cloke, H.L. Floodwater Detection in Urban Areas Using Sentinel-1 and WorldDEM Data. *J. Appl. Remote Sens.* **2021**, *15*, 32003. [CrossRef]
4. Jonkman, S.N.; Vrijling, J.K. Loss of Life Due to Floods. *J. Flood Risk Manag.* **2008**, *1*, 43–56. [CrossRef]
5. Zhao, G.; Pang, B.; Xu, Z.; Yue, J.; Tu, T. Mapping Flood Susceptibility in Mountainous Areas on a National Scale in China. *Sci. Total Environ.* **2018**, *615*, 1133–1142. [CrossRef]
6. Huang, M.; Jin, S. Rapid Flood Mapping and Evaluation with a Supervised Classifier and Change Detection in Shouguang Using Sentinel-1 SAR and Sentinel-2 Optical Data. *Remote Sens.* **2020**, *12*, 2073. [CrossRef]
7. Jonkman, S.N. Global Perspectives on Loss of Human Life Caused by Floods. *Nat. Hazards* **2005**, *34*, 151–175. [CrossRef]
8. Li, Y.; Martinis, S.; Wieland, M. Urban Flood Mapping with an Active Self-Learning Convolutional Neural Network Based on TerraSAR-X Intensity and Interferometric Coherence. *ISPRS J. Photogramm. Remote Sens.* **2019**, *152*, 178–191. [CrossRef]
9. Zhang, H.; Qi, Z.; Li, X.; Chen, Y.; Wang, X.; He, Y. An Urban Flooding Index for Unsupervised Inundated Urban Area Detection Using Sentinel-1 Polarimetric SAR Images. *Remote Sens.* **2021**, *13*, 4511. [CrossRef]
10. Mignot, E.; Li, X.; Dewals, B. Experimental Modelling of Urban Flooding: A Review. *J. Hydrol.* **2019**, *568*, 334–342. [CrossRef]
11. Tavus, B.; Kocaman, S.; Gokceoglu, C.; Nefeslioglu, H.A. Considerations on the Use Of Sentinel-1 Data in Flood Mapping in Urban Areas: Ankara (Turkey) 2018 Floods. *Int. Arch. Photogramm. Remote Sens. Spat. Inf. Sci.* **2018**, *XLII-5*, 575–581. [CrossRef]
12. Chini, M.; Pelich, R.; Pulvirenti, L.; Pierdicca, N.; Hostache, R.; Matgen, P. Sentinel-1 InSAR Coherence to Detect Floodwater in Urban Areas: Houston and Hurricane Harvey as a Test Case. *Remote Sens.* **2019**, *11*, 107. [CrossRef]

13. Tanim, A.H.; McRae, C.B.; Tavakol-Davani, H.; Goharian, E. Flood Detection in Urban Areas Using Satellite Imagery and Machine Learning. *Water* **2022**, *14*, 1140. [CrossRef]
14. Li, C.; Sun, N.; Lu, Y.; Guo, B.; Wang, Y.; Sun, X.; Yao, Y. Review on Urban Flood Risk Assessment. *Sustainability* **2023**, *15*, 765. [CrossRef]
15. Tanaka, T.; Kiyohara, K.; Tachikawa, Y. Comparison of Fluvial and Pluvial Flood Risk Curves in Urban Cities Derived from a Large Ensemble Climate Simulation Dataset: A Case Study in Nagoya, Japan. *J. Hydrol.* **2020**, *584*, 124706. [CrossRef]
16. Peng, J.; Zhang, J. Urban Flooding Risk Assessment Based on GIS-Game Theory Combination Weight: A Case Study of Zhengzhou City. *Int. J. Disaster Risk Reduct.* **2022**, *77*, 103080. [CrossRef]
17. Parvin, F.; Ali, S.A.; Calka, B.; Bielecka, E.; Linh, N.T.T.; Pham, Q.B. Urban Flood Vulnerability Assessment in a Densely Urbanized City Using Multi-Factor Analysis and Machine Learning Algorithms. *Theor. Appl. Climatol.* **2022**, *149*, 639–659. [CrossRef]
18. Ye, C.; Xu, Z.; Lei, X.; Liao, W.; Ding, X.; Liang, Y. Assessment of Urban Flood Risk Based on Data-Driven Models: A Case Study in Fuzhou City, China. *Int. J. Disaster Risk Reduct.* **2022**, *82*, 103318. [CrossRef]
19. Wu, Z.; Xue, W.; Xu, H.; Yan, D.; Wang, H.; Qi, W. Urban Flood Risk Assessment in Zhengzhou, China, Based on a D-Number-Improved Analytic Hierarchy Process and a Self-Organizing Map Algorithm. *Remote Sens.* **2022**, *14*, 4777. [CrossRef]
20. Li, J.; Mahalov, A.; Hyde, P. Effects of Urbanization on Extreme Rainfall in an Arid/Semiarid Region. *Atmos. Sci. Lett.* **2020**, *21*, e966. [CrossRef]
21. Fowler, H.J.; Ali, H.; Allan, R.P.; Ban, N.; Barbero, R.; Berg, P.; Blenkinsop, S.; Cabi, N.S.; Chan, S.; Dale, M. Towards Advancing Scientific Knowledge of Climate Change Impacts on Short-Duration Rainfall Extremes. *Philos. Trans. R. Soc. A* **2021**, *379*, 20190542. [CrossRef] [PubMed]
22. Ahmad, M.N.; Shao, Z.; Javed, A. Mapping Impervious Surface Area Increase and Urban Pluvial Flooding Using Sentinel Application Platform (SNAP) and Remote Sensing Data. *Environ. Sci. Pollut. Res.* **2023**, *30*, 125741–125758. [CrossRef] [PubMed]
23. Bentivoglio, R.; Isufi, E.; Jonkman, S.N.; Taormina, R. Deep Learning Methods for Flood Mapping: A Review of Existing Applications and Future Research Directions. *Hydrol. Earth Syst. Sci.* **2022**, *26*, 4345–4378. [CrossRef]
24. Bulti, D.T.; Abebe, B.G. A Review of Flood Modeling Methods for Urban Pluvial Flood Application. *Model. Earth Syst. Environ.* **2020**, *6*, 1293–1302. [CrossRef]
25. Chen, J.; Chen, W.; Huang, G. Assessing Urban Pluvial Flood Resilience Based on a Novel Grid-Based Quantification Method That Considers Human Risk Perceptions. *J. Hydrol.* **2021**, *601*, 126601. [CrossRef]
26. Azizi, K.; Meier, C.I. Urban Pluvial Flood Risk Assessment: Challenges and Opportunities for Improvement Using a Community-Based Approach. In Proceedings of the World Environmental and Water Resources Congress 2021, Virtually, 7–11 June 2021; pp. 350–361.
27. Olsen, A.S.; Zhou, Q.; Linde, J.J.; Arnbjerg-Nielsen, K. Comparing Methods of Calculating Expected Annual Damage in Urban Pluvial Flood Risk Assessments. *Water* **2015**, *7*, 255–270. [CrossRef]
28. Löwe, R.; Urich, C.; Domingo, N.S.; Mark, O.; Deletic, A.; Arnbjerg-Nielsen, K. Assessment of Urban Pluvial Flood Risk and Efficiency of Adaptation Options through Simulations—A New Generation of Urban Planning Tools. *J. Hydrol.* **2017**, *550*, 355–367. [CrossRef]
29. Lara-Benítez, P.; Carranza-García, M.; Luna-Romera, J.M.; Riquelme, J.C. Temporal Convolutional Networks Applied to Energy-Related Time Series Forecasting. *Appl. Sci.* **2020**, *10*, 2322. [CrossRef]
30. Faruq, A.; Arsa, H.P.; Hussein, S.F.M.; Razali, C.M.C.; Marto, A.; Abdullah, S.S. Deep Learning-Based Forecast and Warning of Floods in Klang River, Malaysia. *Ingénierie Systèmes Inf.* **2020**, *25*, 365–370. [CrossRef]
31. Bai, S.; Kolter, J.Z.; Koltun, V. An Empirical Evaluation of Generic Convolutional and Recurrent Networks for Sequence Modeling. *arXiv* **2018**, arXiv:1803.01271.
32. Siami-Namini, S.; Namin, A.S. Forecasting Economics and Financial Time Series: ARIMA vs. LSTM. *arXiv* **2018**, arXiv:1803.06386.
33. Gude, V.; Corns, S.; Long, S. Flood Prediction and Uncertainty Estimation Using Deep Learning. *Water* **2020**, *12*, 884. [CrossRef]
34. Song, T.; Ding, W.; Wu, J.; Liu, H.; Zhou, H.; Chu, J. Flash Flood Forecasting Based on Long Short-Term Memory Networks. *Water* **2019**, *12*, 109. [CrossRef]
35. Le, X.-H.; Ho, H.V.; Lee, G.; Jung, S. Application of Long Short-Term Memory (LSTM) Neural Network for Flood Forecasting. *Water* **2019**, *11*, 1387. [CrossRef]

36. Ali, M.H.M.; Asmai, S.A.; Abidin, Z.Z.; Abas, Z.A.; Emran, N.A. Flood Prediction Using Deep Learning Models. *Int. J. Adv. Comput. Sci. Appl.* **2022**, *13*. [CrossRef]
37. Goodfellow, I.; Pouget-Abadie, J.; Mirza, M.; Xu, B.; Warde-Farley, D.; Ozair, S.; Courville, A.; Bengio, Y. Generative Adversarial Networks. *Commun. ACM* **2020**, *63*, 139–144. [CrossRef]

sustainability

MDPI

Article

Long-Term NDVI Trends and Vegetation Resilience in a Seismically Active Debris Flow Watershed: A Case Study from the Wenchuan Earthquake Zone

Wen Zhang [1,2], Zelin Wang [3,*], Minghui Meng [4], Tiantao Li [3], Jian Guo [5], Dong Sun [4], Liang Qin [4], Xiaoya Xu [3] and Xiaoyu Shen [3]

1 Technology Innovation Center for Risk Prevention and Mitigation of Geohazard, Ministry of Natural Resources, Chengdu 611734, China; zhangwen@mail.cgs.gov.cn
2 Institute of Exploration Technology, Chinese Academy of Geological Sciences, Chengdu 611734, China
3 State Key Laboratory of Geohazard Prevention and Geo-Environment Protection, Chengdu University of Technology, Chengdu 610059, China; litiantao18@cdut.edu.cn (T.L.); 13881718927@163.com (X.X.); 18599070257@163.com (X.S.)
4 Sichuan Geological Environment Survey and Research Center, Chengdu 610081, China; mengmh1990@foxmail.com (M.M.); sundong2003@126.com (D.S.); qinliang1988.good@163.com (L.Q.)
5 Department of Civil Engineering, Panzhihua University, Panzhihua 617000, China; 17844627712@163.com
* Correspondence: wjywcbl1@gmail.com

Abstract: Vegetation restoration in seismically active regions involves complex interactions between geological hazards and ecological processes. Understanding the spatiotemporal patterns of vegetation recovery is critical for assessing disaster evolution, evaluating mitigation effectiveness, and guiding ecological resilience planning. This study investigates post-earthquake vegetation dynamics in the Chutou Gully watershed, located in the 12 May 2008 Wenchuan earthquake zone, using NDVI data from 2000 to 2022. Results reveal a sharp decline in vegetation cover following the earthquake, followed by a steady recovery trend, with NDVI values projected to return to pre-earthquake levels by 2030. Degradation was concentrated in debris flow channels, while more stable adjacent slopes exhibited stronger recovery. Over time, the area of poorly restored vegetation significantly declined, indicating increased ecosystem resilience. The findings highlight the need for site-specific ecological restoration strategies tailored to localized recovery conditions. This study provides valuable insights for disaster mitigation agencies, ecological planners, and local governments working in mountainous hazard-prone regions, and contributes to the long-term sustainability of ecosystems in disaster-prone areas.

Keywords: NDVI; spatiotemporal dynamics; vegetation recovery; trend analysis; geological hazards

1. Introduction

Geological hazards represent a major global threat to ecological security, often initiating cascading secondary disasters such as vegetation degradation, soil erosion, and hydrological disruption [1,2]. These processes can cause persistent instability in regional ecosystems. Common geological hazards include earthquakes, landslides, debris flows, and rockfalls, all of which are prevalent in tectonically active and topographically complex mountain regions [3–5]. In southwestern China, the interplay of active fault systems, steep slopes, and concentrated monsoonal precipitation fosters a high incidence of such hazards [6,7].

Among these, debris flows—a type of rapid mass movement involving saturated soil, rock, and organic material—are particularly destructive due to their sudden onset, high mobility, and capacity for repeated occurrence. They are often triggered by intense rainfall events, particularly in landscapes already destabilized by prior seismic activity or human disturbance [8]. Following the 2008 Wenchuan earthquake, vast quantities of loose deposits were generated in the alpine gorge zones of western Sichuan. These unstable materials, under repeated rainfall stimulation, have formed a typical "earthquake–landslide–debris flow" hazard chain [9]. This compounded hazard regime has led to complex, nonlinear patterns of vegetation destruction and recovery, posing challenges for long-term ecological rehabilitation [10].

In such high-risk areas, monitoring vegetation recovery through spatiotemporal analyses is critical. Understanding vegetation dynamics not only helps reveal ecosystem resilience, but it is also crucial for designing sustainable land-use strategies and ecological restoration in hazard-prone regions [11,12].

Remote sensing technology has played an increasingly important role in vegetation monitoring in recent years. Owing to its sensitivity to vegetation cover and biomass, the normalized difference vegetation index (NDVI) has been widely used to quantify vegetation dynamics across various ecosystems [13,14]. Numerous studies have explored post-disaster vegetation recovery. For example, Yang et al. [15] identified linkages between vegetation change and landslide activity in the Wenchuan earthquake zone; Zhang et al. [16] analyzed the long-term influence of debris flow evolution on vegetation restoration via a decade of observations; and Wang et al. [17] investigated the feedback of vegetation growth on slope erosion induced by debris flows. However, existing research has focused largely on the spatiotemporal features of hazards or short-term ecological responses, with insufficient attention given to long-term vegetation recovery via multi-scale and integrative approaches. Moreover, many studies rely on single statistical methods (e.g., linear regression or basic trend analysis), which often fail to capture the nonlinear characteristics and spatial heterogeneity of vegetation dynamics. In regions with frequent cascading hazards, the coupling mechanisms between secondary disaster disturbances and natural vegetation recovery remain poorly understood, limiting the precision of ecological restoration strategies [18].

The Chutou Gully watershed, located in the core rupture zone of the 2008 Wenchuan earthquake, has experienced repeated seismic activity and frequent debris flows, making it a representative area for studying vegetation response under compound hazard conditions [19,20]. Despite this, systematic research linking vegetation recovery with the spatial distribution of geological hazards and the role of mitigation engineering remains limited. In light of the limitations in existing studies—particularly the lack of long-term, multi-scale, and integrative analyses—this study conducts a comprehensive assessment of vegetation recovery using Landsat NDVI data from 2000 to 2022. Compared to previous work, our study offers the following three key contributions: (1) it quantifies long-term vegetation recovery rates in a typical high-risk watershed; (2) it captures nonlinear and spatially heterogeneous recovery patterns through advanced spatiotemporal analysis; and (3) it examines the coupled effects of secondary hazard activity and protective engineering interventions on vegetation dynamics. By addressing these underexplored dimensions, the study aims to improve understanding of ecosystem resilience in seismically active regions and provide scientific evidence to inform ecological restoration, disaster risk reduction, and spatial planning in mountainous environments [21].

2. Study Area and Data Sources

2.1. Overview of the Study Area

The study area is located in Miansi town, Wenchuan County, Sichuan Province, China, within the geographic coordinates of 103°25′–103°29′ E and 31°20′–31°23′ N, and it is situated along the northern bank of the Minjiang River, within the seismically active Maowen Fault Zone (Figure 1). The watershed covers a total area of approximately 21.8 km^2 and features a typical gully morphology. The terrain generally inclines from west–north to east–south, with the main gully extending 7.8 km and exhibiting a steep average longitudinal slope of 260‰. The elevation ranges from 1152 to 4012 m, creating significant vertical relief. The area is characterized by well-developed tributary gullies and steep valley side slopes [22,23].

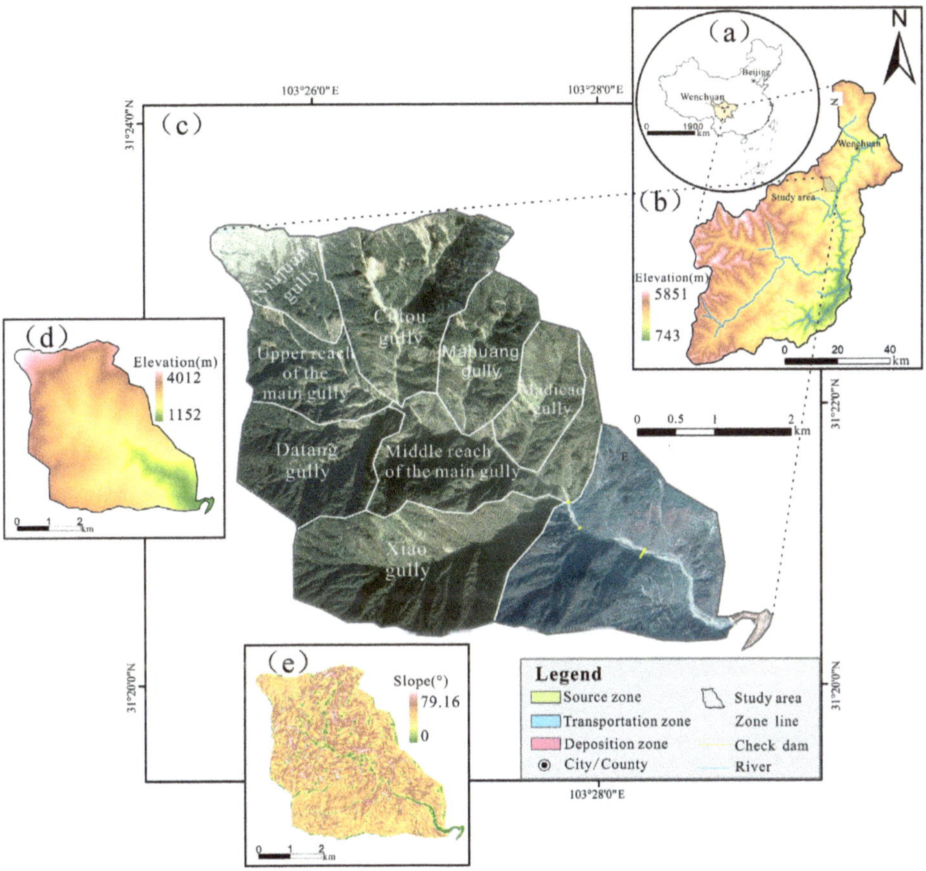

Figure 1. Location and environmental setting of the Chutou Gully watershed: (**a**) location in Wenchuan County; (**b**) location of the study area; (**c**) subregions and related information within the study area; (**d**) elevation map of the study area; (**e**) slope map of the study area.

Geologically, the region is primarily composed of granitic and dioritic lithologies. On 12 May 2008, the Wenchuan earthquake triggered widespread slope failures throughout the watershed, generating large volumes of coseismic loose deposits that subsequently provided abundant material sources for debris flow activity. The area experiences a subtropical monsoon climate, with pronounced seasonal and spatial variations in precipitation.

Approximately 60% of the annual rainfall occurs between May and September, during which the coupling of intense rainfall events with widespread loose deposits significantly amplifies debris flow hazards [24].

Under favorable geomorphic conditions and frequent high-intensity rainfall, abundant loose materials have led to pronounced post-seismic debris flow effects throughout the watershed [25]. According to field investigations and monitoring data, four large-scale debris flow events occurred between 2013 and 2020 (Figure 2), all of which were concentrated in the rainy season and exhibited strong characteristics of coseismic source material reactivation [26–28].

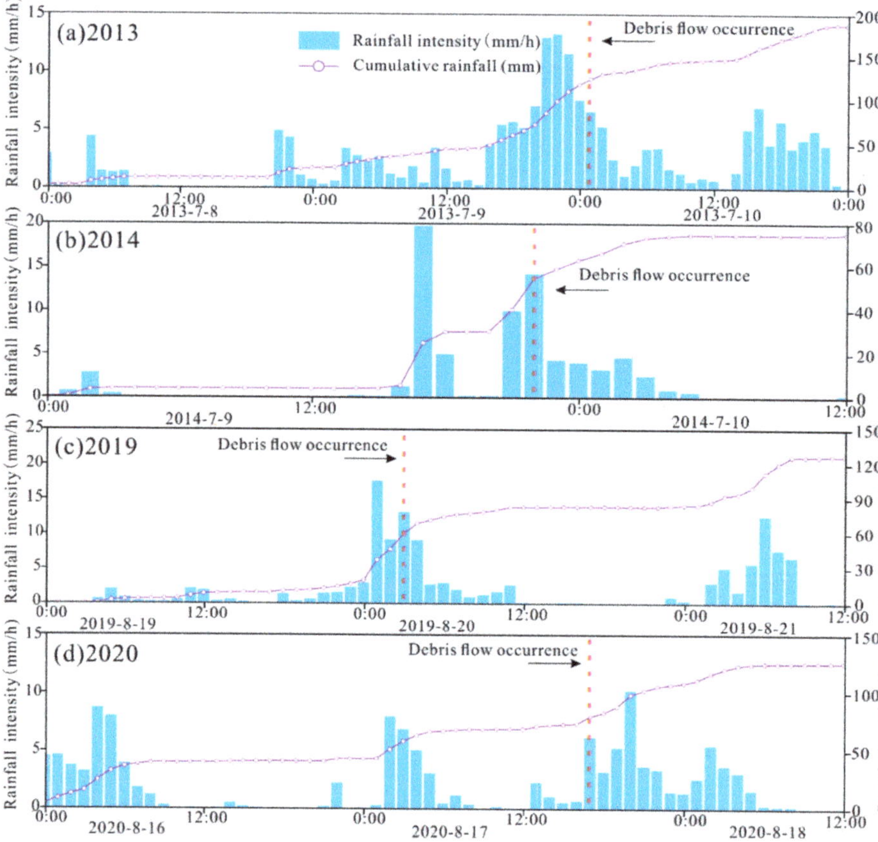

Figure 2. Hourly and cumulative rainfall during debris flow events in the Chutou Gully watershed: (**a**) 2013 event; (**b**) 2014 event; (**c**) 2019 event; (**d**) 2020 event.

2.2. Data Sources

Table 1 summarizes the spatial resolution, temporal coverage, and data sources used in this study.

A combination of remote sensing, topographic, and meteorological datasets was integrated to support spatiotemporal analysis of vegetation dynamics in the study area. (1) The NDVI data are as follows: The annual maximum NDVI at 30-m resolution (2000–2022) was obtained from the National Science & Technology Infrastructure of China. To address striping and missing values, supplementary NDVI layers were generated using the Google Earth Engine based on full-year Landsat 5/7/8/9 imagery, which were processed with cloud masking, radiometric calibration, and temporal smoothing. (2) The remote sensing imagery is as follows: Landsat 5/7/8/9 images (30 m resolution, 0–20% cloud cover) were sourced from the Geospatial Data Cloud platform and preprocessed using ENVI software for atmospheric and radiometric correction. (3) Topographic data are as follows: Elevation data from the ALOS Global Digital Surface Model (12.5 m resolution) were provided by NASA and processed in ArcGIS to generate slope information. (4) Meteorological data are as follows: Hourly precipitation data were collected from the Sichuan Provincial Meteorological Bureau.

Table 1. Remote sensing data and other relevant data used in this study.

Data	Acquisition Time	Resolution	Data Source
NDVI data	2000–2022	30 m	https://www.nesdc.org.cn/ (accessed on 21 December 2024)
High-resolution remote sensing imagery			https://www.gscloud.cn/ (accessed on 11 February 2025)
	2000–2012	30 m	Landsat5/Landsat7 (accessed on 15 January 2025)
	2013–2022	30 m	Landsat7/Landsat8/Landsat9 (accessed on 15 January 2025)
Elevation data	2022	12.5 m	https://www.earthdata.nasa.gov/ (accessed on 22 March 2025)
Slope data	2022	12.5 m	https://www.earthdata.nasa.gov/ (accessed on 22 March 2025)
Rainfall data	8–10 July 2013 9–10 July 2014 19–21 August 2019 16–18 August 2020	Hourly Hourly Hourly Hourly	Sichuan Provincial Meteorological Service (accessed on 14 January 2025)

All spatial datasets were resampled as needed, aligned to the WGS84 coordinate system, and harmonized to ensure spatial consistency throughout the analysis.

3. Research Methods

3.1. NDVI Data Processing

The NDVI is widely recognized as an effective indicator of vegetation growth status and canopy coverage, exhibiting a linear or near-linear relationship with green leaf density, photosynthetically active radiation, vegetation productivity, and plant biomass [29]. In this study, NDVI values were calculated from the selected Landsat satellite imagery using ENVI 5.6 software, and the vegetation activity levels were subsequently classified. The NDVI was calculated using the standard formula, as presented in Equation (1)

$$NDVI = \frac{NIR - \mathrm{R}}{NIR + \mathrm{R}} \tag{1}$$

where NIR represents the reflectance in the near-infrared band and R denotes the reflectance in the red band. $NDVI$ values range from -1 to 1, with values close to 1 indicating high green vegetation density, whereas values near 0 or below typically correspond to non-vegetated surfaces.

3.2. Theil–Sen Median Trend Analysis and the Mann–Kendall Test

The Theil–Sen median trend analysis method calculates the trend on the basis of clustering and rank-order statistics [30]. Compared to ordinary least squares linear regression, it offers greater robustness and reliability, especially in the presence of outliers or nonnormal data distributions. The corresponding calculation is shown in Equation (2), as follows:

$$\beta_{ij} = mean\left(\frac{NDVI_j - NDVI_i}{j - i}\right), \forall j > i \tag{2}$$

where $NDVI_j$ and $NDVI_i$ are the $NDVI$ values at time points j and i, respectively; $\beta_{ij} > 0$ indicates an increasing trend, whereas $\beta_{ij} < 0$ indicates a decreasing trend over time.

To determine whether the detected trend is statistically significant, the Mann–Kendall test is employed. This nonparametric test does not require the data to follow a normal distribution, nor does it assume a linear trend. It is also resistant to missing values and outliers, making it widely applicable in long-term trend detection in the environmental time series [31]. The test statistics are computed using the following calculation method:

$$Z = \begin{cases} \frac{S}{\sqrt{Var(S)}} (S > 0) \\ 0\ (S < 0) \\ \frac{S}{\sqrt{Var(S)}} (S < 0) \end{cases} \tag{3}$$

$$S = \sum_{i=1}^{n-1} \sum_{j=i+1}^{n} sgn(x_j - x_i) \tag{4}$$

$$sgn(x_j - x_i) = \begin{cases} +1\ x_j - x_i > 0 \\ 0\ x_j - x_i = 0 \\ -1\ x_j - x_i < 0 \end{cases} \tag{5}$$

$$Var(S) = \frac{n(n-1)(2n+5)}{18} \tag{6}$$

where Z denotes the standardized statistic derived from the Mann–Kendall trend test; where S represents the Mann–Kendall test value. The variables x_i and x_j refer to the NDVI values of a specific pixel in year i and year j, respectively. The variable n stands for the total number of years in the time series. The Sgn function is used to identify the sign of the difference between two data values.

Under the standard normal distribution, the significance of the trend is assessed via the Z statistic. When $|Z|>Z_{1-\alpha/2}$, the trend is considered statistically significant at the confidence level α. Here, $Z_{1-\alpha/2}$ represents the critical value corresponding to the desired confidence level in the standard normal distribution table. For a significance level of $\alpha = 0.05$, the change trend is significant at the 95% and 99% confidence levels when $|Z|$ exceeds the critical values of 1.96 and 2.58, respectively.

On the basis of the combination of the β and Z values, the trends are classified into different levels, as summarized in Table 2.

Table 2. Trend classification based on Theil–Sen Median trend analysis and Mann–Kendall Test.

β	Z	Trend Category	Trend Characteristics
$\beta > 0$	$Z > 2.58$	4	Extremely significant increase
	$1.96 < Z \leq 2.58$	3	Significant increase
	$1.65 < Z \leq 1.96$	2	Slightly significant increase
	$Z \leq 1.65$	1	Nonsignificant significant increase
$\beta = 0$		0	No change
$\beta < 0$	$Z \leq 1.65$	−1	Nonsignificant significant decrease
	$1.65 < Z \leq 1.96$	−2	Slightly significant decrease
	$1.96 < Z \leq 2.58$	−3	Significant decrease
	$Z > 2.58$	−4	Extremely significant decrease

3.3. Spatial Autocorrelation Analysis

Global and local spatial autocorrelation analyses are important methods for examining the spatial distribution characteristics of the NDVI [32]. Global spatial autocorrelation is typically assessed via Moran's I index, which reflects the overall spatial dependence of the NDVI across a study area. A significantly positive Moran's I value indicates positive spatial autocorrelation, suggesting the presence of clusters with high or low NDVI values. Conversely, a significantly negative Moran's I implies spatial negative autocorrelation, reflecting a trend toward spatial heterogeneity in the NDVI distribution.

In contrast, local spatial autocorrelation is analyzed via local Moran's I, which provides a deeper insight into spatial heterogeneity [33]. This method identifies specific areas with significant clustering, such as high–high (H-H) clusters (high NDVI values surrounded by high values) and low–low (L-L) clusters (low values surrounded by low values). Additionally, it can detect spatial outliers or anomalies, including high–low (H-L) regions (high values surrounded by low values) and low–high (L-H) regions (low values surrounded by high values), thereby revealing localized patterns of spatial discontinuity. The formula for the local Moran's I_i statistic is expressed in Equation (7), as follows:

$$I_i = \frac{n(x_i - \overline{x})}{\sum_{i=1}^{n} (x_i - \overline{x})^2} \sum_j w_{ij}(x_j - \overline{x}) \tag{7}$$

where x_i and x_j represent the attribute values of spatial units i and j, respectively; $\overline{x}$ denotes the mean of all attribute values; w_{ij} is the spatial weight between units i and j; and n is the total number of spatial units.

This index quantifies the strength of the association between each spatial unit and its neighboring units, effectively distinguishing different patterns of spatial correlation.

3.4. Vegetation Recovery Rate

The vegetation recovery status of landslide areas is expressed via the vegetation recovery rate (VRR), which is calculated via Equation (8), as follows:

$$VRR = \frac{NDVI_0 - NDVI_1}{NDVI_0 - NDVI_2} \tag{8}$$

where $NDVI_0$ represents the post-earthquake NDVI, specifically the NDVI value for the year 2008 in this study. $NDVI_1$ refers to the NDVI value for a selected year after the earthquake, and $NDVI_2$ denotes the pre-earthquake NDVI, which was calculated as the mean NDVI of the study area from 2000–2007.

According to previous research, the VRR values are classified into the following six categories [34]: Type I (VRR < 0), Type II (0 $\leq$ VRR < 25), Type III (25 $\leq$ VRR < 50), Type IV (50 $\leq$ VRR < 75), Type V (75 $\leq$ VRR < 100), and Type VI (VRR $\geq$ 100).

3.5. Hurst Exponent

The Hurst exponent (H) quantifies the long-term dependence of a time series and can reveal the persistence characteristics of NDVI variations. Its value ranges from 0 to 1. When H = 0.5, the series exhibits random walk behavior. H > 0.5 indicates positive persistence, suggesting that future trends are likely to follow the historical direction. In contrast, H < 0.5 implies anti-persistence, meaning that the future trend is likely to reverse [35].

In this study, the persistence characteristics derived from the Hurst exponent were integrated with Theil–Sen Median trend analysis to construct a two-parameter matrix model. On the basis of this framework, future NDVI trends were classified into several types, as shown in Table 3.

Table 3. Classification of trend levels based on the Hurst exponent.

β	H	Trend Category	Trend Characteristics
$\beta > 0$	$0.65 < H < 1$	4	Strongly persistent improvement
	$0.5 < H \leq 0.65$	3	Weakly persistent improvement
$\beta < 0$	$0 < H \leq 0.35$	2	Anti-strongly persistent improvement
	$0.35 < H \leq 0.5$	1	Anti-weakly persistent improvement
$\beta = 0$		0	No change
$\beta > 0$	$0.35 < H \leq 0.5$	−1	Anti-weakly persistent degradation
	$0 < H \leq 0.35$	−2	Anti-strongly persistent degradation
$\beta < 0$	$0.5 < H \leq 0.65$	−3	Weakly persistent degradation
	$0.65 < H < 1$	−4	Strongly persistent degradation

4. Results and Analysis

4.1. Spatiotemporal Evolution of the NDVI in the Chutou Gully Watershed

Through data collection and batch downloading of satellite imagery, the annual maximum NDVI values for the Chutou Gully watershed from 2000–2022 were derived via band processing and maximum value composite methods. Selecting the annual maximum NDVI helps to minimize the impact of cloud contamination, snow cover, or other atmospheric effects that may cause artificially low NDVI values on certain dates, thereby enabling a more accurate assessment of vegetation changes in the study area. Specifically, the 2008 maximum NDVI was extracted from imagery acquired after the "5.12" Wenchuan earthquake to better capture vegetation changes following seismic disturbances. The spatial distributions of the annual NDVI values in the Chutou Gully watershed from 2000–2022 are shown in Figure 3.

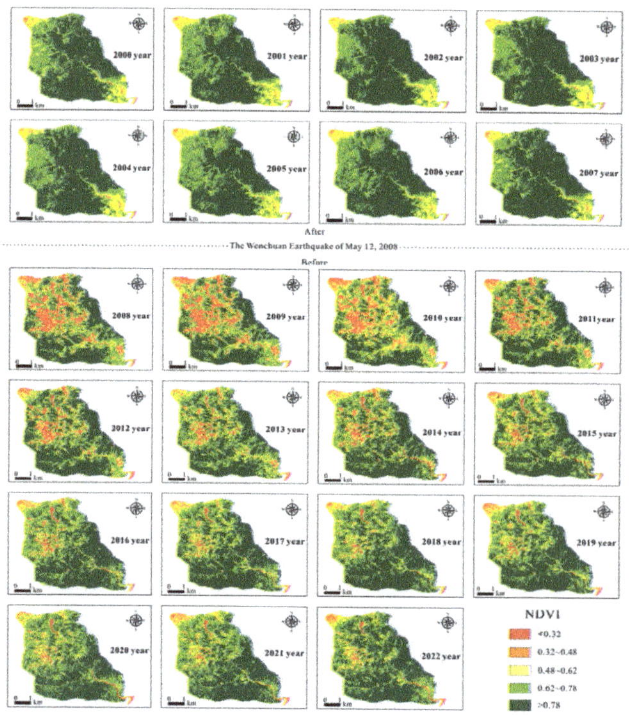

Figure 3. Annual spatial distribution of the NDVI in the Chutou Gully watershed from 2000–2022.

4.1.1. Temporal Variation Characteristics of the NDVI in the Chutou Gully Watershed

The NDVI values in the Chutou Gully watershed exhibited phase-based fluctuation patterns over the multiyear period. Prior to the earthquake, the average NDVI was 0.793, with a maximum value of 0.804. However, coseismic landslides triggered by the 2008 Wenchuan earthquake caused a sharp decline in the NDVI to 0.575. In the early post-earthquake stage, the vegetation gradually recovered, with the NDVI values continuously increasing. A slight decrease occurred in 2014 due to a debris flow event, followed by a rapid rebound until another decline in 2019, which was again associated with debris flow activity. The area subsequently entered a slow recovery phase. The results suggest that geological hazard events have a significant disturbance effect on the NDVI dynamics, and that the vegetation restoration process after major disturbances follows a typical "disturbance-recovery" alternation pattern.

A Slogistic1 regression model was fitted to the NDVI data from 2008–2022, yielding a high coefficient of determination ($R^2 = 0.953$). The model results indicate that the NDVI in the Chutou Gully watershed has been in a general state of recovery since the earthquake, with a recovery rate (β) of 0.119 per year. The NDVI is projected to return to the pre-earthquake average level between 2030 and 2031 (Figure 4). In the long term, vegetation recovered rapidly during the early post-seismic phase, followed by a prolonged period of slow recovery.

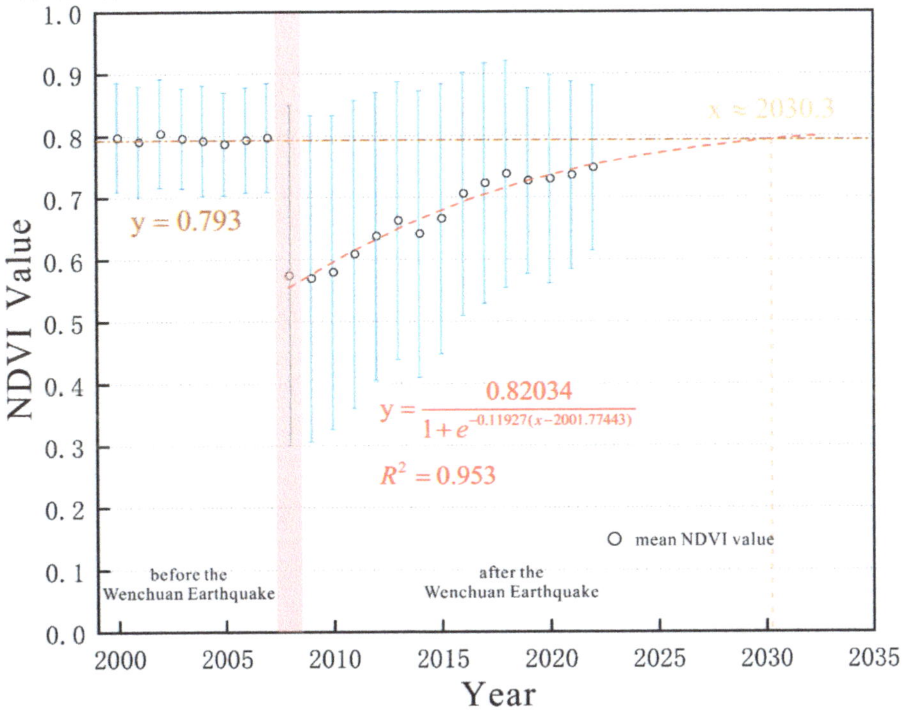

Figure 4. NDVI variation and trend prediction in the Chutou Gully watershed from 2000–2022 (error bars represent the standard deviation of the NDVI values).

The mean NDVI serves as an indicator of the regional vegetation recovery trend, while its standard deviation reflects the stability of the ecosystem [36]. A larger standard deviation indicates greater instability in the ecological system. As illustrated in the figure, prior to the Wenchuan earthquake, the average standard deviation of the NDVI in the Chutou Gully watershed was 0.0863. After the earthquake, it rose sharply to 0.274, highlighting a significant increase in ecosystem instability due to seismic disturbance. In the following years, as vegetation gradually re-established on slope surfaces, the mean NDVI increased, and the standard deviation correspondingly declined, suggesting a partial improvement in ecological stability. Nevertheless, the current standard deviation of the NDVI remains above the pre-earthquake level, indicating that full vegetation recovery has not yet been achieved. Some areas within the watershed still exhibit limited vegetation cover or sparse growth, reflecting ongoing ecological fragility in localized zones.

4.1.2. Spatial Distribution and Variation Characteristics of the NDVI in the Chutou Gully Watershed

To examine the spatial pattern of the NDVI in the Chutou Gully watershed, a pixel-based analysis was conducted to generate a multiyear NDVI distribution map from 2000–2022. The NDVI values were categorized into the following five vegetation activity levels: low (<0.32), moderately low (0.32–0.48), moderate (0.48–0.62), moderately high (0.62–0.78), and high vegetation activity zone (>0.78). The spatial distributions of the mean NDVI during the pre-earthquake period (2000–2007) and post-earthquake period (2008–2022) are shown in Figure 5a,b. The pre-seismic average NDVI was 0.793, whereas the post-seismic average declined to 0.657.

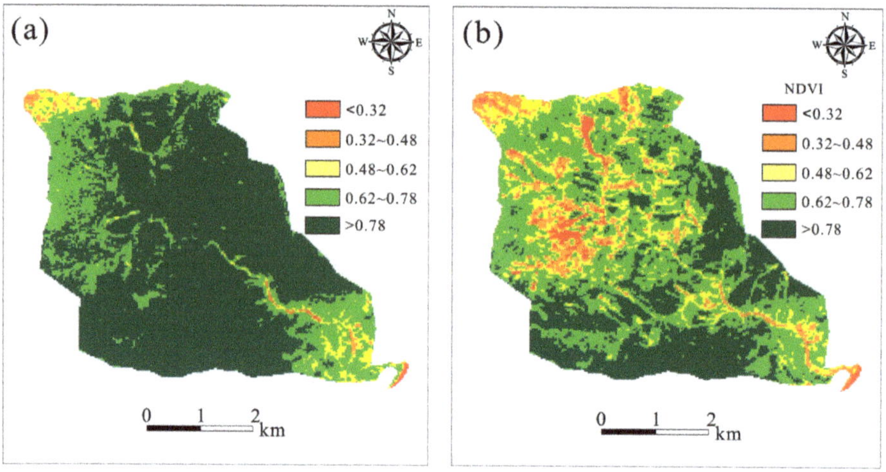

Figure 5. Spatial distribution of the average NDVI in the Chutou Gully watershed: (**a**) pre-earthquake period; (**b**) post-earthquake period.

The vegetation activity zoning in the Chutou Gully watershed exhibited a three-phase transition pattern in response to the Wenchuan earthquake (Figure 6). Prior to the earthquake, the high vegetation activity zone dominated the landscape, accounting for an average of 60%, with a peak of 66.2%, whereas the low and moderately low zones were negligible. Immediately following the seismic event, coseismic landslides and other disturbances triggered an abrupt shift in the vegetation activity structure, as follows: the high zone sharply declined from 64.6% to 35.0%, whereas the low and moderately low zones increased to 11.0% and 11.3%, respectively.

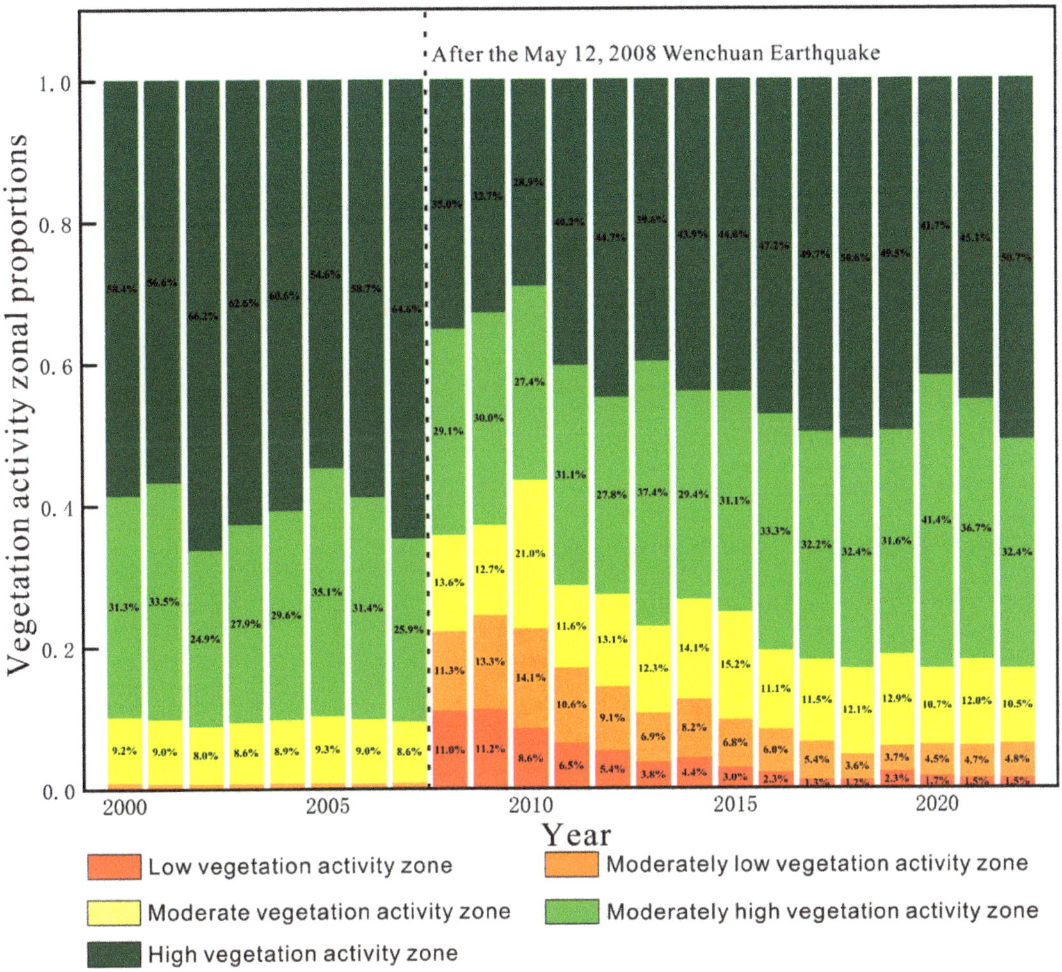

Figure 6. Proportions of vegetation activity levels in the Chutou Gully watershed from 2000–2022.

During the recovery phase, the vegetation exhibited a fluctuating recovery trend. Although the proportions of low and moderately low zones gradually decreased, the trajectory was interrupted by secondary disasters such as debris flows, resulting in periodic fluctuations. Moreover, the high-vegetation activity zone exhibited a slow upward trend, with occasional stagnation or slight regression in certain years. These oscillatory dynamics indicate that the post-seismic ecological restoration process in the Chutou Gully watershed is nonlinear but maintains an overall positive recovery trajectory.

To better examine the long-term spatial variation in the NDVI, Theil–Sen Median trend analysis and the Mann–Kendall test were applied to the annual NDVI data of the Chutou Gully watershed for the following two distinct periods: 2000–2022 and 2008–2022 (Figure 7).

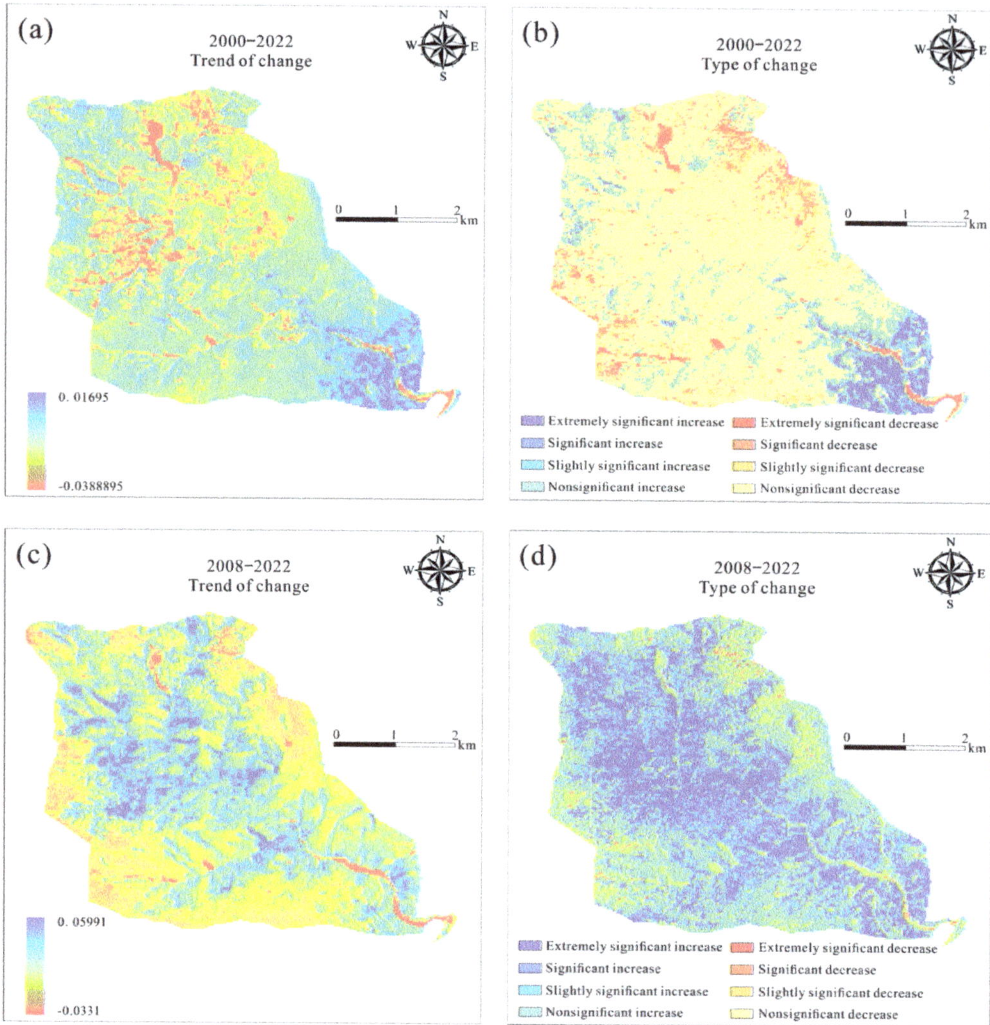

Figure 7. NDVI change trends and types in the Chutou Gully watershed over two distinct periods: (**a**) NDVI change trend from 2000 to 2022; (**b**) NDVI change types from 2000 to 2022; (**c**) NDVI change trend from 2008 to 2022; (**d**) NDVI change types from 2008 to 2022.

According to the results of Theil–Sen Median trend analysis and the MK significance test, the NDVI in the Chutou Gully watershed exhibited an overall declining trend from 2000–2022. Specifically, 80.73% of the area showed a decreasing trend, with 2.10% showing an extremely significant decrease, 4.22% showing a significant decrease, 5.34% showing a slightly significant decrease, and 69.07% showing a nonsignificant decrease (Table 4). These results suggest that the region experienced substantial vegetation loss due to disturbances from geological hazards such as earthquakes and debris flows. However, the large proportion of nonsignificant decreases implies that, despite the disturbances, vegetation has been gradually recovering over the 23-year period.

Table 4. NDVI change trends in the Chutou Gully watershed during the two time periods.

Period	Proportion%							
	Extremely Significant Decrease	Significant Decrease	Slightly Significant Decrease	Nonsignificant Significant Decrease	Nonsignificant Significant Increase	Slightly Significant Increase	Significant Increase	Extremely Significant Increase
2000–2022	2.10	4.22	5.34	69.07	11.50	1.68	2.18	3.91
2008–2022	0.14	0.41	0.44	9.94	35.02	8.90	22.10	23.05

Further analysis of the 2008–2022 period reveals a reversal in the trend, with NDVI generally increasing across the watershed. An increase was observed in 89.02% of the area, including 23.05% with an extremely significant increase, 22.10% with a significant increase, 8.90% with a slightly significant increase, and 35.02% with a nonsignificant increase. This trend reflects the post-disaster vegetation recovery that has occurred since the 2008 Wenchuan earthquake. Notably, the slopes on both sides of the transport zone exhibited marked increases in NDVI, which aligns with observed patterns of vegetation regeneration. In contrast, the river channels within the transport zone showed a decreasing trend in NDVI, primarily due to sediment deposition from debris flows after 2008, which resulted in a decline in vegetation cover and NDVI values in these zones.

4.2. Spatial Clustering Characteristics of the NDVI in the Chutou Gully Watershed

To investigate the spatial clustering characteristics of the NDVI, spatial autocorrelation analysis was performed for the following two time periods: 2000–2007 (pre-earthquake) and 2008–2022 (post-earthquake). The global spatial autocorrelation results revealed that the global Moran's I values were 0.89 and 0.87, respectively, indicating a strong and significant positive spatial correlation in both periods. The results of the local spatial autocorrelation analysis are presented in Figure 8.

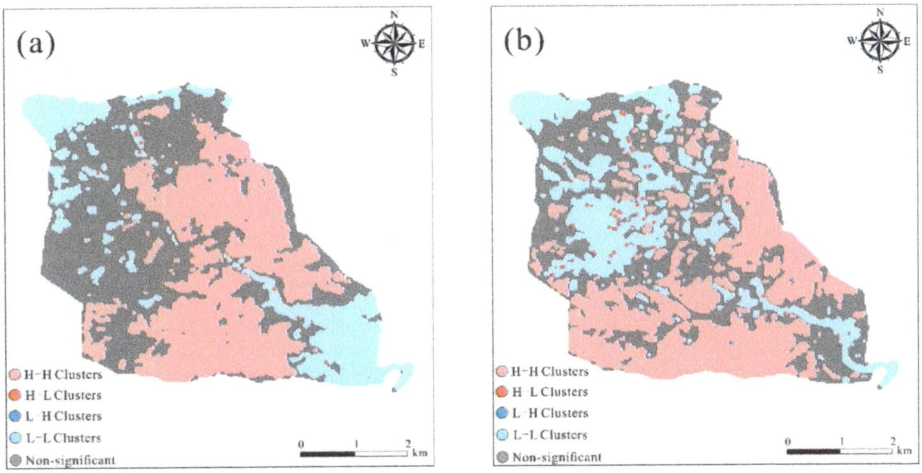

Figure 8. Local spatial autocorrelation analysis of the changes in the NDVI in the Chutou Gully watershed: (**a**) pre-earthquake period; (**b**) post-earthquake period.

A comparison of the NDVI clustering patterns between the two periods revealed significant dynamics in the spatial distribution of vegetation in the Chutou Gully watershed. During the pre-earthquake period (2000–2007), H-H clusters were concentrated along both sides of the gully in the source and deposition zones, indicating stable vegetation growth with minimal geological disturbance. The L-L clusters displayed a bimodal spatial distribution. In the Niujuan Gully area, the NDVI remained consistently low because of permanent snow cover in the high-altitude permafrost zone (elevation > 4000 m). In downstream residential areas within the source zone, reduced vegetation cover was attributed mainly to human disturbance.

Following the earthquake, a noticeable shift in clustering patterns was observed. H-H clusters were relocated to high-altitude mountain areas within the source and transport zones, where vegetation was largely unaffected by geological hazards and remained similar to pre-seismic conditions. Moreover, L-L clusters expanded, and new hazard-sensitive areas emerged, including the upper main gully, Datang Gully, Chutou Gully, and river channels within the transport zone. In these regions, secondary debris flows deposit large volumes of material, leading to a decline in the NDVI across the gully areas. The permafrost zone in Niujuan Gully has continued to exhibit persistently low vegetation cover.

Overall, the spatial clustering pattern of the NDVI was dominated by H-H and L-L clusters. Prior to the earthquake, the distribution was spatially differentiated, whereas in the post-earthquake period, clustering exhibited a patchy mosaic structure, with some areas forming large, contiguous zones of high or low vegetation activity [37].

4.3. Vegetation Recovery Rate in the Chutou Gully Watershed

Figure 9 shows the trend of the vegetation recovery rate (VRR) in the Chutou Gully watershed following the earthquake disturbance. On the basis of recovery levels, VRR was classified into six types (Type I–Type VI). In this classification, Type I and Type II represent poor recovery, Type III and Type IV indicate moderate recovery, and Type V and Type VI represent good recovery. Given the upward trend of the NDVI since 2010, this study focuses primarily on vegetation recovery patterns from 2010–2022.

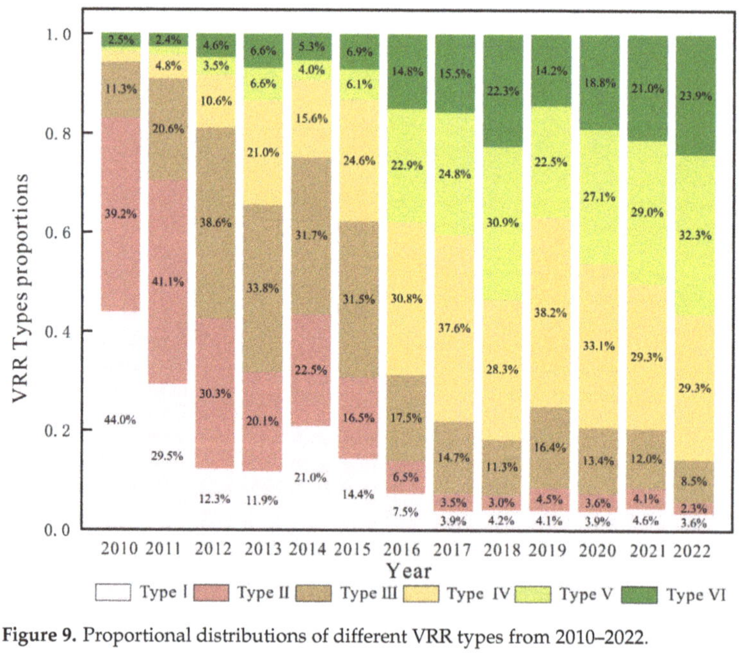

Figure 9. Proportional distributions of different VRR types from 2010–2022.

Analysis of VRR data from 2010–2022 revealed significant spatiotemporal heterogeneity in vegetation recovery across the study area. In the early post-seismic stage (2010–2013), the proportion of poorly recovered areas decreased markedly from 83.2% to 32%, indicating a general shift toward ecosystem recovery. Notably, in 2010, low-recovery zones were mainly concentrated in high-altitude regions, which may have experienced less initial earthquake damage (Figure 10a). However, the debris flow event in 2014 caused a rebound in poorly recovered areas to 43.5%, highlighting the strong disruptive impact of secondary disasters on the recovery process (Figure 10b,c).

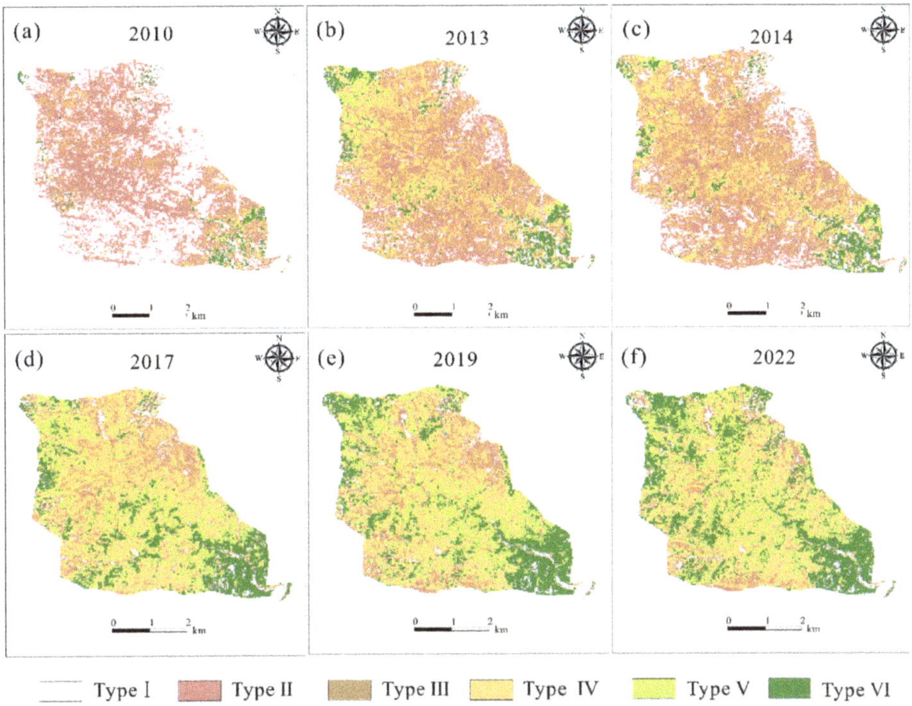

Figure 10. Spatial distribution of VRR-based vegetation recovery in the Chutou Gully watershed for typical years: (**a**) 2010; (**b**) 2013; (**c**) 2014; (**d**) 2017; (**e**) 2019; (**f**) 2022.

Thereafter, the watershed entered a fluctuating recovery phase. From 2015–2018 (Figure 10d), vegetation recovery improved significantly, yet consecutive debris flows in 2019 and 2020 led to another decline in recovery levels, especially in the upper reaches of Chutou Gully, where intense debris flow erosion likely damaged the vegetation structure, as this area serves as a primary sediment source zone (Figure 10e). By 2022, the proportion of well-recovered areas increased to 56.2%. Natural recovery was most evident on stable terrain along both sides of the main river channel, whereas collapsing debris accumulation zones and exposed bedrock areas remained in low-recovery states due to harsh site conditions. These spatiotemporal patterns reflect the cumulative effects of geological hazards and the heterogeneity in vegetation recovery capacity across different zones (Figure 10f).

4.4. Future Trends in Vegetation Change in the Chutou Gully Watershed

On the basis of the integrated analysis of the Theil–Sen Median trend analysis, the Mann–Kendall test, and the Hurst exponent for the period of 2008–2022, the vegetation change in the Chutou Gully watershed exhibited the following characteristics: the overall trend indicated persistent improvement, with strong persistence of improvement (32.26%) and weak persistence of improvement (48.08%) predominantly distributed along both sides of the gully. This suggests that vegetation recovery in these areas has high spatiotemporal stability.

Notably, degraded zones account for only a small proportion of the watershed, with strong and weak persistence of degradation accounting for just 0.78% and 0.87%, respectively (Table 5). In the source area, exposed collapse deposits with limited material mobility act as major barriers to vegetation recovery. In the downstream sections of the transport zone, the combined effects of debris flow deposition and road construction led to a sustained decline in the NDVI values (Figure 11).

Table 5. Future change trends in the NDVI in the Chutou Gully watershed.

Change Type	Trend Characteristics	Proportion
Significant	Strongly persistent improvement	32.26%
Slightly significant	Weakly persistent improvement	48.08%
Nonsignificant	Anti-strongly persistent degradation	0.01%
	Anti-weakly persistent degradation	0.19%
	Anti-weakly persistent improvement	16.80%
	Anti-strongly persistent improvement	1.00%
Slightly significant	Weakly persistent degradation	0.78%
Significant	Strongly persistent degradation	0.87%

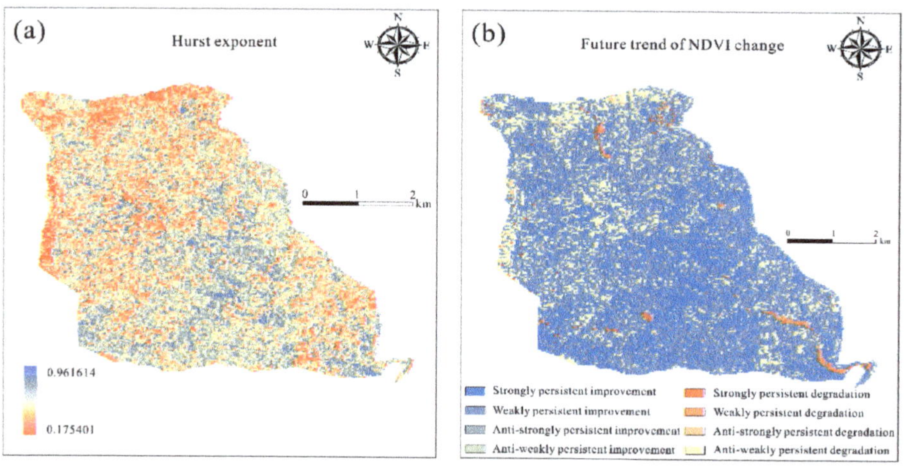

Figure 11. (**a**) Hurst exponent and (**b**) predicted NDVI change trend in the Chutou Gully watershed.

Based on the temporal variation of NDVI and the results of the Hurst exponent analysis, the NDVI in the Chutou Gully watershed is projected to exhibit an overall increasing trend in the future. However, the steep gorge terrain and remoteness of the source area pose significant challenges to debris management, while infrastructure development, such as road construction in the downstream transport zone, is a key driver of anticipated vegetation degradation. Overall, vegetation dynamics in the watershed are jointly driven by geological conditions and human activities, underscoring the need for targeted and differentiated ecological restoration strategies in degradation-sensitive areas.

5. Debris Flow Impact Extent and Its Influence on the NDVI

5.1. Spatiotemporal Extent of Debris Flow Impact

To assess the ecological impacts of four post-earthquake debris flow events in the Chutou Gully watershed, this study employed a multitemporal remote sensing comparative analysis approach to systematically reveal the evolution mechanisms of surface ecological disturbances caused by secondary debris flow hazards. On the basis of the temporal sequence of debris flow occurrences, satellite images captured before and after each of the four events were selected. Using visual interpretation and spatial overlay techniques, the following three key elements were extracted: (1) the spatial distribution of pre-event exposed collapse deposits; (2) the incremental distribution of newly formed collapse deposits following debris flows; and (3) the spatial evolution of debris flow transport zones.

On this basis, a set of dynamic impact maps was developed to illustrate the changing spatial extent of each debris flow event. Specifically, the pre-disaster image from 21 May 2013 was used to visually extract exposed slope failures as potential debris sources (Figure 12a). The image from 9 August 2013 was then spatially overlaid to identify newly exposed deposits and the impact zone of the 2013 debris flow (Figure 12b). The image from 23 October 2014 captured a subsequent stage close to the 2013 event; on the basis of previous interpretation results, newly affected areas and fresh debris deposits were extracted through comparison (Figure 12c). Against the background of natural recovery, the image from 22 May 2019 was used to delineate residual debris sources potentially triggering the 2019 event (Figure 12d). This analytical method was extended to the 27 September 2019 and 7 October 2020 images, in which debris flow boundaries and new deposit zones for 2019 and 2020 were determined via spatial overlay with prior interpretation results (Figure 12e,f).

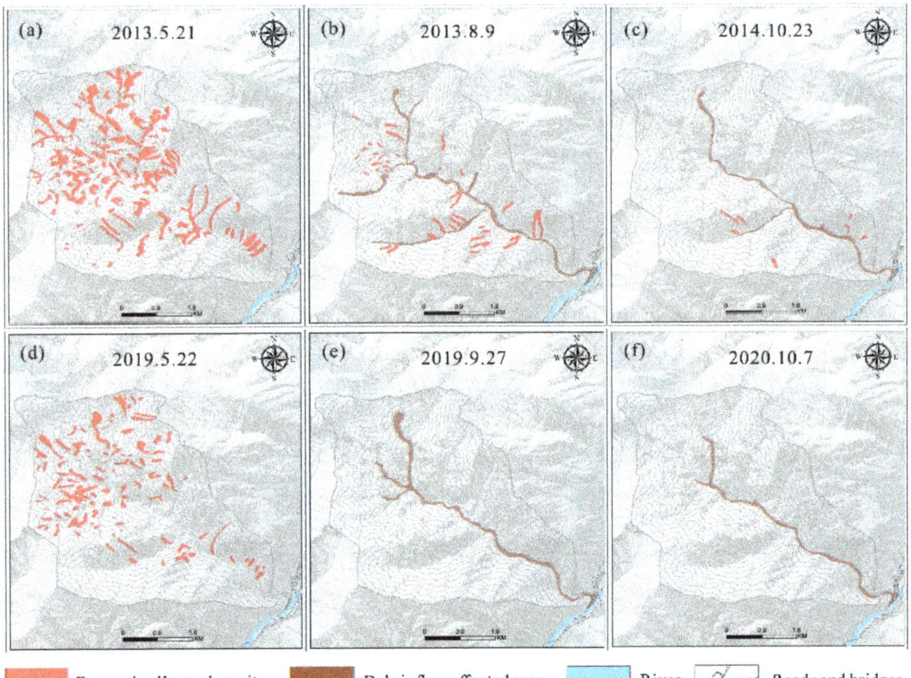

Figure 12. Remote sensing interpretation of exposed collapse deposits and debris flow impact extent on typical dates: (**a**) 21 May 2013; (**b**) 9 August 2013; (**c**) 23 October 2014; (**d**) 22 May 2019; (**e**) 27 September 2019; (**f**) 7 October 2020.

The 2013–2020 debris flow sequence analysis revealed that the study area experienced dynamic contraction in the impact range. The initiation zones of debris flows were widespread in Chutou Gully, Datang Gully, Xiao Gully, and the upper reaches of the main gully in 2013. By 2014, activity was concentrated in the Chutou and Xiao gullies. In 2019, debris flows reappeared in the Chutou, Datang, and upper main gully areas, but by 2020, only a single tributary in Chutou Gully remained active, reflecting a clear trend of spatial adjustment. The evolution of collapse deposits closely corresponded with vegetation recovery. In 2013, owing to unstable vegetation cover and debris flow impacts, large areas of exposed slope failures emerged along both sides of tributaries. By 2014, collapse activity weakened, with only small deposits appearing near ridgelines. After 2019, as the vegetation continued to regenerate, the slopes stabilized, and no new exposed deposits were detected for two consecutive years.

These findings suggest a dual mechanism for the attenuation of debris flow activity. On the one hand, river incision and successive post-seismic debris flows may have transported and depleted the available loose sediment over time; on the other hand, vegetation recovery effectively suppressed the formation of new collapse deposits and progressively limited the extent of the debris flow. This provides compelling evidence for the long-term regulatory role of ecological restoration in mitigating geological hazards [38,39].

5.2. *Effects of Debris Flow Activity on the NDVI*

On the basis of the selected imagery dates from the previous section, the NDVI values were extracted and analyzed via multitemporal remote sensing data to generate NDVI spatial distribution maps for the Chutou Gully watershed on representative dates (Figure 13). On 21 May 2013, the NDVI was 0.658. In August 2013, the first debris flow event caused the expansion of channel deposits and destruction of vegetation cover, resulting in a decrease to 0.651. In October 2014, a second debris flow introduced a small amount of newly exposed bedrock, and the vegetation in the previously impacted areas had not yet recovered. The cumulative disturbance led to a further decline in the NDVI to 0.645. After five years of natural recovery, the NDVI significantly increased to 0.723 in May 2019. The third debris flow, which occurred in September 2019, did not create new exposed areas and caused only a slight decrease to 0.720. The October 2020 event, dominated by loose material transport rather than fresh erosion, did not cause substantial vegetation loss, with the NDVI slightly decreasing to 0.716.

The NDVI values for these selected dates were categorized into vegetation activity levels following the classification method outlined previously. As shown in Figure 14, the consecutive debris flows from 2013–2014 resulted in significant ecological changes. In addition to newly exposed surfaces and increased channel sediment, the proportion of zones with low and moderately low vegetation activity increased from 14.4% to 18.5%, whereas the proportion of zones with high and moderately high vegetation activity decreased from 70.1% to 64.7%. Over the subsequent five-year natural recovery period, the ecosystem demonstrated strong resilience, with low-activity zones declining steadily to 8.8%, and high-activity zones rising to 78.8%. However, disturbances from 2019–2020 interrupted this trend, with the proportion of low-activity zones rebounding to 11.0% and that of high-activity zones slightly declining to 75.2%.

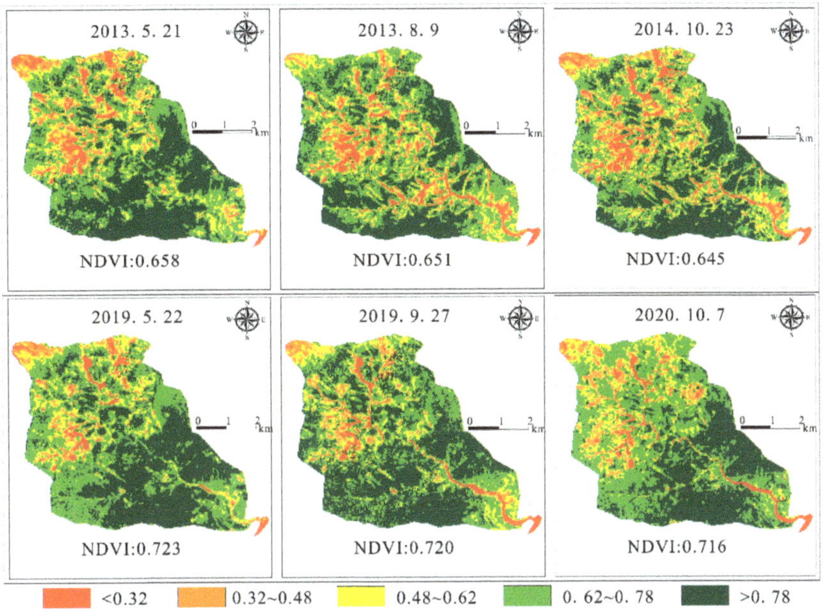

Figure 13. Spatial distribution of the NDVI in the Chutou Gully watershed on typical dates.

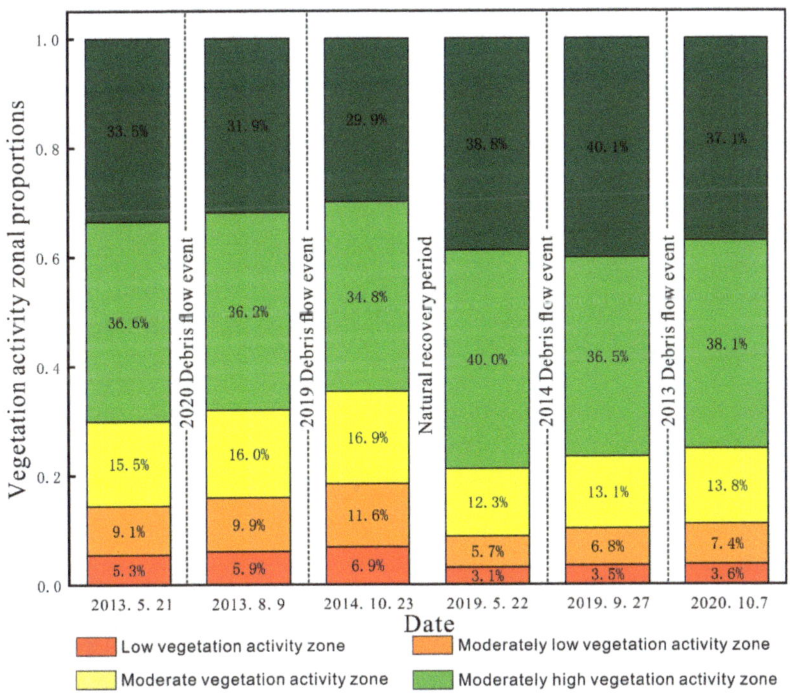

Figure 14. Proportions of vegetation activity levels in the Chutou Gully watershed before and after debris flow events.

This evolution process highlights the destructive effects of debris flow disturbances on vegetation activity, with impact intensity inversely correlated with post-disaster recovery time. During the early post-disaster stage (2013–2014), weak ecological foundations resulted in high sensitivity of vegetation, as follows: the proportion of low-activity zones increased by 4.1%, whereas high-activity zones decreased by 5.4%, reflecting the vulnerability of newly regenerated vegetation. In contrast, from 2019–2020, after several years of ecological recovery, the system exhibited enhanced stability through self-repair mechanisms. Even under higher-intensity debris flow events, only minor changes were observed, as follows: low-activity zones increased by 2.2%, and high-activity zones declined by 3.6%, indicating a significant reduction in vegetation loss compared to that in the early phase. This differentiated response suggests a progressive increase in ecosystem resistance over time, where vegetation communities build stronger disaster resilience mechanisms through structural reorganization and functional optimization [40,41].

6. Discussion

This study systematically revealed the post-seismic vegetation recovery dynamics and future trends in the Chutou Gully watershed of the Wenchuan earthquake zone, where cascading geological hazards, such as debris flows, exert persistent ecological pressure. By integrating multi-scale remote sensing and spatial analysis techniques, the findings enrich the theoretical framework of post-earthquake ecosystem succession and provide insight into vegetation resilience under complex disturbance regimes. These insights are also essential for informing sustainable ecological planning and land management in mountainous hazard-prone regions.

The results show that the annual NDVI recovery rate in the study area is 0.119, which is notably lower than the recovery rates observed in areas undergoing natural succession without recurrent secondary hazards [42]. This suggests that frequent disturbances, such as debris flows, significantly inhibit the self-recovery of ecosystems [43]. Compared to findings from more geomorphologically stable Wenchuan-affected regions—such as those reported by Lv et al. [44]—the lower recovery rate observed here underscores the cumulative suppressive effect of repeated debris flow events. This highlights the importance of considering not only the initial seismic impact, but also the prolonged influence of secondary hazards when evaluating recovery potential. Such long-term limitations must be addressed to support the sustainability of vegetation restoration in earthquake-impacted environments.

Further analysis revealed that high-frequency hazard events caused temporary declines in NDVI, while extended intervals between events allowed for increased ecosystem resistance, resulting in a cyclical "disturbance–recovery" pattern. This observation is consistent with established theories of ecosystem resilience [10]. Spatially, low-NDVI clusters shifted progressively toward debris flow channels, indicating degradation of site conditions due to sediment deposition. In contrast, slope areas—benefiting from relative geomorphic stability—showed better recovery outcomes. These patterns confirm previous insights regarding the role of terrain in controlling post-disturbance vegetation heterogeneity [45]. Understanding these spatial dynamics is fundamental for designing sustainability-oriented restoration strategies adapted to varied topographic conditions.

In terms of methodology, this study constructed a dual-parameter NDVI trend identification and prediction model by combining Theil–Sen trend estimation, Mann–Kendall significance testing, and Hurst exponent analysis. This approach improved the accuracy of long-term trend classification. The Hurst exponent results show that 80.34% of the watershed exhibited a persistent improvement trend, with these areas mainly located along stable slopes. While the direct field validation data were limited, the consistency of these trends with observed protective engineering zones and slope-based recovery areas—

as also found in studies such as that by Sun et al. [46]—provides indirect validation of model reliability. Moreover, the model's capacity to capture the spatial heterogeneity of resilience is supported by its alignment with geomorphological gradients and historical debris flow pathways.

However, model performance was less accurate in some degradation-prone areas. This contrasts with findings from flatter terrains where prediction accuracy tends to be higher, such as in the work by Jiao et al. [47]. One likely reason is the complex microtopography of the Chutou Gully watershed, which alters sediment transport patterns and introduces small-scale variability that is not fully captured by the current dataset resolution. Additionally, unquantified anthropogenic activities—such as informal road expansion and construction—may have affected local vegetation dynamics. These discrepancies point to the need for incorporating high-resolution terrain indices and human disturbance indicators in future modeling efforts to better explain localized degradation processes.

From a practical perspective, the destructive influence of debris flows on vegetation appears to diminish over time, indicating that the first five years post-disaster constitute a critical intervention window. During this period, artificial restoration measures—such as seeding and improvement of soil conditions—can effectively accelerate early-stage vegetation recovery [48]. In the long term, the following distinct spatial trends have emerged: areas with strongly persistent improvement (32.26%) exhibit high natural recovery potential and may be preserved through passive restoration strategies, while weakly persistent degradation zones (0.87%) remain highly sensitive to sediment disturbance and require active ecological engineering measures, such as slope stabilization and erosion control. These findings underscore the importance of tailoring restoration strategies to local vegetation dynamics and disturbance sensitivity within debris flow-affected landscapes.

Despite these contributions, this study has several limitations. First, the 30-m resolution of the NDVI dataset may be insufficient in areas with complex microtopography, potentially leading to pixel mixing and reduced accuracy in characterizing fine-scale vegetation patterns. Incorporating higher-resolution remote sensing data (e.g., UAV or LiDAR imagery) in future research would help overcome this limitation. Second, NDVI alone cannot fully represent changes in vegetation type or community structure. Integrating field investigation data and species-specific remote sensing indicators could enhance ecological interpretation. Finally, while the Hurst exponent effectively characterizes trend persistence, it does not reveal underlying drivers. Future studies should consider multi-factor coupling analyses involving terrain, climate, and human activity to deepen understanding of the mechanisms behind vegetation dynamics. Furthermore, model reliability could be improved through validation with long-term field monitoring or high-resolution imagery comparison.

7. Conclusions

Understanding the long-term ecological impact of the Wenchuan earthquake and its associated cascading geological hazards is essential for evaluating vegetation resilience and informing ecological restoration in seismically active mountainous regions. Based on long-term NDVI time series data (2000–2022) and multi-scale spatial analysis methods, this study systematically examined the vegetation recovery dynamics and future trends in the Chutou Gully watershed. The results revealed that the average NDVI dropped sharply from 0.793 to 0.575 after the earthquake, followed by a gradual recovery at an average annual rate of 0.119. Projections indicate a return to pre-earthquake levels by approximately 2030. This recovery generally follows an alternating "disturbance–recovery" cycle influenced by repeated debris flow events.

Spatially, post-earthquake vegetation degradation was most prominent in gully channels affected by debris flow deposition, while slopes on both sides showed significant recovery due to relative geomorphological stability. From 2010 to 2022, the area classified as poorly recovered declined from 83.2% to 5.8%, while well-recovered areas increased to 56.2%, indicating an improved disaster resistance threshold over time. According to Hurst exponent analysis, 80.34% of the study area shows signs of sustained vegetation improvement, with 32.26% exhibiting strong persistence. However, degradation hotspots persist in collapse accumulation zones and areas affected by human disturbances, where targeted restoration strategies are needed. Importantly, vegetation regrowth not only enhanced ecological stability, but also contributed to the long-term suppression of debris flow reactivation by stabilizing loose sediment sources.

This study contributes to the understanding of post-seismic vegetation resilience and offers a practical methodology for trend analysis in other earthquake-impacted watersheds. However, several limitations should be noted. The 30-m NDVI resolution may be inadequate for capturing fine-scale vegetation changes in rugged terrain. Future work should incorporate high-resolution data and field-based vegetation indicators to better assess structural recovery. Furthermore, while the Hurst exponent identifies persistence, it does not explain underlying drivers. Future studies should adopt multi-factor frameworks—integrating terrain, climate, and anthropogenic influences—to deepen our understanding of vegetation dynamics and support resilient and sustainable ecological management in seismically active regions.

Author Contributions: Investigation, X.X., X.S., D.S., L.Q. and J.G.; methodology, Z.W., W.Z. and M.M.; supervision, T.L.; writing—original draft, Z.W.; writing—review and editing, W.Z. All authors have read and agreed to the published version of the manuscript.

Funding: This study was financially supported by the National Natural Science Foundation of China (grant number 42377194), the Financial funds of Sichuan Institute of Geological Survey (SCIGS-CZDZX-2025005), and the Sichuan Province Central Government Guides Local Science and Technology Development Special Project (grant number 2023ZYD0151).

Institutional Review Board Statement: Not applicable.

Informed Consent Statement: Not applicable.

Data Availability Statement: Raw data supporting the findings of this study are available from the corresponding author upon reasonable request.

Conflicts of Interest: The authors declare no conflicts of interest.

References

1. Xiao, H.; Shao, H.; Long, J.; Zhang, S.; He, S.; Wang, D. Spatial-temporal pattern evolution and geological influence factors analysis of ecological vulnerability in Western Sichuan mountain region. *Ecol. Indic.* **2023**, *155*, 110980. [CrossRef]
2. Wang, W.; Song, Y.; Huang, L.; Shi, Y.; Zhang, C. Vulnerability assessment of disaster chains: A case study of rainstorm-landslide disaster chains in the Greater Bay Area. *Int. J. Disaster Risk Reduct.* **2025**, *119*, 105272. [CrossRef]
3. Durlević, U. Assessment of torrential flood and landslide susceptibility of terrain: Case study - Mlava River Basin (Serbia). *Bull. Serbian Geogr. Soc.* **2021**, *101*, 49–75. [CrossRef]
4. Zhang, L.; Zhang, J.; Ming, Z.; Li, H.; Chen, R.; Jia, Y. Effects of changing spatial scale on debris-flow hazard assessment: A case study in the Dadu River basin, China. *Sci. Total Environ.* **2024**, *954*, 176482. [CrossRef]
5. Schovanec, H.; Walton, G.; Kromer, R.; Malsam, A. Development of Improved Semi-Automated Processing Algorithms for the Creation of Rockfall Databases. *Remote Sens.* **2021**, *13*, 1479. [CrossRef]
6. Zheng, H.; Ding, M. Spatiotemporal changes of landslide susceptibility in response to rainfall and its future prediction—A case study of Sichuan Province, China. *Ecol. Inform.* **2024**, *84*, 102862. [CrossRef]
7. Zhang, L.; Guo, Z.; Qi, S.; Zhao, T.; Wu, B.; Li, P. Landslide susceptibility evaluation and determination of critical influencing factors in eastern Sichuan mountainous area, China. *Ecol. Indic.* **2024**, *169*, 112911. [CrossRef]

8. Cui, P.; Lin, Y.-M.; Chen, C. Destruction of vegetation due to geo-hazards and its environmental impacts in the Wenchuan earthquake areas. *Ecol. Eng.* **2012**, *44*, 61–69. [CrossRef]
9. Zhang, X.; Tang, C.; Tie, Y.; Li, X.; Tang, C.; Xiong, J.; Chen, M.; Gong, L. Dynamic susceptibility assessment of debris flow hazard after a strong earthquake, Wenchuan County, Sichuan, China. *Landslides* **2024**, *21*, 1915–1928. [CrossRef]
10. Chang, M.; Dou, X.; Luo, C.; Su, Y. Multi-scenario simulations for quantitative assessment of debris flow chain hazards in southwestern China. *Catena* **2025**, *253*, 108900. [CrossRef]
11. Sun, M.; Zhang, L.; Yang, R.; Li, X.; Zhang, Y.; Lu, Y. Construction of an integrated framework for assessing ecological security and its application in Southwest China. *Ecol. Indic.* **2023**, *148*, 110074. [CrossRef]
12. Zhou, Y.; Yue, D.; Li, S.; Wang, Y.; Meng, X.; Xu, X. Coupling mechanism of the eco-geological environment in debris flow prone area: A case study of the Bailong River basin. *Sci. Total Environ.* **2024**, *955*, 177230. [CrossRef]
13. Gao, W.; Zheng, C.; Liu, X.; Lu, Y.; Chen, Y.; Wei, Y.; Ma, Y. NDVI-based vegetation dynamics and their responses to climate change and human activities from 1982 to 2020: A case study in the Mu Us Sandy Land, China. *Ecol. Indic.* **2022**, *137*, 108745. [CrossRef]
14. Xu, Y.; Dai, Q.-Y.; Lu, Y.-G.; Zhao, C.; Huang, W.-T.; Xu, M.; Feng, Y.-X. Identification of ecologically sensitive zones affected by climate change and anthropogenic activities in Southwest China through a NDVI-based spatial-temporal model. *Ecol. Indic.* **2024**, *158*, 111482. [CrossRef]
15. Yang, W.; Qi, W.; Zhou, J. Decreased post-seismic landslides linked to vegetation recovery after the 2008 Wenchuan earthquake. *Ecol. Indic.* **2018**, *89*, 438–444. [CrossRef]
16. Zhang, S.; Peng, J.Y.; Zhang, M.P.; Chen, Y.B.; Han, Y.Y.; Su, C.X.; Zhuang, D.Y. Evolution of debris flow activities in the epicentral area, 10 years after the 2008 Wenchuan earthquake. *Eng. Geol.* **2023**, *319*, 107118. [CrossRef]
17. Wang, X.A.; Chen, X.; Chen, H.; Zhao, W. Influence mechanism of herbaceous plants on debris flow bank erosion. *Catena* **2024**, *245*, 108308. [CrossRef]
18. Zhang, J.; Zhang, Y.; Dannenberg, M.P.; Guo, Q.; Atkins, J.W.; Li, W.; Sun, G. Eco-hydrological recovery following large vegetation disturbances from a mega earthquake on the eastern Tibetan plateau. *J. Hydrol.* **2025**, *651*, 132595. [CrossRef]
19. Fan, R.L.; Zhang, L.M.; Wang, H.J.; Fan, X.M. Evolution of debris flow activities in Gaojiagou Ravine during 2008–2016 after the Wenchuan earthquake. *Eng. Geol.* **2018**, *235*, 1–10. [CrossRef]
20. Yang, Y.; Tang, C.; Tang, C.; Chen, M.; Cai, Y.; Bu, X.; Liu, C. Spatial and temporal evolution of long-term debris flow activity and the dynamic influence of condition factors in the Wenchuan earthquake-affected area, Sichuan, China. *Geomorphology* **2023**, *435*, 108755. [CrossRef]
21. Zhai, L.; Cheng, S.; Sang, H.; Xie, W.; Gan, L.; Wang, T. Remote sensing evaluation of ecological restoration engineering effect: A case study of the Yongding River Watershed, China. *Ecol. Eng.* **2022**, *182*, 106724. [CrossRef]
22. Chen, M.; Tang, C.; Zhang, X.; Xiong, J.; Chang, M.; Shi, Q.; Wang, F.; Li, M. Quantitative assessment of physical fragility of buildings to the debris flow on 20 August 2019 in the Cutou gully, Wenchuan, southwestern China. *Eng. Geol.* **2021**, *293*, 106319. [CrossRef]
23. Wang, T.; Deng, M. Comparison of Relationship Between Debris-Flow Volume and Peak Discharge in Different Regions. In *Understanding and Reducing Landslide Disaster Risk: Volume 6 Specific Topics in Landslide Science and Applications*; Arbanas, Ž., Bobrowsky, P.T., Konagai, K., Sassa, K., Takara, K., Eds.; Springer International Publishing: Cham, Switzerland, 2021; pp. 149–154.
24. Yang, F.; Fan, X.; Wei, Z.; Subramanian, S.S.; Van Asch, T.W.J.; Xu, Q. Modelling the evolution of debris flows after the 2008 Wenchuan earthquake. *Eng. Geol.* **2023**, *321*, 107152. [CrossRef]
25. Huang, T.; Ding, M.; Gao, Z.; Téllez, R.D. Check dam storage capacity calculation based on high-resolution topogrammetry: Case study of the Cutou Gully, Wenchuan County, China. *Sci. Total Environ.* **2021**, *790*, 148083. [CrossRef]
26. Xu, C.; Xu, X.; Yao, X.; Dai, F. Three (nearly) complete inventories of landslides triggered by the May 12, 2008 Wenchuan Mw 7.9 earthquake of China and their spatial distribution statistical analysis. *Landslides* **2014**, *11*, 441–461. [CrossRef]
27. Ge, Y.-G.; Cui, P.; Zhang, J.-Q.; Zeng, C.; Su, F.-H. Catastrophic debris flows on July 10th 2013 along the Min River in areas seriously-hit by the Wenchuan earthquake. *J. Mt. Sci.* **2015**, *12*, 186–206. [CrossRef]
28. Yang, F.; Fan, X.; Subramanian, S.S.; Dou, X.; Xiong, J.; Xia, B.; Yu, Z.; Xu, Q. Catastrophic debris flows triggered by the 20 August 2019 rainfall, a decade since the Wenchuan earthquake, China. *Landslides* **2021**, *18*, 3197–3212. [CrossRef]
29. Rouse, J.W.; Haas, R.H.; Schell, J.A.; Deering, D.W. Monitoring Vegetation Systems in the Great Plains with ERTS. In Proceedings of the Third ERTS-1 Symposium NASA, Washington, DC, USA, 10–14 December 1973; pp. 309–317.
30. Sen, P.K. Estimates of the regression coefficient based on Kendall's Tau. *J. Am. Stat. Assoc.* **1968**, *63*, 1379–1389. [CrossRef]
31. Kendall, M.G. *Rank Correlation Methods*; Griffin: London, UK, 1948.
32. Griffith, D.A. Spatial Autocorrelation. In *Encyclopedia of Social Measurement*; Kempf-Leonard, K., Ed.; Elsevier: New York, NY, USA, 2005; pp. 581–590.
33. Chen, T.; Chen, T.; Fu, J. Exploring the spatial and temporal changes of compound disasters: A case study in Gaoping River, Taiwan. *Clim. Risk Manag.* **2024**, *44*, 100617. [CrossRef]

34. Lin, W.-T.; Chou, W.-C.; Lin, C.-Y.; Huang, P.-H.; Tsai, J.-S. Vegetation recovery monitoring and assessment at landslides caused by earthquake in Central Taiwan. *For. Ecol. Manag.* **2005**, *210*, 55–66. [CrossRef]
35. Hurst, H.E. Long-Term Storage Capacity of Reservoirs. *Trans. Am. Soc. Civ. Eng.* **1951**, *116*, 770–799. [CrossRef]
36. Chen, M.; Tang, C.; Li, M.; Xiong, J.; Luo, Y.; Shi, Q.; Zhang, X.; Tie, Y.; Feng, Q. Changes of surface recovery at coseismic landslides and their driving factors in the Wenchuan earthquake-affected area. *Catena* **2022**, *210*, 105871. [CrossRef]
37. Ikhumhen, H.O.; Fang, Q.; Lu, S.; Meilana, L.; Lopes, N.D. R Investigating socio-ecological vulnerability to climate change via remote sensing and a data-driven ranking algorithm. *J. Environ. Manag.* **2023**, *347*, 119254. [CrossRef] [PubMed]
38. Zhang, S.; Zhang, L.M. Impact of the 2008 Wenchuan earthquake in China on subsequent long-term debris flow activities in the epicentral area. *Geomorphology* **2017**, *276*, 86–103. [CrossRef]
39. Chen, M.; Chang, M.; Xu, Q.; Tang, C.; Dong, X.; Li, L. Identifying potential debris flow hazards after the 2022 Mw 6.8 Luding earthquake in southwestern China. *Bull. Eng. Geol. Environ.* **2024**, *83*, 241. [CrossRef]
40. Yin, J.-Z.; He, F.-Q.; Luo, Z.-B. Researching the Relationships between the Environmental Change of Vegetation and the Activity of Debris Flows Based on Remote Sensing and GIS. *Procedia Environ. Sci.* **2011**, *11*, 918–924.
41. Wang, S.; Meng, X.; Chen, G.; Guo, P.; Xiong, M.; Zeng, R. Effects of vegetation on debris flow mitigation: A case study from Gansu province, China. *Geomorphology* **2017**, *282*, 64–73. [CrossRef]
42. Olupot, W. The potential of natural succession to restore degraded areas of a Ugandan rainforest dominated by the exotic paper mulberry Broussonetia papyrifera (L.) L'Hér. ex Vent. *For. Ecol. Manag.* **2022**, *504*, 119861. [CrossRef]
43. Michelini, T.; Bettella, F.; D'Agostino, V. Field investigations of the interaction between debris flows and forest vegetation in two Alpine fans. *Geomorphology* **2017**, *279*, 150–164. [CrossRef]
44. Lv, J.; He, X.; Bao, Y.; Li, H. Spatiotemporal pattern of post-earthquake vegetation recovery in a mountainous catchment in southwestern China. *Nat. Hazards* **2025**, *121*, 3023–3046. [CrossRef]
45. Kang, S.; Lee, S.-R.; Vasu, N.N.; Park, J.-Y.; Lee, D.-H. Development of an initiation criterion for debris flows based on local topographic properties and applicability assessment at a regional scale. *Eng. Geol.* **2017**, *230*, 64–76. [CrossRef]
46. Sun, X.; Yuan, L.; Zhou, Y.; Shao, H.; Li, X.; Zhong, P. Spatiotemporal change of vegetation coverage recovery and its driving factors in the Wenchuan earthquake-hit areas. *J. Mt. Sci.* **2021**, *18*, 2854–2869. [CrossRef]
47. Jiao, Q.; Zhang, B.; Liu, L.; Li, z.; Yue, Y.; Hu, Y. Assessment of spatio-temporal variations in vegetation recovery after the Wenchuan earthquake using Landsat data. *Nat. Hazards* **2014**, *70*, 1309–1326. [CrossRef]
48. He, S.; Wang, D.; Fang, Y.; Lan, H. Guidelines for integrating ecological and biological engineering technologies for control of severe erosion in mountainous areas – A case study of the Xiaojiang River Basin, China. *Int. Soil Water Conserv. Res.* **2017**, *5*, 335–344. [CrossRef]

sustainability

MDPI

Article

Evaluation and Prediction of Ecological Quality Based on Remote Sensing Environmental Index and Cellular Automata-Markov

Weirong Qin [1,2,†], Mohd Hasmadi Ismail [2,†], Mohammad Firuz Ramli [2], Junlin Deng [3,*] and Ning Wu [3,*]

[1] College of Resources and Environment, Beibu Gulf University, Qinzhou 535011, China; qwr71@bbgu.edu.cn
[2] Faculty of Forestry and Environment, Universiti Putra Malaysia (UPM), Serdang 43400, Selangor, Malaysia; mhasmadi@upm.edu.my (M.H.I.); firuz@upm.edu.my (M.F.R.)
[3] Key Laboratory of Beibu Gulf Offshore Engineering Equipment and Technology, Beibu Gulf University, Qinzhou 535011, China
* Correspondence: junlin.deng@163.com (J.D.); n.wu@bbgu.edu.cn (N.W.)
† These authors contributed equally to this work.

Abstract: The evaluation and prediction of ecological environmental quality are essential for sustainable development and environmental management, particularly in regions experiencing rapid urbanization and industrial growth like Johor in southern Peninsular Malaysia. This study evaluates the temporal and spatial changes in ecological environmental quality in Johor from 1990 to 2020 using the Remote Sensing Environmental Index (RSEI) and Cellular Automata-Markov (CA-Markov). A CA-Markov model was employed to predict ecological environmental quality for the next 12 months based on historical data. The results reveal significant changes over the 30 years, highlighting the dynamic nature of ecological conditions. The prediction results indicate that areas with excellent ecological quality are primarily focused in the central and northern regions, while the southern and eastern edges show mixed ecological conditions. The western region, characterized by intensive land use, shows significant environmental degradation. The poorest ecological points are mainly distributed in urban and semiurban areas with frequent human activities, such as cities, ports, and villages. These findings highlight the need for targeted environmental policies and management strategies to mitigate ecological degradation and promote sustainable development in Johor State of Peninsular Malaysia.

Keywords: ecological environmental quality; Remote Sensing Environmental Index (RSEI); Cellular Automata-Markov (CA-Markov); prediction; sustainable development

1. Introduction

The evaluation of ecological environmental quality plays a crucial role in supporting ecosystem restoration and sustainable development. As ecosystems face increasing pressures from human activities, climate change, and land use intensification, it has become essential to conduct scientific and systematic monitoring and assessment of ecological conditions. Such evaluations provide critical insights into the current state and dynamic changes of ecosystems, serving as a foundation for ecological protection and restoration efforts. By identifying areas of environmental degradation and predicting future trends, ecological environmental quality assessments offer valuable scientific evidence for policymakers, enabling informed decision-making and effective management practices.

Johor is a state located in southern Peninsular Malaysia and is a region experiencing rapid economic growth, urbanization, and industrialization. This development, while contributing to economic prosperity, has placed significant pressure on the region's ecological environment, leading to habitat loss, deforestation, and environmental degradation.

As a key area with diverse ecosystems, including forests, wetlands, and coastal zones, Johor's ecological health is vital for maintaining regional biodiversity, supporting local livelihoods, and ensuring sustainable development. The evaluation of ecological environmental quality in Johor is particularly important for understanding the spatial and temporal impacts of human activities and natural processes on its ecosystems. Such evaluations can help identify critical areas requiring immediate restoration and conservation efforts while providing a scientific basis for managing urban expansion, industrial activities, and resource exploitation.

While ecological environmental quality evaluation is critical for Johor, existing methods often face several limitations that hinder their effectiveness. Traditional evaluation approaches tend to rely on fragmented field surveys and localized data, which can be time-consuming, costly, and insufficient for capturing large-scale spatial and temporal changes. Additionally, these methods often cannot integrate multiple environmental indicators, leading to incomplete or inaccurate assessments of ecological conditions. In Johor, the rapid pace of urbanization and industrialization exacerbates these challenges, as conventional methods struggle to keep up with the dynamic changes in land use and ecosystem health. Furthermore, predictive analysis is rarely incorporated into existing evaluations, limiting the ability to foresee future environmental trends and plan proactive conservation strategies. To address these issues, there is a need for advanced, integrated approaches that utilize remote sensing technology and predictive models to provide comprehensive, efficient, and reliable assessments of ecological environmental quality in Johor.

In previous studies, remote sensing technology has played a significant role in the evaluation of ecological environmental quality, providing an effective means for regional ecological monitoring and dynamic analysis. For instance, vegetation index models, such as the Normalized Difference Vegetation Index (NDVI) and Enhanced Vegetation Index (EVI), have been widely used to monitor changes in vegetation coverage, assess forest health, and estimate agricultural yields [1–3]. Land cover and land use change detection techniques leverage multi-temporal remote sensing data to analyze dynamic changes such as urban expansion and deforestation [4]. Ecosystem productivity assessment methods combine remote sensing data with photosynthesis models to study carbon cycles and the impacts of climate change on ecosystems [5,6]. In water quality assessments, remote sensing technology extracts parameters such as water transparency and chlorophyll content through spectral analysis to monitor water quality conditions [7]. High-resolution remote sensing imagery, integrated with ground survey data, has also been utilized for biodiversity evaluation and wetland monitoring, supporting ecological restoration planning through the construction of habitat distribution models [8–10].

The Remote Sensing Ecological Index (RSEI) is a comprehensive ecological evaluation method that integrates four key ecological factors, such as greenness, humidity, dryness, and heat. By employing principal component analysis (PCA) to construct a composite index, RSEI effectively eliminates the subjectivity associated with weight assignment in traditional methods [11–15]. In recent years, researchers have optimized RSEI to suit the characteristics of different regions, leading to the development of variants such as the Multi-Indicator Remote Sensing Ecological Index (MSRE) and the Integrated Remote Sensing Ecological Index (IRSEI), which are tailored to the specific features of various ecosystems [16–19]. RSEI has been widely applied in ecological quality assessments across urbanized areas, wetland conservation regions, and desertification-prone zones [20–24].

In order to enhance the predictive capabilities of ecological environment assessments [25–30], the CA-Markov model has been widely used to simulate land use changes and their impacts on ecological environments [18,26,31–35]. By combining the strengths of Cellular Automata (CA) and Markov Chain, this model efficiently simulates dynamic ecosystem changes across temporal and spatial scales [36–41]. For instance, Li et al. utilized the CA-Markov model to analyze future ecological quality changes in urban clusters in central Yunnan [36]; Yang et al. compared the impacts of urbanization on ecological environments across different regions in China [38]; and Wang et al. simulated land use changes in the Heihe River Basin to evaluate ecological quality variations under different scenarios [41].

Despite significant progress in the application of remote sensing technology and dynamic models in ecological environment assessments, existing research still has several limitations. First, traditional methods often rely on scattered field surveys or single indicators, making it difficult to comprehensively reflect the complexity and dynamic changes of ecosystems. Second, the predictive capabilities of current studies are limited, and they cannot provide strong support for the development of forward-looking conservation strategies. Moreover, in Johor, a region with a complex and diverse ecological environment, existing assessment methods lack the ability to effectively integrate multi-scale analyses and adapt models to the region's specific characteristics.

This study focuses on Johor State, Peninsular Malaysia, a rapidly developing region, to evaluate and predict its ecological environmental quality using the Remote Sensing Environmental Index (RSEI) and CA-Markov model. This study aims to support targeted ecological restoration strategies, ensuring balanced development and environmental sustainability. This study utilizes the Google Earth Engine (GEE) cloud platform to process multi-temporal remote sensing data, using Landsat 5 satellite images recorded from 1990 to 2013 and Landsat 8 satellite images recorded from 2013 to 2023. The RSEI values for 34 years are calculated to comprehensively analyze the dynamic changes in the ecological quality of Johor. By integrating remote sensing technology and predictive models, this study reveals the spatial distribution, temporal trends, and potential driving factors of Johor's ecological environment.

The contributions of this study are as follows:

(i) A comprehensive method for assessing and predicting ecological environment quality is proposed, which integrates the RSEI with the CA-Markov model. By leveraging the spatial characteristics of remote sensing data and the dynamic spatiotemporal simulation capabilities of the CA-Markov model, this approach achieved integrated ecological environment analysis in the Johor region, addressing the research gap in large-scale, long-term time series prediction of ecological environment dynamics.

(ii) Quantitative spatial distribution and temporal trend analysis reveal long-term ecological change patterns and the relationship between regional differences in ecological quality and spatial aggregation characteristics, providing a theoretical foundation for formulating ecological conservation and restoration strategies.

(iii) Spatial autocorrelation analysis is introduced to identify spatiotemporal distribution patterns of priority ecological conservation zones, vulnerable areas, and transitional regions through autocorrelation analysis of ecological quality data across different periods. This significantly enhances the understanding of spatial heterogeneity in regional ecological environments.

(iv) Based on spatiotemporal analysis results, this study proposed targeted ecological protection and management strategies for rapidly urbanizing and industrializing zones in Johor, along with concrete recommendations for balancing economic development and ecological conservation. The policy implications offer valuable references for ecological environment management in similar regions.

2. Materials and Methods

2.1. Materials

2.1.1. Study Area

Johor is located at the southernmost tip of Peninsular Malaysia, with geographical coordinates ranging from 1°20′ N to 2°35′ N latitude and 103°20′ E to 104°20′ E longitude, as shown in Figure 1. The region has a typical tropical rainforest climate, with annual rainfall ranging between 2000 mm and 3000 mm, concentrated mainly during the northeast monsoon period from November to March. The temperature remains relatively high throughout the year, with an average range between 25 °C and 30 °C. The hottest months are May and June, when temperatures can exceed 30 °C, while December and January are relatively cooler, with temperatures ranging from 24 °C to 26 °C. Johor is home to diverse wetland ecosystems, including coastal wetlands, inland wetlands, swamps, and estuarine areas. These wetland ecosystems play a crucial role in maintaining biodiversity, providing wildlife habitats, regulating the water cycle, and controlling flooding. The eastern and southern coastlines of Johor, particularly around Johor National Park and Pulau Kukup National Park, are characterized by extensive mangrove forests. These mangroves are vital in preventing coastal erosion, protecting coastal ecosystems, and supporting fisheries production. In aquaculture, Johor's activities are mainly concentrated in the coastal and estuarine areas, particularly around Batu Pahat and Mersing. The main aquaculture species in these areas include tilapia, prawns, and various shellfish. Wetlands and mangrove forests provide essential ecological conditions for aquaculture while playing an irreplaceable role in maintaining ecological balance and protecting biodiversity.

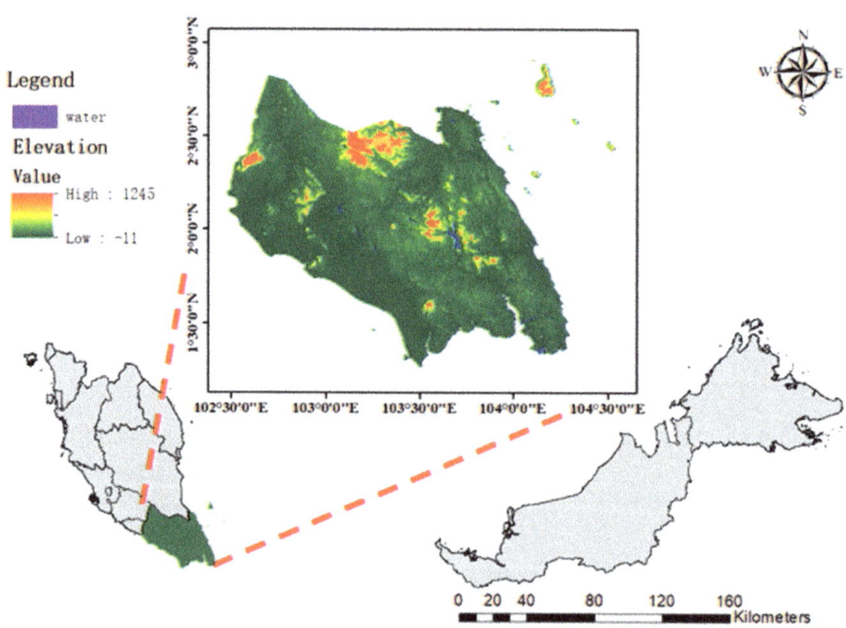

Figure 1. Study area Johor, Malaysia.

2.1.2. Data Sources

The remote sensing data collected for this study were collected from cloud-free images captured by the Landsat 5 and Landsat 8 satellites, as shown in Table 1. Landsat data were primarily used to calculate Normalized Difference Vegetation Index (NDVI), Wetness Index (WET), Normalized Difference Built-up Space Index (NDBSI), and Land

Surface Temperature (LST), Principal Component Analysis (PCA), and Remote Sensing Environmental Index (RSEI), providing the foundation for analyzing trends in ecological landscape patterns, conducting spatial autocorrelation analysis, and preparing predictive models. MODIS data were used as a substitute when the ST_10 band for a specific year was damaged and unable to compute LST, and the data fusion method is given in Equation (4) listed in Table 1. The JRC/GSW1_3 Yearly Water Classification History dataset was employed to perform water masking.

Table 1. Data sources used in this study.

Data	Time	ee.ImageCollection	Resolution (m)
Landsat 5	1990–2012	("LANDSAT/LT05/C02/T1_L2")	30
Landsat 8	2013–2023	("LANDSAT/LC08/C02/T2_L2")	30
MODIS		("MODIS/061/MOD11A1")	LST_1000
Water(mask)		JRC/GSW1_3/Yearly History	30

2.2. Methods

2.2.1. Remote Sensing Ecological Index (RSEI)

The Remote Sensing Ecological Index (RSEI) is a comprehensive indicator designed to effectively assess regional ecological environmental quality, providing a holistic reflection of ecosystem health. The calculation of RSEI integrates multiple ecological factors, including the Normalized Difference Vegetation Index (NDVI), Wetness Index (WET), Normalized Difference Built-up Space Index (NDBSI), and Land Surface Temperature (LST), as outlined in Table 2. The RSEI methodology is particularly well-suited for evaluating Johor's ecological environment. Its inclusion of the Wetness Index enables precise characterization of the dynamic changes in Johor's wetland and mangrove ecosystems, while the Land Surface Temperature and Built-up Space Index provide critical support for quantifying the impacts of urbanization. Through Principal Component Analysis (PCA), the primary information (PC1) from these four indicators is extracted, enabling a dimensional reduction and comprehensive analysis of ecological environmental quality. Standardization and normalization of the data further allow the classification of RSEI into five levels (from Poor to Excellent), laying the groundwork for spatiotemporal change analysis and spatial clustering studies.

Johor, with its diverse ecosystems, relies heavily on the ecological services provided by its wetlands, mangroves, and coastal areas for regional biodiversity conservation and environmental management. Based on Johor's ecological characteristics, this study utilizes remote sensing imagery from 1990 to 2023 to calculate RSEI values for different years, identifying the spatial distribution patterns and temporal trends in ecological environmental quality. By conducting spatial autocorrelation analysis of RSEI levels, this study uncovers the spatial clustering patterns and dynamic changes in ecological quality. Additionally, the CA-Markov model is applied to predict future ecological quality, simulating the potential impacts of urbanization and land-use changes on the ecosystem. The RSEI assessment framework designed in this study aims to provide a scientific basis for ecological conservation and land-use optimization in Johor, while also offering a theoretical model and practical approach for ecological monitoring and management in similar regions.

Table 2. The calculation methods for RSEI.

Index	Sources	No.
$NDVI = \frac{(B_{NIR} - B_{RED})}{(B_{NIR} - B_{RED})}$	Xu et al., 2013 [11]; Crist, 1985 [42]; Chen et al., 2022 [43].	(1)
$L5\_Wet = 0.0315 * B1 + 0.2021 * B2 + 0.3012 * B3 + 0.1594 * B4 - 0.6806 * B5 - 0.6109 * B7$ $L8\_Wet = 0.1509 * B2 + 0.1973 * B3 + 0.3279 * B4 + 0.3406 * B5 - 0.7122 * B6 - 0.4572 * B7$	Xu et al., 2013 [11]; Crist, 1985 [42]; Baig et al., 2014 [44]	(2)
$IBI = \frac{2\frac{B_{SWIR1}}{B_{SWIR1}+B_{NIR}} - \frac{B_{NIR}}{B_{NIR}+B_{RED}} + \frac{B_{GREEN}}{B_{GREEN}+B_{SWIR1}}}{2\frac{B_{SWIR1}}{B_{SWIR1}+B_{NIR}} + \frac{B_{NIR}}{B_{NIR}+B_{RED}} + \frac{B_{GREEN}}{B_{GREEN}+B_{SWIR1}}}$ $SI = \frac{(B_{SWIR1}+B_{RED})-(B_{NIR}+B_{BLUE})}{(B_{SWIR1}+B_{RED})+(B_{NIR}+B_{BLUE})}$ $NDBSI = \frac{IBI+SI}{2}$	Xu et al., 2013 [11]; Hu et al., 2018 [45]; Jiménez-Muñoz et al., 2014 [46]	(3)
$L_{\lambda} = Gain * DN + bias$ $T_{BT} = \frac{K_2}{\text{In}\left(\frac{K_1}{L_{\lambda}}+1\right)}$ $pv = \left(\frac{NVDI-NDVI_{min}}{NDVI_{max}}^2\right)$ $\varepsilon = 0.004 * pv + 0.986$ $LST = \frac{T_{BT}}{1+(\lambda T_{BT}/\rho) * \text{In}(\varepsilon)}$	Xu et al., 2013 [11]; Hu et al., 2018 [45];	(4)
$RESI_0 = 1 - \beta 1 * NDVI + \beta 2 * WET + \beta 3 * NDBSI + \beta 4 * LST$ β1, β2, β3, and β4 are the main components of PCA		(5)
$RSEI = \frac{RSEI_0 - RESI_{0-min}}{RESI_{0-max} - RESI_{0-min}}$		(6)

$\rho = 1.438 * 10^{-2}$	DN	λ	gain	bias	K1	K2
Landsat5	$B_{SWIR1=B6}$	11.450 μm	0.0552	1.2378	607.76	1260.56
Landsat8	$B_{SWIR1=B10}$	10.895 μm	-	-	774.89	1321.08

2.2.2. Spatial Auto-Correlation Analysis

The diversity and regional characteristics of Johor's ecological environment highlight the importance of local spatial autocorrelation analysis. By integrating local Moran's I, Z-values, and *p*-values, it is possible to accurately identify ecological priority zones, vulnerable areas, and transitional zones within Johor's complex ecological landscape.

Spatial autocorrelation methods, particularly through global indicators such as Moran's I, are highly effective in evaluating the spatial relationships of ecological quality across different regions in Johor. These methods provide insights into whether the spatial distribution of ecological quality is uniform [47,48]. Johor's ecosystems, including wetlands, mangroves, and coastal zones, often exhibit significant ecological gradients when juxtaposed with adjacent urban and industrial areas. Using Moran's I, as expressed in Equation (7), the spatial clustering effect of these ecological differences can be quantified, along with the strength of the spatial relationships between high-quality and low-quality ecological areas [49]. A Moran's I value approaching +1 indicates that high-quality ecological areas are spatially clustered, potentially forming critical ecological conservation zones or corridors. Conversely, a value nearing −1 may reflect pronounced ecological quality differentiation, such as the negative impact of industrial activities on surrounding wetland ecosystems.

The analyses for the Johor area are instrumental in identifying high-priority areas for ecological conservation as well as ecological vulnerability zones or areas requiring restoration efforts. The application of spatial autocorrelation methods thus deepens the understanding of the spatial dependencies of ecological quality in Johor, providing a robust foundation for designing targeted ecological protection and management policies [47–49].

$$I = \frac{n}{\sum_{i=1}^{n}\sum_{j=1}^{n} w_{ij}} \frac{\sum_{i=1}^{n}\sum_{j=1}^{n} w_{ij}(x_i - \overline{x})(x_j - \overline{x})}{\sum_{i=1}^{n}(x_i - \overline{x})^2} \tag{7}$$

where n represents the number of observations; x_i and x_j are the attribute values of spatial units i and j, respectively; $\overline{x}$ is the mean of all attribute values across spatial units; and w_{ij} is the weight between units i and j in the spatial weight matrix.

The index of local indicators of spatial association (LISA) is shown in Equation (8), which calculates the relationship between each grid cell and the ecological environment quality of its neighboring areas, providing a more detailed perspective on the spatial heterogeneity of ecological quality within Johor [50]. For instance, wetland and mangrove areas may exhibit H-H clustering types (high–high), indicating regions of ecological importance and good conditions for conservation. Conversely, areas undergoing rapid urbanization, such as Johor Bahru and its surroundings, may form L-L clustering types (low–low), reflecting the negative impacts of urbanization on the ecological environment. Additionally, L-H or H-L types can identify ecological boundaries or transition zones, such as the potential threats posed by urban expansion to wetland ecosystems.

$$\text{LISA} = \frac{x_i - \overline{x}}{\sum_{i=1}^{n}(x_i - \overline{x})^2/n} \sum_{j=1}^{n} w_{ij}(x_j - \overline{x}) \tag{8}$$

LISA cluster maps categorize local spatial autocorrelation of different regions in Johor into the following five types: high–high (H-H), low–low (L-L), low–high (L-H), high–low (H-L), and not significant. These classifications describe the relative ecological quality of a specific area compared to its neighboring areas. H-H indicates that both the selected area and its neighboring regions exhibit high ecological quality; L-L signifies low ecological quality for both; L-H implies that the selected area has low ecological quality while its neighbors are relatively better; and H-L represents the opposite, where the selected area has high ecological quality, but the surrounding areas are relatively poorer [51].

A positive LISA index value (typically indicating positive autocorrelation) corresponds to high–high or low–low cluster regions, whereas a negative LISA index value (typically indicating negative autocorrelation) corresponds to high–low or low–high neighbors. For instance, aquaculture areas in Batu Pahat and Mersing may fall under the L-H category, suggesting low ecological quality in these regions but relatively better conditions in adjacent wetlands or mangroves. Such classifications assist policymakers in identifying key areas where low ecological quality significantly impacts neighboring regions, enabling the prioritization of ecological restoration measures.

In analyzing the ecological quality of Johor, the Z-value can be calculated using Equation (9), while the p-value can be computed using Equation (10). When the Z-value exceeds 1.96, it indicates the presence of spatial autocorrelation in certain regions of Johor, such as H-H clustering patterns in coastal mangroves or wetland conservation areas. Conversely, urbanized areas may exhibit L-L clustering. If the p-value falls below typical significance levels, such as 0.05 or 5%, the observed pattern is considered significant, confirming the existence of spatial autocorrelation.

$$Z(I) = \frac{I - E(I)}{\sqrt{\text{VAR}(I)}} \tag{9}$$

where $E(I)$ represents the mathematical expectation and VAR(I) denotes the variance.

The *p*-value is calculated based on the Z score, assuming that the Z score follows a standard normal distribution, as follows:

$$p = 2 \times \Phi(-\mid Z \mid) \tag{10}$$

where Φ is the cumulative distribution function (CDF) of the standard normal distribution. For a one-sided test, when the *p*-value is less than 0.01, it indicates that the Moran's I index has strong statistical significance.

2.2.3. Cellular Automata-Markov (CA-Markov)

The CA-Markov model is an integrated model combining Cellular Automata (CA) and Markov Chain methodologies. It is widely used to simulate and predict the spatiotemporal evolution of spatial data, particularly in land use change studies within the field of Geographic Information Systems (GIS) [52]. By integrating the state transition probabilities of Markov Chains with the spatial neighborhood rules of CA, the model enables accurate prediction and simulation of dynamic change processes in both temporal and spatial dimensions [53,54].

CA is a discrete model consisting of a regular grid and simple rules, where each grid cell (referred to as a cell) updates its state at each time step based on predetermined rules and the states of its neighboring cells. This spatial dependency allows CA to reflect the spatial relationships and neighborhood effects in geographic regions. The Markov Chain, however, is a statistical model that describes the probabilities of transitioning from one state to another depending only on its current state without any memory of the previous states. The CA-Markov model combines these two features, utilizing the Markov Chain to predict the transition probabilities for each cell's state while using the CA rules to simulate the spatial change process.

In this study, the ecological quality levels derived from RSEI are used as the cell states for the CA-Markov model. The RSEI is categorized into the following five levels with corresponding labels: Poor (1), Fair (2), Moderate (3), Good (4), and Excellent (5). The model leverages the state changes among these levels to build a Transition Probability Matrix (TPM), which is used to predict future trends in ecological quality.

The first step involves classifying the RSEI results for Johor from 1990 to 2020 into five levels (Poor to Excellent) and assigning a cell state to each year based on these categories. At each time step, the total number of transitions between different states is counted—for example, transitions from Poor to Poor (1→1), Poor to Fair (1→2), and so on. The transition probability for a particular state can be calculated as follows:

$$P = \begin{bmatrix} p11 & p12 & p13 & p14 & p15 \\ p21 & p22 & p23 & p24 & p25 \\ p31 & p32 & p33 & p34 & p35 \\ p41 & p42 & p43 & p44 & p45 \\ p51 & p52 & p53 & p54 & p55 \end{bmatrix}$$
$$P_{ij} \in [0,\ 1),\ \sum_{i=1}^{n} P_{ij} = 1,\ i,\ j = 1,\ 2,\ 3\ldots\ldots n. \tag{11}$$

$$\text{PRE\_RESI}_{2024} = \text{RSEI}_{2023} \bullet P \tag{12}$$

where P is a probability matrix for the entire map, and the prediction probability p in the Markov model represents the probability P^{ij} for each pixel.

As shown in Equation (11), the prediction of the ecological index for the next year can be given by the ecological index of the current year multiplied by *P*, as follows:

$$\text{PRE\_RESI}_{2025} = \text{RSEI}_{2024} \bullet P_{\text{new}} \quad (13)$$

$$\lim_{n \to \infty} \pi = \pi \bullet P_{1990-2020} \quad (14)$$

where Pnew is an updated probability matrix *P*, RSEI2024 is the RSEI in the year 2024, PRE_RSEI2025 is the predicted RSEI in the year 2025, and P1990–2025 is the transfer probability matrix obtained from the years 1990~2020. The probability matrix *P* is calculated using the fixed ecological indices from 1990 to 2023 by statistically analyzing the transition probabilities of each pixel. After predicting the ecological data for 2024 using Equation (12), the transition probabilities from 1990 to 2024 are recalculated to obtain a new probability matrix P_{new}, which is then used to predict the ecological index for 2025, as shown in Equation (13).

Markov predictions cannot be extended indefinitely, as they are subject to a limit in predictive capability. It is necessary to obtain new data for the study area to continue making accurate forecasts, and therefore the actual data from 1990 to 2024 must be obtained to ensure the accuracy and reliability of the predictions. Consequently, the CA-Markov model effectively simulates the spatiotemporal changes in ecological quality in Johor from 1990 to 2020 and predicts future ecological states, providing scientific decision support for ecological conservation and land use planning in the region [52–54]. The CA-Markov model plays a significant role in the evaluation of Johor's ecological quality, particularly in dynamic land-use change simulation, future ecological quality prediction, and multi-scenario analysis for decision-making support.

The CA-Markov model precisely captures the spatiotemporal dynamics of land-use changes in Johor, simulating the transformation processes between different land-use types such as urban land, agricultural land, forests, and wetlands. This capability helps to reveal the impacts of land-use changes on ecological quality. By analyzing historical data and current patterns, the CA-Markov model can predict future land-use change trends in Johor, providing a scientific basis for regional ecological planning and policymaking. For example, the model can predict the impact of urban expansion on critical ecosystems like wetlands and forests, supporting the delineation of priority conservation areas.

Moreover, the CA-Markov model enables simulations under various development scenarios, such as economically driven, ecologically prioritized, or integrated development approaches. This allows the assessment of the potential impacts of land-use decisions on ecological quality, providing valuable insights for optimizing land-use strategies and balancing development with environmental sustainability.

2.2.4. Rolling Forecast Validation

Rolling forecast validation is a commonly used method for evaluating time series forecasting models, primarily aimed at assessing the model's performance and stability through a gradual expansion of the training dataset. The core concept of this method is to use all available data from the starting point to the current time point for model training at each prediction step, followed by forecasting the data for the next time point. The actual values are then compared with the predicted values to evaluate the model's forecasting accuracy [55,56].

In this study, the model is first trained using data from 1990 to 2018 to predict the ecological quality values for 2019. The training data is then extended to include data from 1990 to 2019 to predict the values for 2020, and so on until the data from 1990 to 2020 is used to predict the values for 2023. The predicted values are compared with the actual values to calculate the mean squared error (MSE) and the coefficient of determination (R^2) for each year, evaluating the model's stability and accuracy during the rolling forecast process [55,56] such that

$$\mathrm{MSE} = \frac{1}{n}\sum_{i=1}^{n}\left(y_i - \hat{y}_i\right)^2 \tag{15}$$

$$R^2 = 1 - \frac{\sum_{i=1}^{n}(y_i-\hat{y}_i)^2}{\sum_{i=1}^{n}(y_i-\overline{y})^2} \quad i,\ j = 1,\ 2,\ 3\ldots\ldots,\mathrm{n}. \tag{16}$$

where $\hat{y}_i$ is the predicted value, $\overline{y}$ is the average value, and y_i is the true value. A smaller MSE value and an R^2 value closer to one indicate a higher precision.

The advantage of this validation method lies in its ability to dynamically reflect the model's forecasting performance over an expanding time series. Specifically, in forecasting the ecological quality of Johor, rolling forecast validation not only effectively tests the stability of the CA-Markov model but also identifies potential biases in different periods. This provides a basis for model optimization and parameter adjustments, ensuring more accurate and reliable predictions.

3. Results and Analysis

3.1. Factor Attributes

After calculating the NDVI, WET, NDBSI, and LST data for the period from 1990 to 2023 using Equations (1)–(4), this study conducted Principal Component Analysis (PCA) on the results for four different time spots, such as 1990, 2000, 2010, and 2020. The PCA analysis decomposed the covariance matrix of the standardized ecological indicators to extract the principal components that explain the variance in the data. Principal components (e.g., PC1, PC2, etc.) were selected based on their eigenvalues, and their loadings and contribution rates across the indicators were calculated. Table 3 shows the calculation results with the contribution rates of the principal components and the variable loadings for each time spot.

As shown in Table 3, for the imagery from 1990, 2000, 2010, and 2020, the proportion of variance explained by the first principal component (PC1) exceeded or equaled 80.01%, with the highest being 84.17% in 2020. Therefore, this study selected the loadings of PC1 (β_1, β_2, β_3, and β_4) as weight coefficients to calculate the Remote Sensing Ecological Index (RSEI) for each corresponding year using Equation (5). Furthermore, the results were normalized according to Equation (6) to obtain standardized RSEI values (Normalized RSEI, NRSEI). Finally, the Normalized RSEI values were used to generate spatial landscape maps for the respective years, visually depicting the spatial distribution characteristics and trends in ecological environment quality in Johor for 1990, 2000, 2010, and 2020.

PCA was employed to extract the most explanatory components from ecological indicators, including the NDVI, WET, NDBSI, and LST. The contribution weights of these indicators exhibited significant variations across the distinct years of 1990, 2000, 2010, and 2020, reflecting dynamic shifts in regional ecological conditions and the progressive impact of human activities.

Around 1990, the primary drivers of ecological quality were low urbanization and limited industrialization. NDVI held a relatively high weight, indicating that vegetation coverage and green spaces played a dominant role in sustaining ecological health. Meanwhile, WET and NDBSI contributed minimally to the principal components, suggesting negligible impacts from moisture fluctuations or urban expansion at this stage.

Table 3. Results of the PCA of the indices in the years 1990, 2000, 2010, and 2020.

Year	Principal Component	NDVI	WET	NDBSI	LST	Percent Correlation Eigenvalue (%)	Eigenvalue
		β1	β2	β3	β4		
1990	**PC1**	**0.171**	**0.249**	**−0.033**	**−0.952**	**80.01**	**0.0064**
	PC2	0.372	0.301	−0.859	−0.175	12.49	0.0049
	PC3	0.329	−0.081	−0.467	−0.247	7.35	0.0018
	PC4	0.850	0.485	−0.202	0.017	0.15	0.0011
2000	**PC1**	**0.818**	**−0.074**	**−0.425**	**0.088**	**81.55**	**0.033**
	PC2	0.669	0.030	−0.713	−0.203	10.23	0.004
	PC3	0.702	0.213	−0.530	−0.423	7.81	0.003
	PC4	0.147	−0.097	−0.171	0.032	0.41	0.001
2010	**PC1**	**0.103**	**0.450**	**−0.319**	**−0.930**	**81.25**	**0.033**
	PC2	0.708	−0.361	−0.361	−0.180	9.61	0.004
	PC3	0.678	−0.129	−0.649	−0.317	8.64	0.003
	PC4	0.168	−0.091	−0.374	−0.036	0.50	0.001
2020	**PC1**	**0.136**	**0.343**	**−0.076**	**−0.925**	**84.17**	**0.026**
	PC2	0.683	−0.696	−0.141	−0.169	10.89	0.003
	PC3	0.705	−0.624	−0.061	−0.331	4.57	0.001
	PC4	0.131	0.088	−0.985	0.068	0.37	0.001

By 2000, accelerated urbanization and industrialization, particularly in coastal regions, had marked this period. As human activities intensified, the weight of LST gradually increased, highlighting the growing influence of temperature variations on ecological quality. Concurrently, NDVI's weight slightly declined, likely due to reduced green spaces caused by urban sprawl and agricultural expansion.

Around 2010, the urbanization and industrialization in regions like Johor reached a mature phase. Rising temperatures, exacerbated by industrial emissions and transportation activities in urban areas, became more pronounced. LST's weight surged further, dominating the principal components. NDBSI also began to show increased significance, reflecting the escalating ecological pressure from urban infrastructure development, particularly in industrialized zones.

In 2020, the implementation of ecological conservation policies, such as protections for northwestern wetlands and mangrove reserves, led to slight rebounds in NDVI and WET weights. This signaled improved roles of vegetation and moisture conditions in ecological restoration. However, LST retained a high weight, underscoring persistent temperature impacts from climate change and industrial activities.

These temporal shifts in PCA weights illustrate the interplay between natural ecological processes and anthropogenic forces, offering critical insights for sustainable environmental management.

3.2. Spatiotemporal Distribution and Changes in Ecological Environmental Quality

Based on the five RSEI classification standards mentioned earlier, the RSEI indices for the study area were classified in ArcGIS, producing spatiotemporal landscape pattern maps for 1990, 2000, 2010, 2020, and the 30-year average, as shown in Figure 2. These maps clearly illustrate the spatial distribution and changing trends of RSEI in Johor during the study period. Overall, regions with "Excellent" and "Good" ecological status are primarily located in the central, northern, western, and eastern areas, while "Poor" RSEI values are mainly concentrated in the southern urban areas and their surroundings.

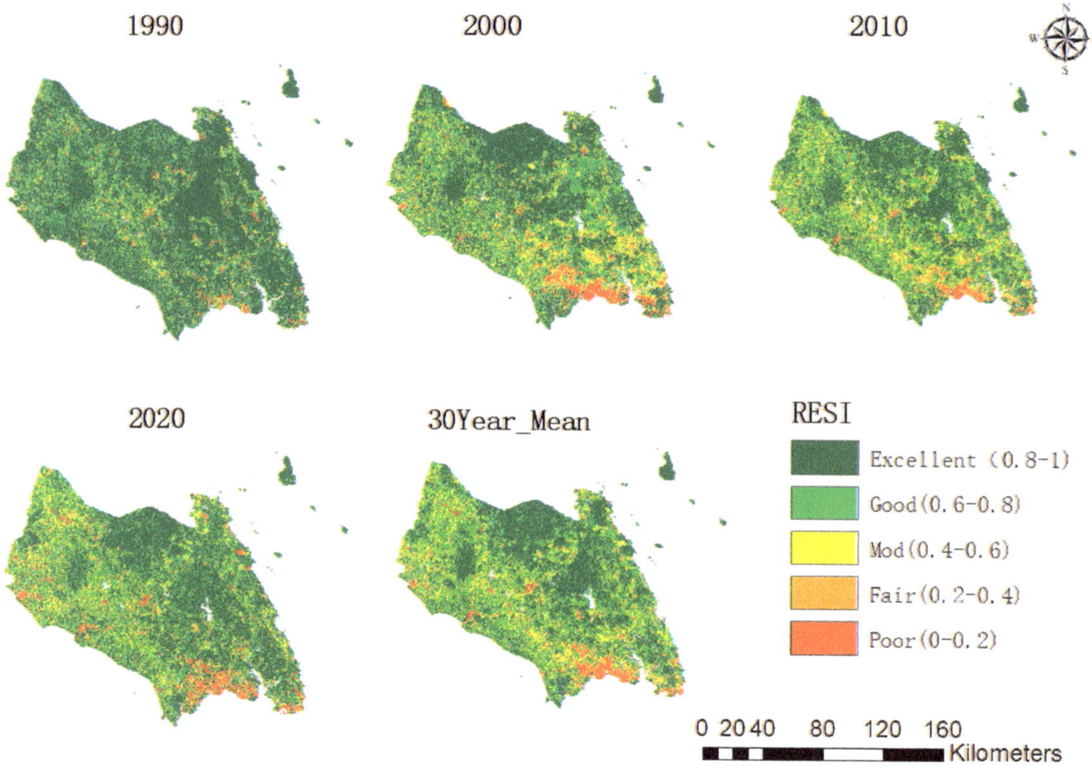

Figure 2. Spatiotemporal landscape pattern maps for 1990, 2000, 2010 and 2020, and the 30-year mean RSEI of Johor state.

Specifically, in 1990, the areas with "Poor" RSEI values were concentrated in the southern urban regions, reflecting that urban planning and environmental protection measures at the time were primarily focused on the city and its surrounding areas. By 2000, the "Poor" RSEI value areas gradually expanded from the southern urban areas to the surrounding towns and rural regions, indicating the broader impact of urbanization and the increased ecological pressure in rural areas. In 2010, the "Poor" RSEI values further extended to include water bodies and their riverbank areas, such as reservoirs and rivers, indicating significant human interference with water resources and aquatic ecosystems. By 2020, the "Poor" RSEI values covered more wetland parks and mangrove protection areas, reflecting increased disturbance in the watershed's surrounding areas, while also suggesting that ecological protection zones and nature parks may have been established and expanded in an attempt to mitigate this trend.

Table 4 presents the area and proportion changes of different RSEI levels in Johor from 1990 to 2020. The data indicates that over the past three decades, the spatial distribution of ecological environment quality in Johor has undergone some changes. First, the area of the highest quality (RSEI 0.8–1) decreased from 11,476 square kilometers in 1990 to 10,376 km^2 in 2020, a reduction of 1100 km^2. This suggests that the highest-quality ecological areas have shrunk over the past 30 years, possibly due to urbanization, land use changes, and ecosystem degradation. In contrast, the area of the "Good" category (RSEI 0.6–0.8) gradually increased from 5024 km^2 in 1990 to 5776 km^2 in 2020, an increase of 752 km^2, which reflects improvements in some medium-quality ecological areas at certain times. The area of the "Moderate" category (RSEI 0.4–0.6) increased by 268 km^2 between 1990 and 2000 but then decreased each year, showing only a 28 km^2 increase by 2020 compared to 1990, indicating a more fluctuating change in ecological quality at this level. The area of the "Fair" category (RSEI 0.2–0.4) showed a consistent annual increase, growing from 621 km^2 in 1990 to 731 km^2 in 2020, an increase of 110 km^2. Furthermore, the area of the lowest quality category (RSEI 0–0.2) significantly increased from 412 km^2 in 1990 to 612 square kilometers in 2020, a total increase of 200 km^2. This indicates a clear deterioration in the ecological environment quality in some areas, especially those more heavily impacted by human activities and land use.

Table 4. Ecological level area statistics of Johor.

Year	1990		2000		2010		2020	
RESI Level	**Area (km^2)**	**Pct. (%)**	**Area (km^2)**	**Pct. (%)**	**Area (km^2)**	**Pct. (%)**	**Area (km^2)**	**Pct. (%)**
Excellent (0.8–1)	11,476	60.16	10,772	56.47	10543	55.27	10,376	54.39
Good (0.6–0.8)	5024	26.34	5288	27.72	5627	29.50	5776	30.28
Mod (0.4–0.6)	1552	8.14	1820	9.54	1736	9.10	1580	8.28
Fair (0.2–0.4)	612	3.21	704	3.69	721	3.78	731	3.84
Poor (0–0.2)	412	2.16	492	2.58	449	2.35	613	3.21
	19,076	100.00	19,076	100.00	19,076	100.00	19,076	100.00

Figure 3 illustrates the percentage change in the area of different ecological environment quality levels in Johor from 1990 to 2020. The proportion of excellent quality (RSEI 0.8–1) decreased from 60.16% in 1990 to 54.39% in 2020, with a decrease of 5.77%. A reduction in the share of high-quality ecological areas in Johor's total area indicates that high-quality ecological regions are gradually shrinking due to the impacts of rapid urbanization and land use changes. The proportion of good quality (RSEI 0.6–0.8) increased from 26.34% in 1990 to 30.28% in 2020. An increase of 3.94% suggests that areas with better environmental quality have gained a larger share, likely due to ecological protection and restoration measures in some regions. The proportion of the lowest quality (RSEI 0–0.2) increased from 2.16% in 1990 to 3.21% in 2020. A rise of 1.05% shows an increase in the proportion of areas with the worst environmental quality, reflecting the ecological degradation issue in certain areas, which is a concerning negative trend.

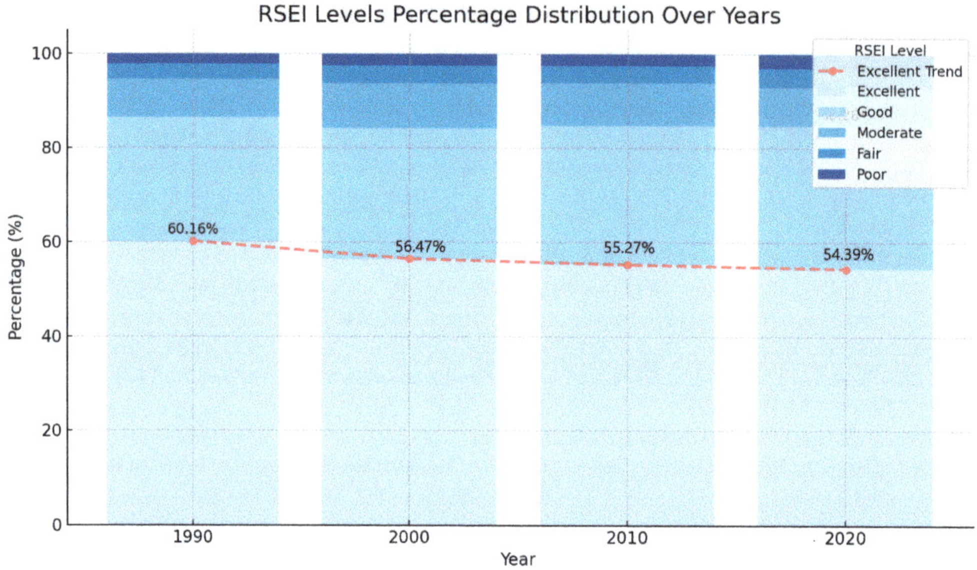

Figure 3. Area distribution of ecological environment quality levels in Johor from 1990 to 2020.

3.3. Spatial Autocorrelation Analysis of Ecological Quality

By calculating the Normalized RSEI average values for Johor from 1990 to 2020, this study uses the UTM-Malaysia-48N projection, resampling the 30 m resolution imagery to a 1 km resolution. The spatial autocorrelation analysis function of GIS is then used to compute the Moran's I value in Equation (7) and the Z value in Equation (9). It can be seen that the Moran's I value is 0.40, the Z value is 38.11, and the p-value is less than 0.001, suggesting a significant positive spatial autocorrelation in the ecological quality distribution, as shown in Figure 4. The ecological conditions exhibit a clustered spatial distribution rather than a random one. From the spatial perspective, as shown in Figure 5, forested areas, urban regions, and plantations present significant clustering characteristics, providing a foundation for subsequent clustering and outlier analysis.

Through the ArcGIS spatial analysis of the RSEI data for 1990, 2000, 2010, and 2020, the results reveal the spatiotemporal evolution characteristics of the ecological quality in Johor, as shown in Figure 6. From the analysis, it can be seen that HH and LL clusters exhibit significant spatial clustering characteristics, reflecting the polarization of ecological conditions within the region. Among them, the HH clusters are mainly concentrated in urbanized areas with intense economic activity, while the LL clusters are primarily found in areas with relatively untouched natural environments or lower levels of development. Furthermore, the HL and LH outlier distributions show a certain spatial dispersion, indicating that these areas may have special ecological or resource utilization features. Further scatter plot analysis in Figure 7 presents the spatial correlation of the RSEI values, particularly highlighting the dominance of HH and LL clusters, which underscores the ecological quality disparities across different regions of Johor.

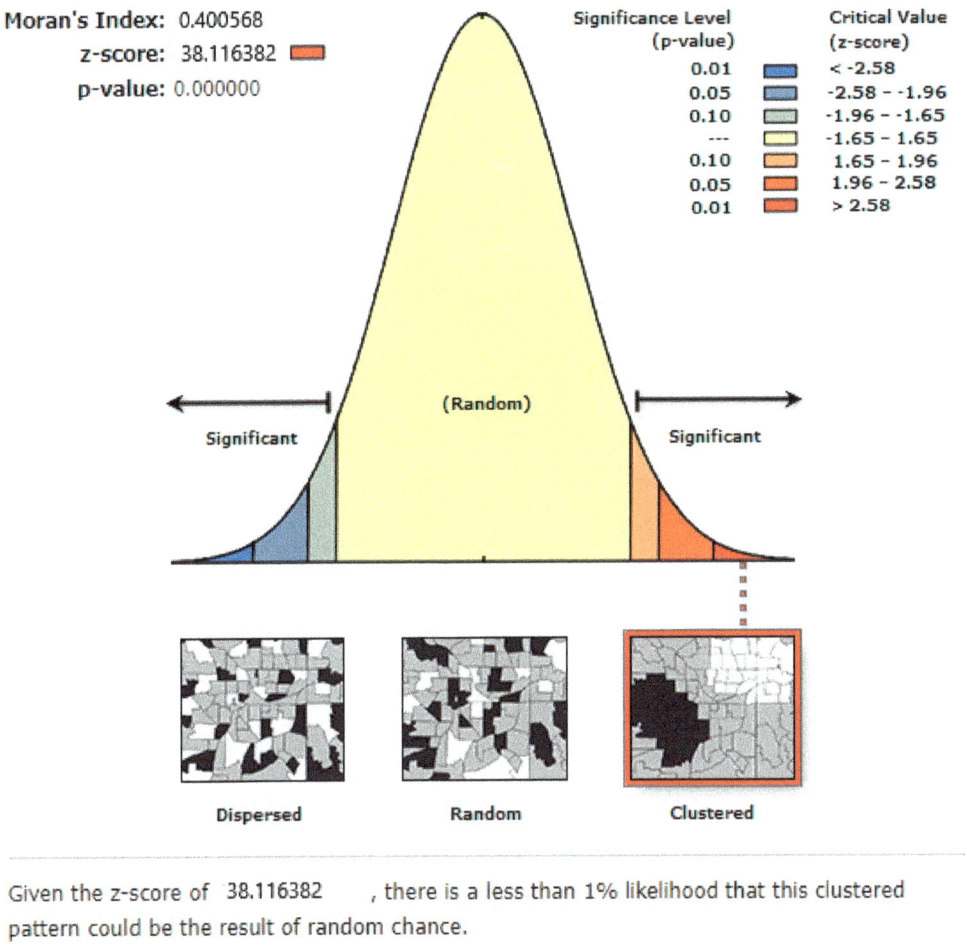

Figure 4. The probability of spatial perspective forested areas, urban regions, and plantations clustering characteristics in Moran I provides a foundation for subsequent clustering and outlier analysis.

The spatial pattern changes in Johor from 1990 to 2020, as shown in Figure 6, clearly illustrate the dynamic distribution characteristics of LL and HH clusters, as well as HL and LH outliers. The LL clusters were initially concentrated in the central region of Johor in 1990, and over the past thirty years, they have gradually expanded towards the northern and eastern coastal areas, forming a large-scale aggregation in the northern region by 2020. The central region of Johor serves as an important link between the economically developed southern region and the less urbanized northern region, primarily characterized by agricultural activities and medium-scale industrial development. The northern region, bordering Pahang, Melaka, and Negeri Sembilan, has relatively low urbanization but plays a significant role in agricultural production, providing essential support for Johor's economic development.

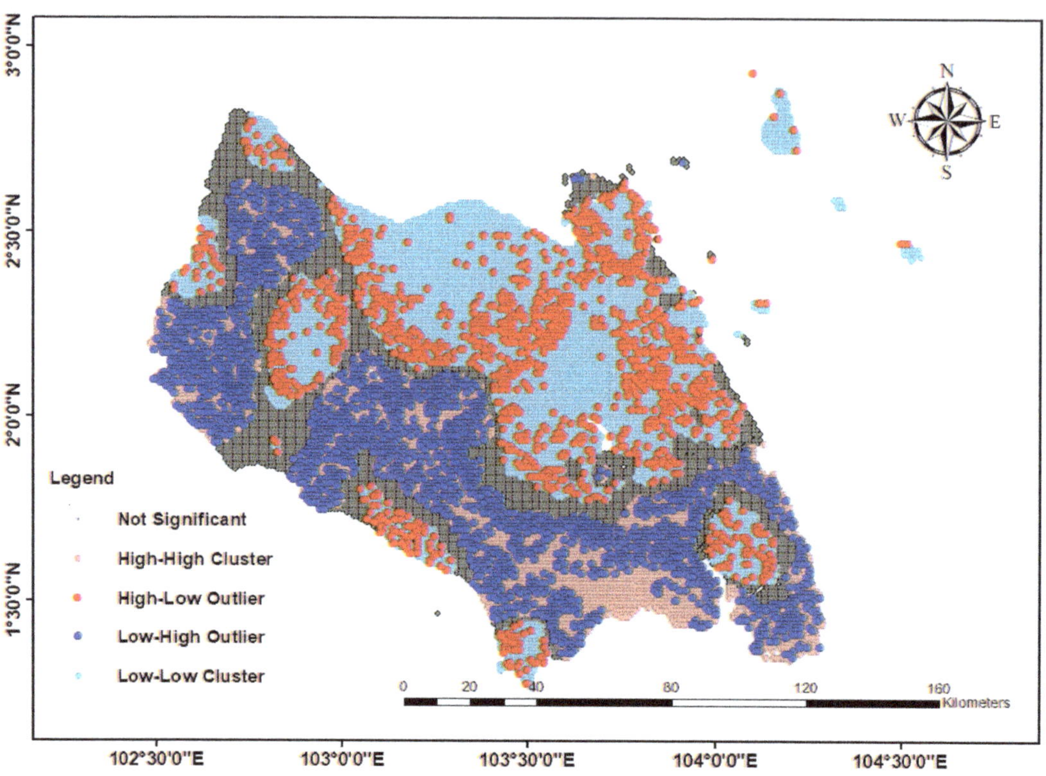

Figure 5. HH\HL\LH\LL spatial landscape patterns.

The HH clusters in 1990 were primarily distributed in the southern and central-western regions of Johor. Over the past 30 years, these clusters have gradually concentrated in the southern, western coastal, and northwestern areas. The southern region, centered around Johor Bahru, is the most economically active area in Johor. The western coastal region, bordering the Strait of Malacca, has become a key economic hub due to its important port facilities and industrial parks, such as large-scale development projects like the Forest City. The eastern region is renowned for its beautiful coastline and vibrant maritime activities, particularly in the Johor River Basin, where the government has actively promoted tourism and fisheries industries by leveraging its natural landscapes and geographical advantages.

It can also be seen from Figure 7 that the distribution of HL and LH outliers is more scattered. The HL areas are mainly found in the southwestern part of Johor, while the LH areas are concentrated in the northeastern part. These regions do not show significant clustering characteristics. The spatial characteristics of these areas reflect a more dispersed distribution of villages or lower levels of resources.

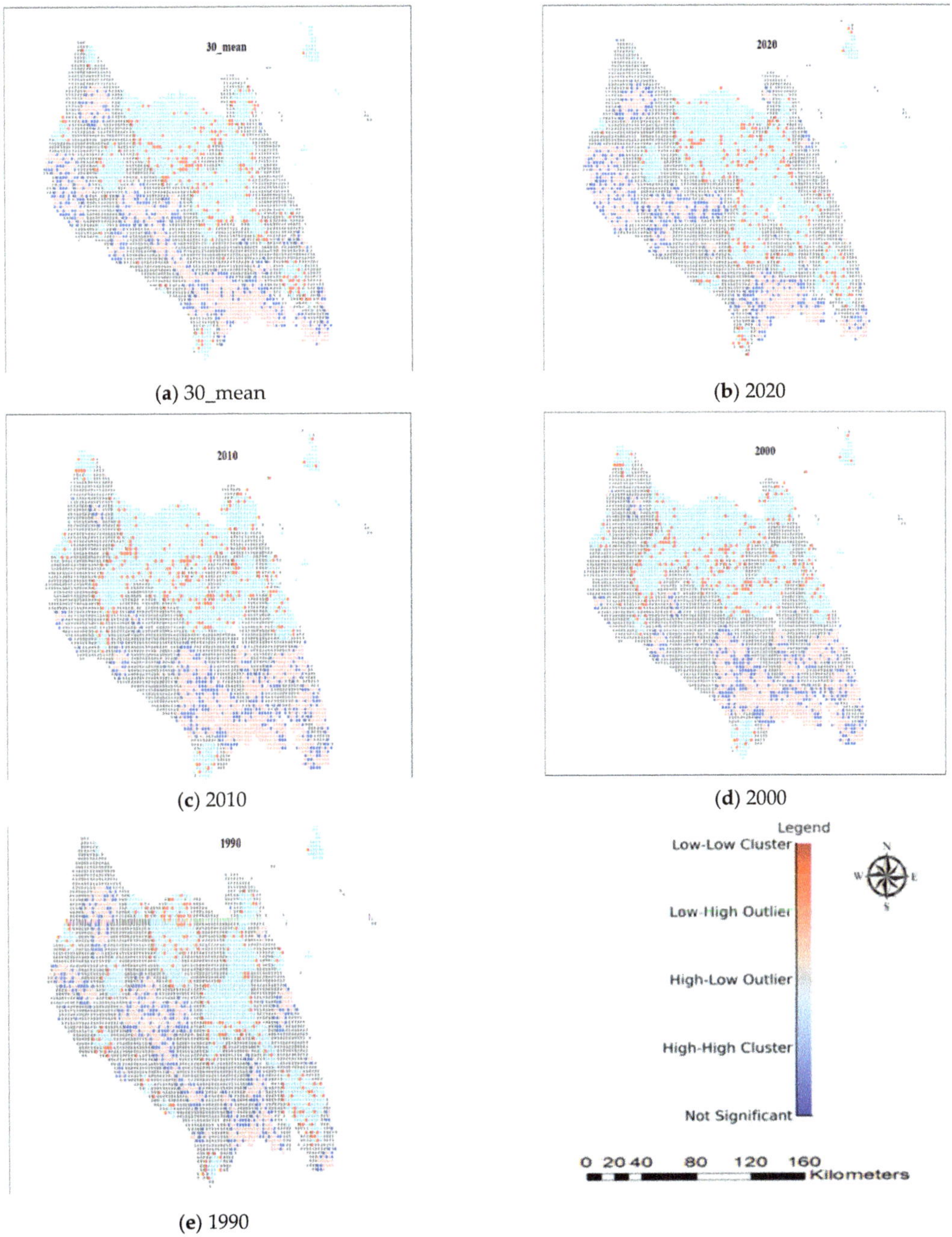

Figure 6. LISA cluster map of the RSEI in Johor for 1990, 2000, 2010, and 2020 and the mean over 30 years.

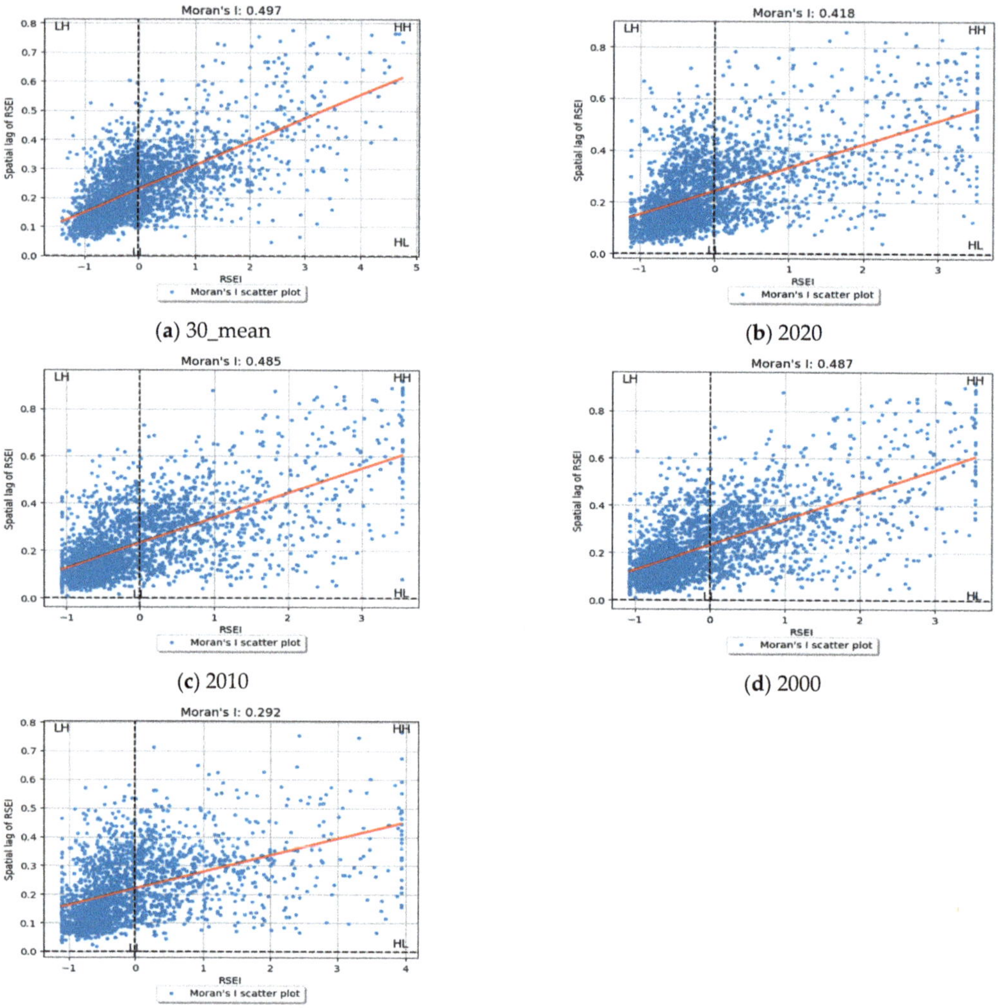

(**a**) 30_mean (**b**) 2020 (**c**) 2010 (**d**) 2000 (**e**) 1990

Figure 7. Moran scatter plots of the RSEI in Johor in 1990, 2000, 2010, and 2020 and the mean value over 30 years. The first, second, third, and fourth quadrants represent the top-right (HH), top-left (LH), bottom-left (LL), and bottom-right (HL) corners of the scatter plot, respectively.

Figure 7 shows the scatter plot of the Moran's I, and it can be seen that the data points from 1990 to 2020 are primarily concentrated in the first quadrant (HH) and the third quadrant (LL), indicating the presence of spatial clusters where geographically adjacent areas have similar attribute values. There are more data points distributed in the third quadrant (LL) than in the first quadrant (HH), reflecting that LL areas are mostly natural or undisturbed, while HH clusters represent urbanized or economically developed regions. Additionally, the density of data points in the second quadrant (LH) is higher than in the fourth quadrant (HL), suggesting that low-value areas are more commonly surrounded by high-value areas. This phenomenon indicates that the geographical characteristics where urban green spaces, parks, and other natural areas are surrounded by highly developed regions, or areas with scarce resources are encircled by agricultural land or plantations.

3.4. One-Year Distribution of RSEI for Johor Area

In the experiment, the CA-Markov model was applied to calculate the dynamic changes in the ecological environment distribution of Johor, based on Equations (12), (13), (15), and (16). To evaluate the performance of the CA-Markov model, different amounts of training data were used for calculation. In order to better analyze the relationships within the ecosystem, we sampled data from 1990 to 2020 at varying intervals, including 4-year, 5-year, 2-year, and 1-year intervals. The detailed sampling strategy and corresponding results are presented in Table 5.

Table 5. Evaluation of the CA-Markov model for ecological outcomes with varying amounts of data.

Interval of Years	10	5	1	0
No. of years	4	7	16	30
MSE	0.4939	0.2433	0.1638	0.0778
R^2	0.9686	0.8219	0.8792	0.8823

When the CA-Markov model was trained using four years of data, it achieved a high R^2 value of 0.9686; however, due to the limited dataset, the model struggled to effectively capture the long-term dynamic characteristics of Johor's ecological environment. The prediction results exhibited poor stability and signs of overfitting. This indicates that models trained with short-term data may perform well on training datasets but lack reliability for forecasting complex ecological system changes.

As the time span of the training data increased, the model's performance improved progressively. With seven years of data, the model began to capture the long-term trends in the ecological environment. The MSE value significantly decreased, and the R^2 value stabilized at 0.8219, achieving a balance between training and testing data. However, limitations in data volume still constrained the model's ability to fully describe the complexities of the ecosystem, with a slight risk of underfitting or overfitting.

When the dataset was extended to 16 years, the model's performance improved significantly, offering a more comprehensive reflection of Johor's long-term ecological trends and complex dynamics. The MSE value dropped to 0.1638, while the R^2 value increased to 0.8792. At this stage, the model's predictive capacity for new data was greatly enhanced, and the risk of overfitting was substantially reduced, demonstrating that a longer period of data enhances the model's generalization ability and the credibility of its predictions.

Further extending the training data to 30 years yielded the best predictive performance, with an MSE value of 0.0778 and an R^2 value of 0.8823. The ample data volume allowed the model to accurately capture the dynamic changes in Johor's ecological environment, achieving high precision and robustness in prediction results.

By analyzing the RSEI data from 1990 to 2020 and constructing the CA-Markov model based on Equation (14) for final predictions, the geographical distribution map of Johor's ecological quality was generated, as shown in Figure 8.

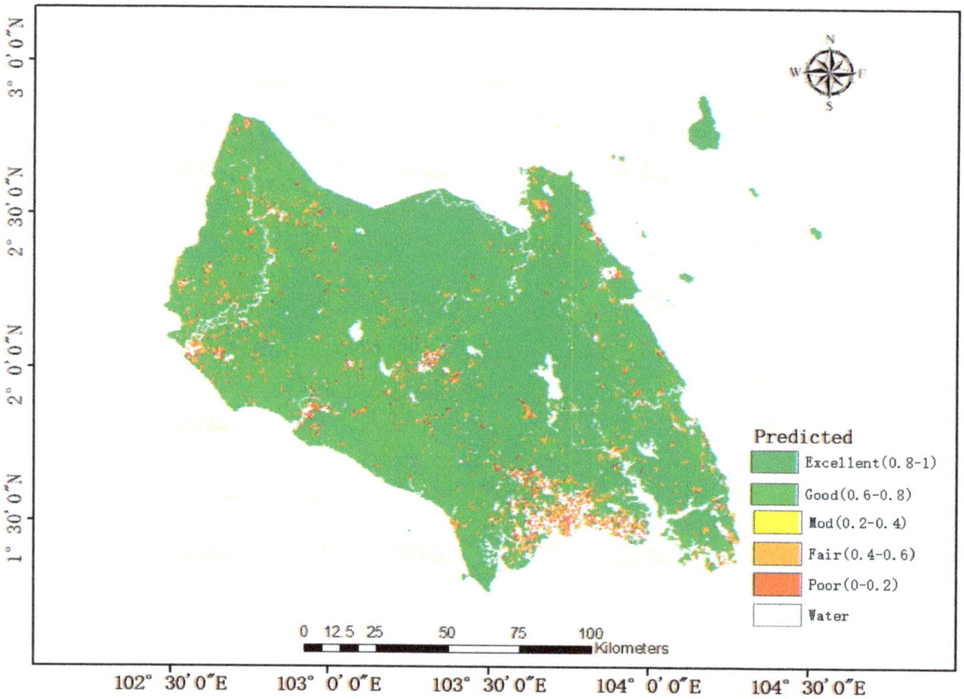

Figure 8. The geographical distribution map of Johor's ecological quality.

It can be seen from Figure 8 that the green areas (Excellent, 0.8–1) dominate most of the study region, primarily concentrated in areas with high forest coverage. These zones exhibit low pollution levels, reflecting the effectiveness of natural resource management and conservation measures. The ecological health in these regions stands out, indicating minimal human interference and high environmental resilience.

Yellow areas (Good, 0.6–0.8) indicate regions with slightly affected but generally stable ecological conditions. These areas might experience limited human activity or have benefited from mitigation measures that prevented further ecological degradation.

Orange areas (Moderate, 0.4–0.6) represent zones with moderate ecological conditions, often associated with light industrial activities, agricultural runoff, or early-stage urbanization. While these areas have not entirely lost their natural landscape, they exhibit signs of ecological stress and potential vulnerability.

Red areas (Fair, 0.2–0.4) and pink areas (poor, 0–0.2) are predominantly located in regions with intense human activity, such as cities, ports, and settlements. These areas face significant ecological challenges, including high pollution levels, habitat degradation, and weakened ecosystem functionality.

Geographically, the ecological quality in Johor's southern and eastern edge regions is highly variable, likely influenced by urban runoff, emissions from aquaculture, or inadequate waste management systems. Central and northern areas are predominantly green, interspersed with small patches of yellow, reflecting low land-use intensity. The western region is mainly yellow, indicating the impact of intensive land use on ecological quality. Additionally, water bodies are often surrounded by red and pink patches, underscoring the environmental pressures faced by these areas.

4. Discussion

In this study, the calculated RSEI values have been influenced by the absence of 2010 Landsat 8 Band 10 data, as shown in Figure 9. To address this issue, MODIS Land Surface Temperature (LST) data were used as a substitute. However, since MODIS data has a resolution of only 1000 m, even after resampling to 30 m, the clustering effect of the data persisted, affecting the spatial accuracy of the distribution. This limitation may have introduced localized errors in the RSEI results, particularly in the densely populated southern and eastern regions, reducing the model's ability to capture subtle ecological changes. Additionally, the data substitution strategy could have introduced instability in the CA-Markov model's prediction results, especially when processing data with a 10-year interval, causing fluctuations. This highlights the need to prioritize higher-resolution and temporally complete datasets in future research to improve the accuracy and consistency of RSEI calculations and related predictive modeling.

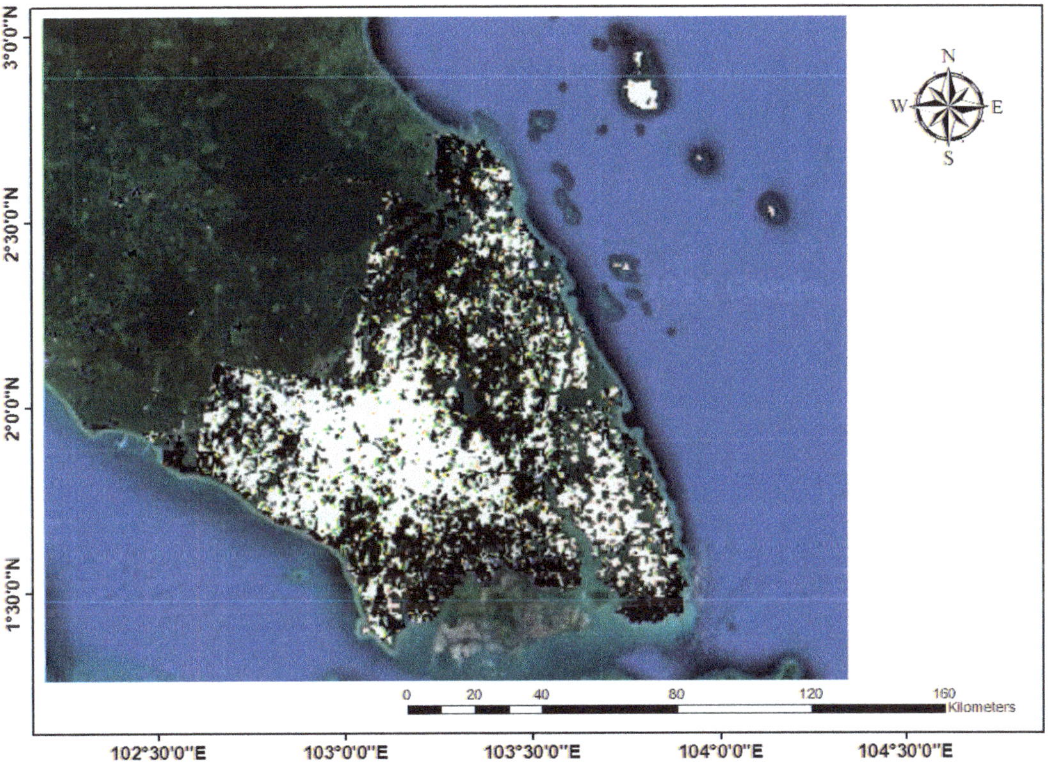

Figure 9. Absence of RSEI values due to the absence of Band 10 in 2010.

Through spatial analysis of Johor's RSEI from 1990 to 2020, significant spatial heterogeneity in ecological quality across the region was observed. The LL clusters were primarily located in the central, northern, and eastern coastal areas, dominated by agriculture and medium-scale industries. These areas exhibited relatively stable ecological conditions with minimal human disturbance. In contrast, the HH clusters were concentrated in the southern urban areas and western coastal industrial zones, reflecting the strong impact of economic activities on regional ecology. The HL and LH outliers were sporadically distributed in the southwestern and northeastern regions, indicating potential for development or localized ecological pressures. This spatial pattern highlights the complex interplay between economic development, land use, and ecological conservation in Johor, underscoring the pronounced regional imbalances in development.

From a temporal perspective, Johor's ecological quality underwent notable changes over the past 30 years. The area classified as having excellent ecological conditions decreased by 1100 km^2, while the area with good conditions increased by 752 km^2. Simultaneously, the area categorized as poor grew from 412 km^2 to 613 km^2, demonstrating the environmental pressures brought by urbanization and industrialization. The southern and eastern edge areas displayed uneven ecological quality, likely influenced by urban runoff, industrial pollution, and inadequate waste management. In contrast, the central and northern regions maintained predominantly high ecological quality, possibly due to lower land use intensity and protective measures. These trends reveal that the dynamic changes in Johor's ecological environment are driven by a combination of factors, including economic activities, land use policies, and resource management practices.

The CA-Markov model's predictions for Johor's future ecological quality indicate that increasing the amount of data significantly enhances the model's accuracy and robustness. However, its limitations remain notable. First, the model relies on a static transition probability matrix, which does not fully account for dynamic factors such as policy adjustments, climate change, and economic activities, potentially leading to insufficient predictions of future complex ecological changes. Second, the use of MODIS data as a substitute in 2010 introduces accuracy issues, particularly in areas with high spatial heterogeneity. Additionally, the computational demands and memory consumption of large-scale datasets limit the model's efficiency when handling high-resolution data. Enhancing the dynamic adaptability of the model's transition probabilities and optimizing computational resources will be critical to improving prediction accuracy.

This study adapted and optimized the RSEI with the CA-Markov model, particularly for wetland and mangrove ecosystems in Johor. The introduction of the WET and NDBSI enhanced the RSEI model's ability to accurately reflect dynamic changes in the local ecological environment, offering a novel methodological framework for similar regions. By integrating remote sensing data with the CA-Markov model, we further improved its spatiotemporal prediction capabilities. The application of a rolling prediction validation method optimized the model's performance in Johor, demonstrating its effectiveness in capturing spatiotemporal patterns of regional ecological changes. Notably, the proposed model successfully revealed future ecological quality trends in long-term time series predictions, providing a scientific basis for ecological conservation in the region.

Through spatial autocorrelation analysis—specifically the combined use of local Moran's I and LISA indices—the spatial aggregation patterns of ecological quality in Johor have been identified. This approach effectively delineated priority ecological conservation zones, vulnerable areas, and transitional regions, offering critical decision-making support for regional ecological management. Based on spatiotemporal analysis results, targeted ecological protection and management policies can be formulated, particularly for urbanized areas and industrial belts with poorer ecological quality. Concrete restoration and conservation measures can be tailored to these findings. These policy recommendations hold significant practical value, providing theoretical support for ecological management in Johor and contributing to sustainable development in similar regions.

5. Conclusions

This study utilized the GEE platform to calculate the RSEI for Johor from 1990 to 2020 and applied the CA-Markov model to predict future ecological environment quality. The analysis revealed the spatiotemporal evolution characteristics and trends of ecological quality in Johor. The results indicated significant regional differences in ecological quality over the past 30 years. High-quality ecological areas decreased by 1100 km^2, while medium-quality areas fluctuated, and poor-quality areas significantly increased, reflecting a notable trend of localized ecological degradation. In spatial dimension, the HH clusters of ecological quality were primarily located in the southern and western regions, characterized by dense urban and industrial activities, while the LL clusters were concentrated in the northern and eastern areas, dominated by low-interference agricultural zones. The scattered HL and LH outliers suggest potential development opportunities in the future. The CA-Markov model predictions demonstrated that the model's robustness and accuracy improved significantly with increased data volume, as indicated by optimized MSE and R^2 values. The forecast showed that future high-quality ecological areas would mainly concentrate in central and northern regions, while poor-quality areas would be concentrated in the southern and western regions, where human activities are most intensive. Remarkably, ecological patches near water sources exhibited a higher risk of degradation, warranting particular attention for conservation efforts.

Author Contributions: Software, W.Q.; validation, W.Q.; writing—original draft, W.Q. and N.W.; formal analysis, W.Q.; conceptualization, W.Q., M.H.I., M.F.R. and J.D.; supervision—investigation, M.H.I., M.F.R., N.W. and J.D.; resources, W.Q.; writing—review and editing, M.H.I., M.F.R. and N.W.; supervision, M.H.I., M.F.R. and N.W.; methodology, W.Q.; project administration: M.H.I., M.F.R. and N.W.; funding acquisition, N.W. and J.D. All authors have read and agreed to the published version of the manuscript.

Funding: This study is supported by the National Natural Science Foundation of China: 42366008, National Natural Science Foundation of China: 52161042; Guangxi Science and Technology Major Program: 2024AA29055; 100 Scholar Plan of the Guangxi Zhuang Autonomous Region of China: 2018.

Institutional Review Board Statement: Not applicable.

Informed Consent Statement: Not applicable.

Data Availability Statement: Data are available upon request.

Acknowledgments: The authors would like to thank all reviewers and editors for their comments on this paper.

Conflicts of Interest: The authors declare that the research was conducted in the absence of any commercial or financial relationships that could be construed as a potential conflict of interest.

Abbreviations

The following abbreviations are used in this manuscript:

RSEI	Remote Sensing Environmental Index
CA-Markov	Cellular Automata-Markov
NDVI	Normalized Difference Vegetation Index
EVI	Enhanced Vegetation Index
PCA	principal component analysis
MSRE	Multi-Indicator Remote Sensing Ecological Index
IRSEI	Integrated Remote Sensing Ecological Index
WET	Wetness Index
NDBSI	Normalized Difference Built-up Space Index
LST	and Land Surface Temperature

References

1. Helali, J.; Asaadi, S.; Jafarie, T.; Habibi, M.; Salimi, S.; Momenpour, S.E.; Shahmoradi, S.; Hosseini, S.A.; Hessari, B.; Saeidi, V. Drought monitoring and its effects on vegetation and water extent change using remote sensing data in Urmia Lake watershed, Iran. *J. Water Clim. Change* **2022**, *13*, 2107–2128. [CrossRef]

2. Zheng, Z.; Wu, Z.; Chen, Y.; Yang, Z.; Marinello, F. Exploration of eco-environment and urbanization changes in coastal zones: A case study in China over the past 20 years. *Ecol. Indic.* **2020**, *119*, 106847. [CrossRef]
3. Huete, A.; Didan, K.; Miura, T.; Rodriguez, E.P.; Gao, X.; Ferreira, L.G. Overview of the radiometric and biophysical performance of the MODIS vegetation indices. *Remote Sens. Environ.* **2002**, *83*, 195–213. [CrossRef]
4. Lu, D.; Mausel, P.; Brondízio, E.; Moran, E. Change detection techniques. *Int. J. Remote Sens.* **2004**, *25*, 2365–2401. [CrossRef]
5. Running, S.W.; Zhao, M. User's Guide for MODIS Land Products (Collection 6): MOD17A2/A3—MODIS GPP and NPP. NASA. 2015. Available online: https://modis.gsfc.nasa.gov/data/dataprod/mod17.php (accessed on 9 April 2014).
6. Wan, Z. *MODIS Land Surface Temperature Products Users' Guide*; Institute for Computational Earth System Science, University of California: Santa Barbara, CA, USA, 2006; p. 805.
7. Dekker, A.G.; Peters, S.W.M. The use of the Thematic Mapper for the analysis of eutrophic lakes: A case study in The Netherlands. *Int. J. Remote Sens.* **1993**, *14*, 799–821. [CrossRef]
8. Turner, W.; Spector, S.; Gardiner, N.; Fladeland, M.; Sterling, E.; Steininger, M. Remote sensing for biodiversity science and conservation. *Trends Ecol. Evol.* **2003**, *18*, 306–314. [CrossRef]
9. Ozesmi, S.L.; Bauer, M.E. Satellite remote sensing of wetlands. *Wetl. Ecol. Manag.* **2002**, *10*, 381–402. [CrossRef]
10. Parmesan, C.; Yohe, G. A globally coherent fingerprint of climate change impacts across natural systems. *Nature* **2003**, *421*, 37–42. [CrossRef] [PubMed]
11. Xu, H.Q. A remote sensing urban ecological index and its application. *Acta Ecol. Sin.* **2013**, *33*, 7853–7862. [CrossRef]
12. Hang, X.; Li, Y.; Luo, X.; Xu, M.; Han, X. Assessing the ecological quality of Nanjing during its urbanization process by using satellite, meteorological, and socioeconomic data. *J. Meteorol. Res.* **2020**, *34*, 280–293. [CrossRef]
13. Li, P. Research on ecoenvironmental quality evaluation system based on big data analysis. *Comput. Intell. Neurosci.* **2022**, *2022*, 5191223. [CrossRef] [PubMed]
14. Shan, W.; Jin, X.; Ren, J.; Wang, Y.; Xu, Z.; Fan, Y.; Gu, Z.; Hong, C.; Lin, J.; Zhou, Y. Ecological environment quality assessment based on remote sensing data for land consolidation. *J. Clean. Prod.* **2019**, *239*, 118126. [CrossRef]
15. Xu, H.; Wang, M.; Shi, T.; Guan, H.; Fang, C.; Lin, Z. Prediction of ecological effects of potential population and impervious surface increases using a remote sensing based ecological index (RSEI). *Ecol. Indic.* **2018**, *93*, 730–740. [CrossRef]
16. Jiang, C.L.; Wu, L.; Liu, D.; Wang, S.M. Dynamic monitoring of eco-environmental quality in arid desert area by remote sensing: Taking the Gurbantunggut Desert China as an example. *Ying Yong Sheng Tai Xue Bao J. Appl. Ecol.* **2019**, *30*, 877–883.
17. Li, S.; Li, X.; Gong, J.; Dang, D.; Dou, H.; Lyu, X. Quantitative analysis of natural and anthropogenic factors influencing vegetation NDVI changes in temperate drylands from a spatial stratified heterogeneity perspective: A case study of Inner Mongolia Grasslands, China. *Remote Sens.* **2022**, *14*, 3320. [CrossRef]
18. Wang, J.; Wang, J.; Xu, J. Spatio-temporal variation and prediction of ecological quality based on remote sensing ecological index–a case study of Zhanjiang City, China. *Front. Ecol. Evol.* **2023**, *11*, 1153342. [CrossRef]
19. Cheng, L.L.; Wang, Z.W.; Tian, S.F.; Liu, Y.T.; Sun, M.Y.; Yang, Y.M. Evaluation of eco-environmental quality in Mentougou District of Beijing based on improved remote sensing ecological index. *Chin. J. Ecol.* **2021**, *40*, 1177.
20. Akbari, M.; Shalamzari, M.J.; Memarian, H.; Gholami, A. Monitoring desertification processes using ecological indicators and providing management programs in arid regions of Iran. *Ecol. Indic.* **2020**, *111*, 106011. [CrossRef]
21. Kosmas, C.; Kairis, O.; Karavitis, C.; Ritsema, C.; Salvati, L.; Acikalin, S.; Alcalá, M.; Alfama, P.; Atlhopheng, J.; Barrera, J.; et al. Evaluation and selection of indicators for land degradation and desertification monitoring: Methodological approach. *Environ. Manag.* **2014**, *54*, 951–970. [CrossRef]
22. Wu, S.; Gao, X.; Lei, J.; Zhou, N.; Guo, Z.; Shang, B. Ecological environment quality evaluation of the Sahel region in Africa based on remote sensing ecological index. *J. Arid. Land* **2022**, *14*, 14–33. [CrossRef]
23. Wang, Y. Evaluation of lake wetland ecotourism resources based on remote sensing ecological index. *Arab. J. Geosci.* **2021**, *14*, 559. [CrossRef]
24. Tang, H.; Fang, J.; Xie, R.; Ji, X.; Li, D.; Yuan, J. Impact of land cover change on a typical mining region and its ecological environment quality evaluation using remote sensing based ecological index (RSEI). *Sustainability* **2022**, *14*, 12694. [CrossRef]
25. Geng, J.; Yu, K.; Xie, Z.; Zhao, G.; Ai, J.; Yang, L.; Yang, H.; Liu, J. Analysis of spatiotemporal variation and drivers of ecological quality in Fuzhou based on RSEI. *Remote Sens.* **2022**, *14*, 4900. [CrossRef]
26. Xiong, Y.; Xu, W.; Lu, N.; Huang, S.; Wu, C.; Wang, L.; Dai, F.; Kou, W. Assessment of spatial–temporal changes of ecological environment quality based on RSEI and GEE: A case study in Erhai Lake Basin, Yunnan province, China. *Ecol. Indic.* **2021**, *125*, 107518. [CrossRef]
27. Shan, Y.; Dai, X.; Li, W.; Yang, Z.; Wang, Y.; Qu, G.; Liu, W.; Ren, J.; Li, C.; Liang, S.; et al. Detecting spatial-temporal changes of urban environment quality by remote sensing-based ecological indices: A case study in Panzhihua city, Sichuan Province, China. *Remote Sens.* **2022**, *14*, 4137. [CrossRef]
28. Yi, S.; Zhou, Y.; Zhang, J.; Li, Q.; Liu, Y.; Guo, Y.; Chen, Y. Spatial-temporal evolution and motivation of ecological vulnerability based on RSEI and GEE in the Jianghan Plain from 2000 to 2020. *Front. Environ. Sci.* **2023**, *11*, 1191532. [CrossRef]

29. Zhang, F.; Wang, Y.; Jim, C.Y.; Chan, N.W.; Tan, M.L.; Kung, H.T.; Shi, J.; Li, X.; He, X. Analysis of Urban Expansion and Human–Land Coordination of Oasis Town Groups in the Core Area of Silk Road Economic Belt, China. *Land* **2023**, *12*, 224. [CrossRef]
30. Zhang, Y.; She, J.; Long, X.; Zhang, M. Spatio-temporal evolution and driving factors of eco-environmental quality based on RSEI in Chang-Zhu-Tan metropolitan circle, central China. *Ecol. Indic.* **2022**, *144*, 109436. [CrossRef]
31. Yao, K.; Halike, A.; Chen, L.; Wei, Q. Spatiotemporal changes of eco-environmental quality based on remote sensing-based ecological index in the Hotan Oasis, Xinjiang. *J. Arid. Land* **2022**, *14*, 262–283. [CrossRef]
32. Luo, M.; Zhang, S.; Huang, L.; Liu, Z.; Yang, L.; Li, R.; Lin, X. Temporal and spatial changes of ecological environment quality based on RSEI: A case study in Ulan Mulun river basin, China. *Sustainability* **2022**, *14*, 13232. [CrossRef]
33. Gao, P.; Kasimu, A.; Zhao, Y.; Lin, B.; Chai, J.; Ruzi, T.; Zhao, H. Evaluation of the temporal and spatial changes of ecological quality in the Hami oasis based on RSEI. *Sustainability* **2020**, *12*, 7716. [CrossRef]
34. Karbalaei Saleh, S.; Amoushahi, S.; Gholipour, M. Spatiotemporal ecological quality assessment of metropolitan cities: A case study of central Iran. *Environ. Monit. Assess.* **2021**, *193*, 305. [CrossRef]
35. Xia, Q.Q.; Chen, Y.N.; Zhang, X.Q.; Ding, J.L. Spatiotemporal changes in ecological quality and its associated driving factors in Central Asia. *Remote Sens.* **2022**, *14*, 3500. [CrossRef]
36. Li, Y.; Li, Y.; Yang, X.; Feng, X.; Lv, S. Evaluation and driving force analysis of ecological quality in Central Yunnan Urban Agglomeration. *Ecol. Indic.* **2024**, *158*, 111598. [CrossRef]
37. Yang, X.; Bai, Y.; Che, L.; Qiao, F.; Xie, L. Incorporating ecological constraints into urban growth boundaries: A case study of ecologically fragile areas in the Upper Yellow River. *Ecol. Indic.* **2021**, *124*, 107436. [CrossRef]
38. Yang, H.; Xu, W.; Yu, J.; Xie, X.; Xie, Z.; Lei, X.; Wu, Z.; Ding, Z. Exploring the impact of changing landscape patterns on ecological quality in different cities: A comparative study among three megacities in eastern and western China. *Ecol. Inform.* **2023**, *77*, 102255. [CrossRef]
39. Zhou, S.; Li, W.; Zhang, W.; Wang, Z. The Assessment of the Spatiotemporal Characteristics of the Eco-Environmental Quality in the Chishui River Basin from 2000 to 2020. *Sustainability* **2023**, *15*, 3695. [CrossRef]
40. Zhang, N.; Xiong, K.; Zhang, J.; Xiao, H. Evaluation and prediction of ecological environment of karst world heritage sites based on google earth engine: A case study of Libo–Huanjiang karst. *Environ. Res. Lett.* **2023**, *18*, 034033. [CrossRef]
41. Wang, S.; Ge, Y. Ecological quality response to multi-scenario land-use changes in the Heihe River Basin. *Sustainability* **2022**, *14*, 2716. [CrossRef]
42. Crist, E.P. A TM tasseled cap equivalent transformation for reflectance factor data. *Remote Sens. Environ.* **1985**, *17*, 301–306. [CrossRef]
43. Chen, C.; Chen, H.; Liang, J.; Huang, W.; Xu, W.; Li, B.; Wang, J. Extraction of water body information from remote sensing imagery while considering greenness and wetness based on Tasseled Cap transformation. *Remote Sens.* **2022**, *14*, 3001. [CrossRef]
44. Baig, M.H.A.; Zhang, L.; Shuai, T.; Tong, Q. Derivation of a tasselled cap transformation based on Landsat 8 at-satellite reflectance. *Remote Sens. Lett.* **2014**, *5*, 423–431. [CrossRef]
45. Hu, X.; Xu, H. A new remote sensing index for assessing the spatial heterogeneity in urban ecological quality: A case from Fuzhou City, China. *Ecol. Indic.* **2018**, *89*, 11–21. [CrossRef]
46. Jiménez-Muñoz, J.C.; Sobrino, J.A.; Skoković, D.; Mattar, C.; Cristóbal, J. Land surface temperature retrieval methods from Landsat-8 thermal infrared sensor data. *IEEE Geosci. Remote Sens. Lett.* **2014**, *11*, 1840–1843. [CrossRef]
47. Fan, G.Y.; Cowley, J.M. Auto-correlation analysis of high-resolution electron micrographs of near-amorphous thin films. *Ultramicroscopy* **1985**, *17*, 345–355. [CrossRef]
48. Martin, D. An assessment of surface and zonal models of population. *Int. J. Geogr. Inf. Syst.* **1996**, *10*, 973–989. [CrossRef]
49. Griffith, D.A. *Spatial Autocorrelation: A Primer Association of American Geographers*; Resource Publications in Geography: Washington, DC, USA, 1987.
50. Anselin, L. Local indicators of spatial association—LISA. *Geogr. Anal.* **1995**, *27*, 93–115. [CrossRef]
51. Jing, Y.; Zhang, F.; He, Y.; Johnson, V.C.; Arikena, M. Assessment of spatial and temporal variation of ecological environment quality in Ebinur Lake Wetland National Nature Reserve, Xinjiang, China. *Ecol. Indic.* **2020**, *110*, 105874. [CrossRef]
52. Hamad, R.; Balzter, H.; Kolo, K. Predicting land use/land cover changes using a CA-Markov model under two different scenarios. *Sustainability* **2018**, *10*, 3421. [CrossRef]
53. Chotchaiwong, P.; Wijitkosum, S. Predicting urban expansion and urban land use changes in Nakhon Ratchasima City using a CA-Markov model under two different scenarios. *Land* **2019**, *8*, 140. [CrossRef]
54. Ruben, G.B.; Zhang, K.; Dong, Z.; Xia, J. Analysis and projection of land-use/land-cover dynamics through scenario-based simulations using the CA-Markov model: A case study in guanting reservoir basin, China. *Sustainability* **2020**, *12*, 3747. [CrossRef]

55. Li, L.; Noorian, F.; Moss, D.J.; Leong, P.H. Rolling window time series prediction using MapReduce. In Proceedings of the 2014 IEEE 15th International Conference on Information Reuse and Integration (IEEE IRI 2014), Redwood City, CA, USA, 13–15 August 2014; pp. 757–764.
56. Inoue, A.; Jin, L.; Rossi, B. Rolling window selection for out-of-sample forecasting with time-varying parameters. *J. Econom.* **2017**, *196*, 55–67. [CrossRef]

sustainability

MDPI

Article

Natural Hazard Assessment in the Southeastern Margin of the Ría de Arosa (Pontevedra, Spain) Using GIS Techniques

Carlos E. Nieto [1,*], Antonio Miguel Martínez-Graña [1] and Leticia Merchán [2]

1 Department of Geology, Faculty of Sciences, University of Salamanca, Merced Square, 37008 Salamanca, Spain; amgranna@usal.es
2 Department of Soil Sciences, Faculty of Agricultural and Environmental Sciences, University of Salamanca, Filiberto Villalobos Avenue, 119, 37007 Salamanca, Spain; leticiamerchan@usal.es
* Correspondence: carlosenriquenm@usal.es

Abstract: The characterization of natural hazards in coastal environments is of great necessity, especially in the current context of global climate change and increasing population concentrations. This research focuses on a multi-hazard analysis of the main geotechnical, geomorphological, hydrological, and lithological risks in the southeastern margin of the Ría de Arosa using Geographic Information System techniques. The integration of geotechnical characterization maps and natural hazard maps has allowed for the identification of areas with a high susceptibility to natural disasters, which is crucial for territorial planning and management in the context of growing urban pressure and global climate change. The results indicate that poorly consolidated surface formations, especially in transitional areas such as dunes and marshes, are particularly vulnerable. Additionally, areas with higher lithological competence have been identified, where slope changes contribute to ground instability. This analysis provides valuable tools for decision-making and the implementation of risk management policies, promoting sustainable development, the protection of coastal ecosystems, and the prevention of risks from urban planning and civil engineering activities in the Ría de Arosa.

Keywords: natural hazard; coastal management; GIS; geomorphology; geotechnical hazard

1. Introduction

The characteristics of coastal areas, from an ecological and economic standpoint, make these territories suitable for the development of human activities [1–4]. These areas currently host 11% of the human population, amounting to approximately 896 million inhabitants [5]. Economic activity, based on urban and industrial development, along with the rise in tourism and recreational activities, makes coastal ecosystems among the most affected and altered ecosystems [1,6,7]. Additionally, these natural environments are threatened by natural processes that annually cause significant socioeconomic damage and, in the most extreme cases, a loss of human lives [8,9]. The current global climate change context leads to a scenario in which extreme events are occurring more frequently [5].

Natural disasters constitute one of the environmental problems that generate significant ecosystem alterations and affect urban areas with high population densities [10,11]. For this reason, there is a need to understand the geological processes associated with these events and delineate the areas that are more vulnerable and have a lower response capacity [12–14]. These coastal environments, where geological hazards play an important role, need to be the subject of effective land-use planning strategies [15,16] and of passive or active measures to mitigate these hazards and ensure sustainable development [17,18]. Furthermore, for the development of urban and civil engineering projects, geotechnical knowledge is crucial to detect potential hazards while planning [19,20].

This vulnerability scenario is clearly observable in the southeastern margin of the Ría de Arosa (ría: an estuary formed by an ancient coastal fluvial valley flooded during the

Quaternary), which is an environment of great ecological importance due to the extensive protected natural areas (PNAs) it encompasses (Ons–O Grove and Umia–O Grove complexes). These environments include intertidal zones (wetlands) with a wide variety of protected fauna (birds, insects) and flora [21,22]. The degradation of wetlands directly affects the regenerative capacity of a coastal area, which in turn directly influences the vulnerability of that territory to the occurrence of a catastrophic event [23]. Additionally, the Ría de Arosa is an area with significant tourism activity, driving the development of infrastructure (hospitality, recreational) that directly increases the vulnerability of the area [6,24].

Therefore, a proper characterization of natural hazards through a multi-risk analysis is needed to evaluate the interactions between individual natural hazards [25–27]. A multi-risk analysis is essential in this context due to the diversity of natural processes studied (coastal flooding, erosion, etc.) that affect the Ría de Arosa [28,29]. The combination of GIS tools, remote sensing, and qualitative or quantitative analysis allows for a comprehensive understanding of these natural risks [30,31]. Numerous studies follow different multi-risk approaches, with the analytical hierarchy process (AHP) and the equal weights method (EWM) standing out [31–37].

In this study, we aim to carry out land-use planning through the development of a natural hazard mapping using ArcGIS 10.8 © (with a 1 m × 1 m resolution), which will serve as a basic tool for decision-making regarding construction processes in the territory [30,31]. This mapping allows for an understanding of the different risks from lithological, geomorphological, hydrogeological, and geotechnical perspectives through the combination of qualitative (geotechnical zoning) [38] and quantitative analyses (the stages of coastal flooding) [28]. A review and an increase in the level of detail of the geotechnical characterization mapping (1:50,000 scale) will be carried out, and the available coastal flooding information for the area will be used [39]. With this detailed mapping, the goal is not only to identify the inherent risks in the area but also to establish a solid foundation for future mitigation and adaptation strategies in the context of vulnerable coastal environments such as the southeastern margin of the Ría de Arosa.

Study Area

This research was conducted in the southeastern margin of the Ría de Arosa, in the province of Pontevedra (Galicia, Spain) (Figure 1). The area has approximately 57,358 inhabitants spread over 9600 hectares. Compared to the national average population density, which is 96 inhabitants per km^2, this area has a high population density (598 inhabitants per km^2) (data from the National Institute of Statistics (INE), https://www.ine.es/, accessed on 30 September 2024). These values exclude the population increase that occurs during the summer season (June, July, and August), because of high tourism activity. The most important municipalities are located along the coastal margin, including Sanjenjo, Portonovo, Cambados, Villanueva de Arosa, El Grove, and La Isla de Arosa. Climatically, the area is considered temperate, with a dry summer compared to the rest of the year. It falls within the Csb type according to the Köppen classification system [40]. Due to the ocean's moderating effect, its average temperatures are mild, ranging from 19.3 °C in summer to 9 °C in winter. The average annual precipitation is 1455 mm, with the rainiest months being October, November, and December. All autumn months experience precipitation above 200 mm (information provided by the Spanish Meteorological Agency (AEMET), https://www.aemet.es/es/serviciosclimaticos/datosclimatologicos, accessed on 30 September 2024).

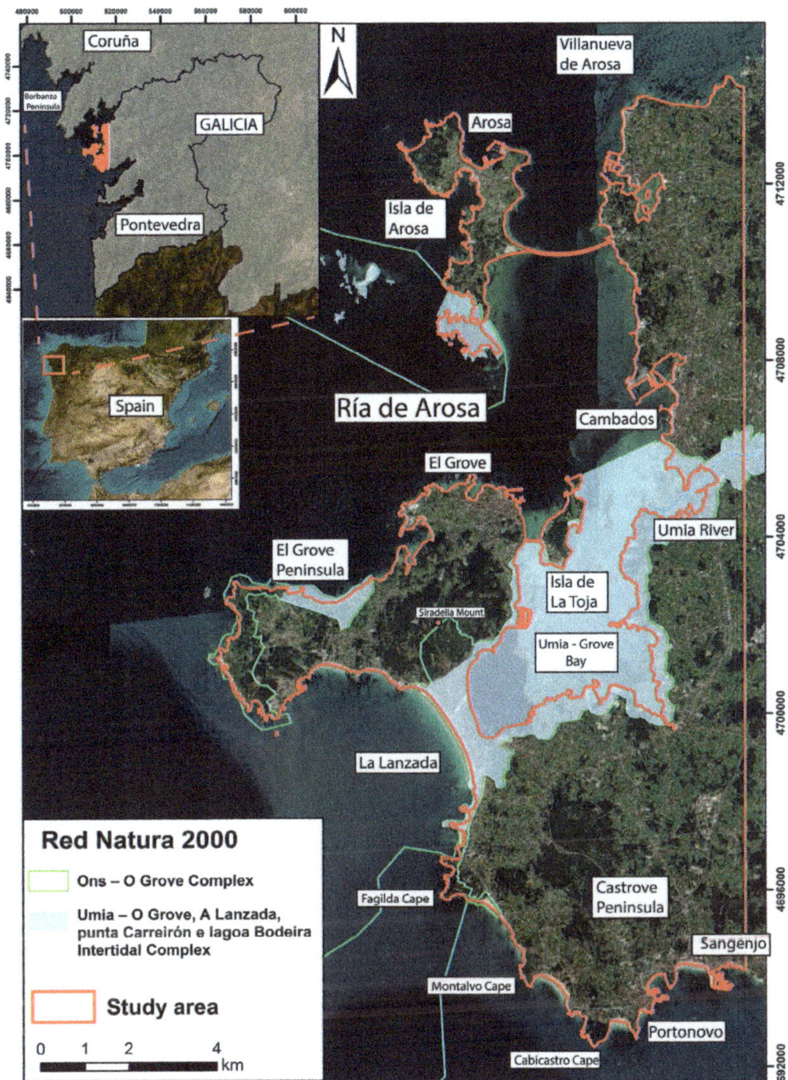

Figure 1. Map of the study area within the province of Pontevedra in Galicia.

The geology and current relief of the area are the result of a long and complex geological history [41]. The area belongs to the northwestern sector of the Iberian Massif, corresponding to the Galicia–Tràs–Os–Montes Zone. Lithologically and structurally, it represents an internal zone of the Variscan Orogen, where allochthonous domains can be recognized in superposition with autochthonous Paleozoic metasediments (d'Home–La Lanzada Complex) and plutonic igneous rocks [42]. The granites and granodiorites exhibit early (synkinematic) and late (post-kinematic) characteristics [43,44]. The area is represented on four geological sheets at 1:50,000 scale from the National Geological Map (MAGNA) (Spanish Geological and Mining Institute (IGME), http://info.igme.es/cartografiadigital/geologica/Magna50.aspx?language=en, accessed on 30 September 2024) [45–48] (Figure 2A).

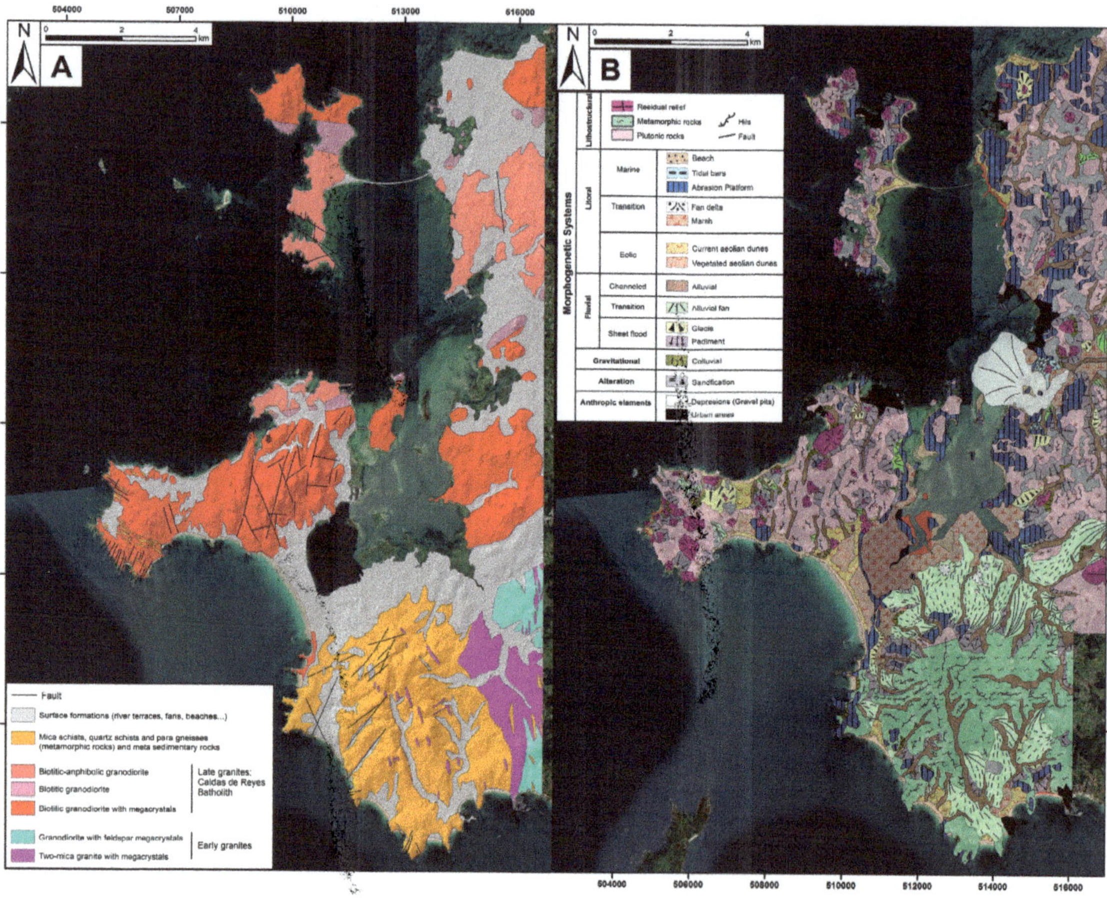

Figure 2. (**A**) Geological map. (**B**) Geomorphological map, modified from [47].

The area's geomorphological characteristics define a highly distinctive relief resulting from an evolutionary process of weathering and differential erosion. Slopes are gentle throughout most of the territory, except in areas dominated by rocky outcrops that reveal residual lithostructural reliefs [49]. The higher granitic areas form peaks, ridges, and hills, often represented by dome-shaped structures, where the evolution is more pronounced; tors; and other minor landforms [50]. The flat areas correspond to the peneplain, which is the result of the differential erosion of granite (mainly the Caldas de Reyes Batholith) and Paleozoic metasediments. Extensive erosion surfaces, known as "rasas" or marine terraces, have developed. In many cases, they are covered by poorly consolidated sandy or conglomerate deposits, resulting from the alteration of the rocky substrate. The strongly transitional marine environment of the area facilitates the formation of coastal landforms through the accumulation of fine deposits by tidal action (marshes and wetlands) and wind action (beaches and dune systems) [49]. In connection with this, two protected natural areas have been designated by the Natura 2000 Network: The Umia–O Grove Intertidal Complex and the Ons–O Grove Complex. The fluvial system is represented by fluvial valley deposits and transitional systems that include active alluvial fans and cones. In more stable settings, extensive floodplain surfaces, such as glacis and pediments, can be identified (Figure 2B).

2. Materials and Methods

The foundation of the work described here is based on fieldwork, satellite image analysis, and data modeling using computer software. First, the study area was initially recorded through sampling, data collection, and the photointerpretation of both current and historical images. All fieldwork and the collection of bibliographic and digital information during this and other previous studies of the area allow for effective planning that addresses the issue in a more detailed and comprehensive way. The selection of strategic sites allows for subsequent validation of the data obtained during digital modeling, which lends greater rigor to the entire study. All digital modeling, both results generated for geotechnical characterization mapping and for natural hazard mapping, was carried out on the Geographic Information System "ArcGIS 10.8 ©". Figure 3 represents the methodological flow followed in this research, from the initial data collection to the generation of the natural hazard mapping. Each stage is described in detail, including the geoprocessing (reclassification, union, or export) used in the layers to obtain the data on the hazard factors (geomorphological, hydrological, lithological, and geotechnical) that make up the natural hazard mapping, which is useful in territorial planning and in preventing associated risks.

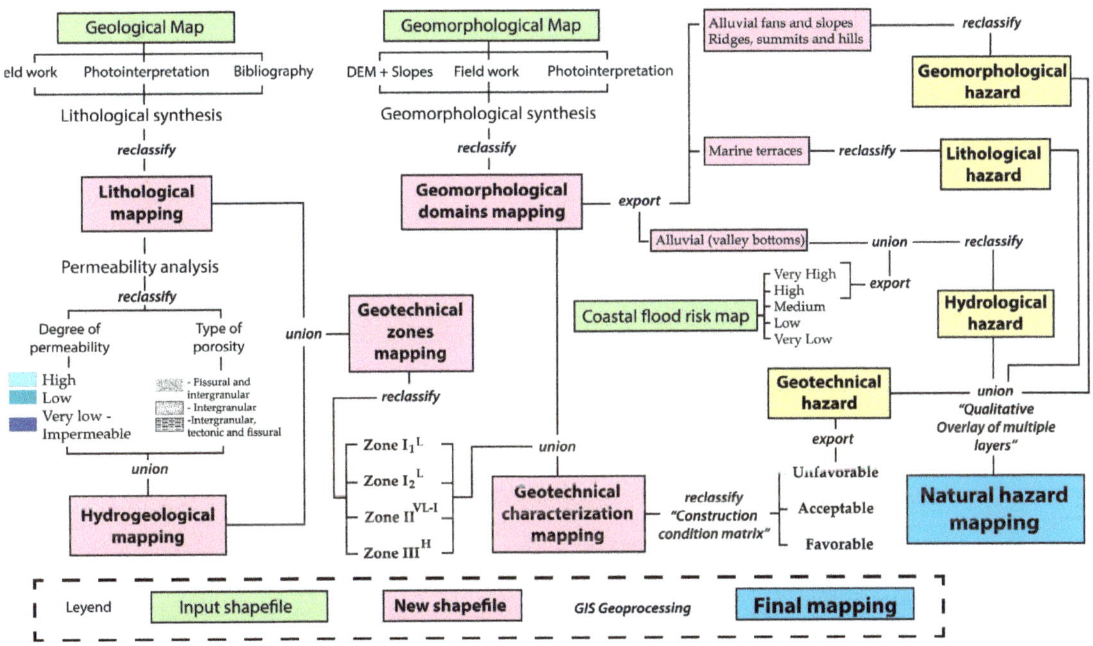

Figure 3. Methodological scheme.

2.1. Geotechnical Characterization Mapping

The creation of the geotechnical characterization mapping relies on three thematic maps, each requiring a preliminary review and synthesis of information to achieve a more representative and easy-to-interpret result. Geotechnical characterization mapping is essential, as it generates a basic information document that must be consulted prior to any decision-making in territorial planning for a specific area, especially for construction projects. It aims, therefore, to prevent any risky situation and generate socioeconomic benefits for the affected civilian population. The zoning of the territory is based on different geological criteria—lithological, hydrogeological, and geomorphological—which overlap and delineate different "geotechnical zones." These various criteria, collectively, aim to show the vulnerability to potential natural or anthropogenic risks of the study area:

1. Lithological Mapping: This map starts with a geological map of the area [45–48], which is simplified by grouping lithologies based on compositional, textural, or geomechanical similarities. Each group defines the homogeneous lithological units that will make up the geotechnical characterization mapping. In this case, lithological groups from Quaternary surface formations are differentiated and reclassified into various lithological domains.
2. Hydrogeological Mapping: Creating the hydrogeological map requires combining data sources and criteria from fieldwork with filtered information from the creation of the lithological map. Hydrogeological data primarily come from official cartography from the Spanish Geological and Mining Institute (IGME) and digital information from the Galician Health Service (https://www.sergas.es/Saude-publica/GIS-Litoloxia-xeoloxia, accessed on 31 October 2024). These data include the hydraulic properties of the different lithological formations. After characterizing lithological units with similar hydraulic properties, hydrogeological units are described. These hydrogeological units are defined based on their permeability degree (from impermeable to very permeable) and porosity type (intergranular, fissure, tectonic, or alteration). These characteristics are intrinsic to each type of rock or substrate. Crystalline rocks, which compose the Variscan basement (igneous and metamorphic), generally exhibit low or no permeability, a characteristic influenced by the low or null effective porosity of these lithologies, allowing water to pass through fractures, joints, or tectonic plates (schistosity). The igneous rocks (granitoids, granodiorites) and metamorphic rocks (slates and schists) found in the study area exhibit these hydrogeological behaviors [45–48]. The substrate belonging to Quaternary surface formations, which shows little or no consolidation, has very different hydrogeological characteristics. Its high effective porosity, which is of an intergranular type, leads to very high permeability values. However, there are no areas with significant thicknesses that would prevent the formation of aquifers with high water-capturing potential in the territory [45–48]. Each hydrogeological unit is assigned a color based on its permeability level for proper differentiation. In this case, the different units are represented in blue tones, with intensity varying according to their permeability level. The darkest color represents very high and impermeable values, which become lighter as permeability increases. In addition, each porosity type is differentiated by a characteristic transparent pattern. Both layers, formed by the reclassification of lithological domains, are combined to create the hydrogeological map.
3. Geotechnical Zone Mapping: The vector polygons from the lithological and hydrogeological maps are merged, and, through a reclassification of the resulting polygons, different geotechnical zones are determined. In this case, the geotechnical zones are numbered with Roman numerals (I, II, III, . . .). Each represents a lithological type with specific hydraulic properties (porosity). This leads to the creation of the geotechnical zone mapping.
4. Geomorphological Domain Mapping: This is derived from the detailed geomorphological map, with a resolution of 1:50,000 for the study area [49]. This type of mapping encompasses the fundamental features of the reliefs, synthesizing them. This simplification requires an in-depth relief analysis, relying on the slope map, DEM, specific fieldwork, and photointerpretation. Its purpose is to facilitate an understanding of the terrain and highlight its physiographic characteristics, which are grouped into morphogenetically coherent domains independent of the substrate type, though the substrate type often influences the morphological appearance of the relief.
5. Geotechnical Characterization Mapping: The combined polygons generated from the geotechnical zone mapping and the geomorphological domain mapping create a 1:50,000 geotechnical characterization map for the southeastern margin of the Ría de Arosa. This map presents both the geotechnical properties of the substrate and the surface formations on it, enabling the identification of areas with limitations or negative triggers for civil engineering, urban, or industrial activities.

2.2. Natural Hazard Mapping

The sectoral analysis of external geodynamic processes, which are responsible for generating potential natural risks, along with their recognition and spatial delimitation, characterizes the natural hazard mapping of a territory. In this area, lithological, geomorphological, geotechnical, and hydrological issues are identified, which can often trigger natural or anthropogenic risks. Geotechnical characterization mapping will serve as a basis for defining and recognizing the various types of issues present:

1. Lithological hazards: These relate to the textural, structural, and compositional characteristics of each lithology. Processes such as karstification and the resulting consequences for carbonate rocks, the planes of weakness seen in some lithologies, or degrees of consolidation represent risks associated with this type of hazard. In our case, the main lithological issues are associated with poorly cemented conglomerate deposits found on terrace or marine "rasas" formations. These polygons are derived from the geomorphological domain map, exported, and reclassified as lithological hazards.
2. Geomorphological hazards: Each geomorphological feature impacts the terrain's construction conditions differently. In these cases, natural or predominantly anthropogenic action can induce instability in certain surface formations. Generally, areas with steeper slopes will have a less favorable response in terms of stability, potentially triggering active gravitational processes like landslides, rockfalls, or soil creep. In the study area, ridges, hills, and summits, with their steep slopes and occasional presence of granite boulders, which may cause instability during construction, are classified as geomorphological hazards. Alluvial fans and gentler slopes also pose risks to stability during construction phases and are classified similarly.
3. Hydrological hazards: The identification of high-risk flood areas is based on integrating coastal flood risk data with a geomorphological analysis to delimit and describe valley floors. Coastal flood risk maps for the southeastern margin of the Ría de Arosa were created using the Flood Hazard Index (FHI) method, which has been successfully applied to areas on the Atlantic margin of the Iberian Peninsula [28]. The parameters analyzed include significant wave height (Fw), annual sea level rise (Fsl), and extreme tidal range (Ftr), which were obtained from public data on PORTUS (Spanish Ports) (https://portus.puertos.es/#/, accessed on 31 October 2024). The rate of sea level rise (mm/year) is considered a variable factor, with values based on extreme scenarios presented by the Intergovernmental Panel on Climate Change (IPCC), where greater increase rates are expected [5]. A raster layer is created from the digital elevation model (DEM), with pixels below the maximum water sheet level for each scenario selected. The higher-risk scenarios Xa and Xb, at 25 and 100 years, respectively, are extracted in order to be included in the flood risk mapping, a part of the natural hazard mapping. Geomorphological data of the area were used to identify alluvial deposits in the valley floors, regarded as flood-prone zones. These deposits, characterized by a low altitude and proximity to fluvial systems, were highlighted as hydrological vulnerability areas. Both maps were combined and used to reclassify hydrological hazards.
4. Geotechnical hazards: The geotechnical characterization map results allow for the identification of favorable, acceptable, or unfavorable areas for active construction processes. Each polygon is evaluated using a construction conditions matrix that crosses geotechnical zones (described in the geotechnical zone mapping) with geomorphological domains. As a result, polygons with unfavorable values are exported and reclassified as geotechnical hazards.

The hazard map enables the combined identification of the different issues previously described. The process of developing this map is based on a Multicriteria Decision-Making (MCDM) methodology, which is used to combine the hazard parameters described and obtained in the previous steps without assigning a hierarchy. Here, the layers are combined through a qualitative overlay of multiple layers (based on union and reclassification) to integrate the analyzed parameters into the resulting polygons. The qualitative overlay allows for the integration and recognition of areas where the different identified risks act individually or jointly. The vector polygon layers of each risk are geoprocessed using the union procedure. In this way, new polygons are created that describe the individual or combined occurrence of a risk. This method aims to provide a descriptive, qualitative approach. No weighting process is applied to the different risks, as this would require a detailed study of the impacts of each risk. This would then provide a solid foundation for the quantitative prioritization of impacts. By avoiding this, subjectivity is prevented in this research, making the natural hazard mapping a preliminary tool for identifying hazard-prone areas in relation to construction processes and their monitoring. The interpretation of the results is based on identifying areas with higher or lower hazard levels according to the number (greater, lesser, or none) of processes that have the potential to pose a hazard in each area.

3. Results and Discussion

3.1. Geotechnical Characterization

The qualitative superposition of the lithological (Figure 4A), hydrogeological (permeability and porosity) (Figure 4B–D), and geomorphological (Figure 4E) maps allows for the differentiation of the various geotechnical areas which are represented in the geotechnical zone and geotechnical characterization mapping (Figure 5A,B).

3.1.1. Lithological Mapping

This mapping synthesizes the lithologies into different domains with lithological, textural, or compositional affinities based on what is reflected in the geological map (Figure 2A). In this case, the superficial formations and the Paleozoic metamorphic rocks remain unchanged. Synthesis occurs within the igneous rocks based on their early or late character (syn- or post-kinematic). The granites and granodiorites with internal deformation (early) are grouped together, while the late ones (Caldas de Reyes batholith) are classified separately (Figure 4A).

3.1.2. Hydrogeological Mapping

Hydrogeologically, a precise differentiation of the various lithological domains present can be made (Figure 4D). The superficial formations exhibit the best hydrogeological characteristics due to their low or nonexistent consolidation. They demonstrate a high degree of permeability and well-connected intergranular porosity. However, the limited thickness of these formations prevents the development of significant aquifers [45–48]. The metamorphic rocks (schists and gneisses) present rather poor hydrogeological characteristics. They show a very low degree of permeability, and in many cases can be considered practically impermeable (<1%) [47,48]. Areas where tectonic weakness planes (schistosity, fractures, jointing) are concentrated and areas with slight surface alterations show an increase in permeability, but in general, this increase is not significant. The two groups of igneous rocks (early and late) show slight variability from a hydrogeological standpoint. Early granites and granodiorites have very low, almost nonexistent permeability (very low–impermeable, Figure 4B,D), which increases as the degree of fracturing and surface weathering rises. The Caldas de Reyes batholith exhibits low permeability, induced by fracturing, and slightly higher permeability where its weathering is better developed [46–48]. The localized development of weathering–fracturing-type aquifers is possible, especially where granitic clay is found [46–48].

3.1.3. Geomorphological Domain Mapping

Geomorphological domain mapping is the grouping of different Quaternary superficial formations found in the study area in a morphogenetically coherent manner (Figure 4E). It is presented according to the morphogenetic systems present and, in turn, based on additional parameters such as slopes or the degree of consolidation of the deposit. Additionally, the main areas of anthropogenic infrastructures (population centers or industrial areas) are represented, which interact with the different domains or are included within them. This map is crucial for understanding the spatial distribution of landforms and their relationship with the evolution of the terrain and with the geotechnical characterization mapping (Figure 5B).

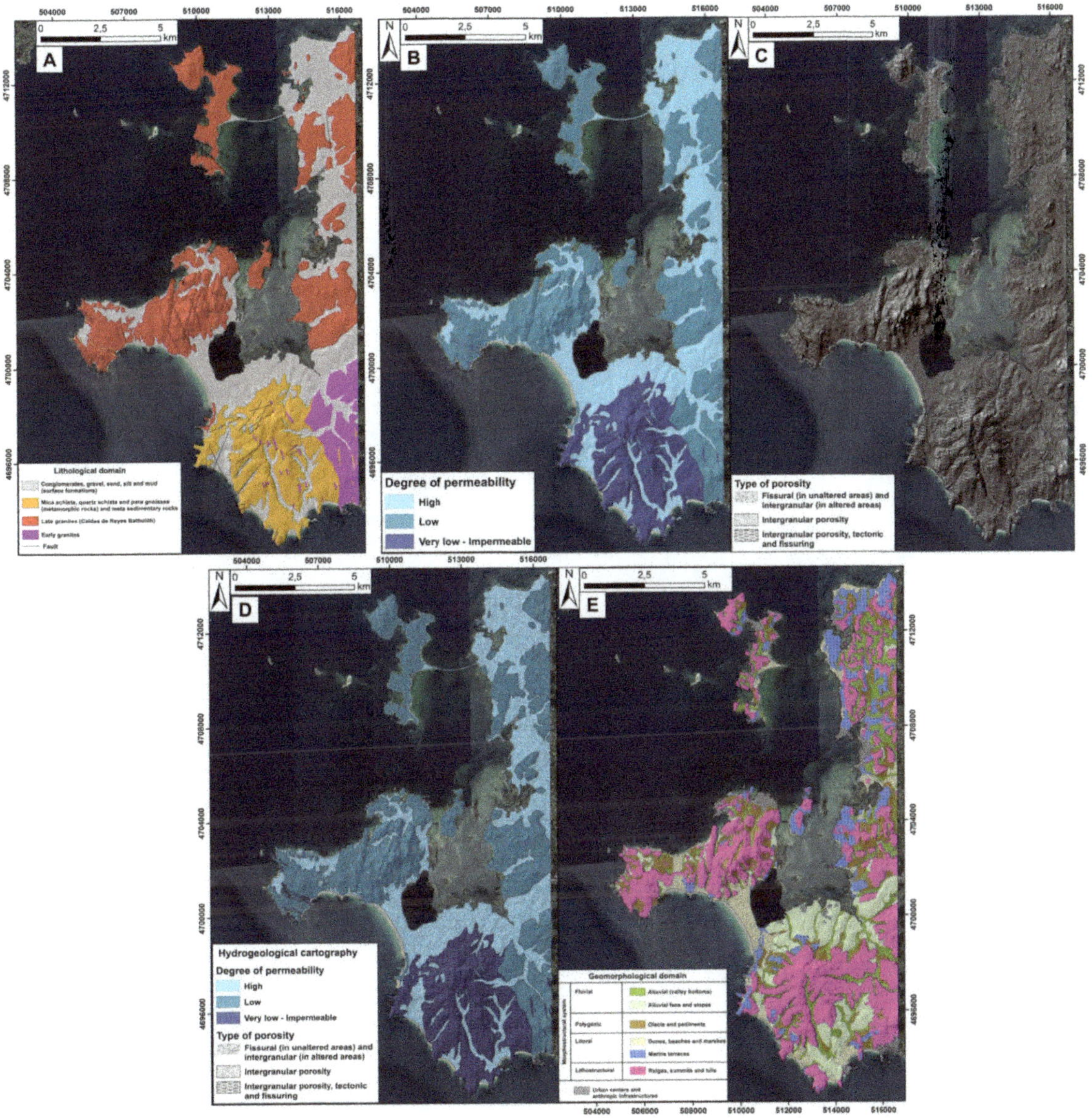

Figure 4. (**A**) Lithological mapping. (**B**) Degree of permeability. (**C**) Type of porosity. (**D**) Hydrogeological mapping. (**E**) Geotechnical domain mapping.

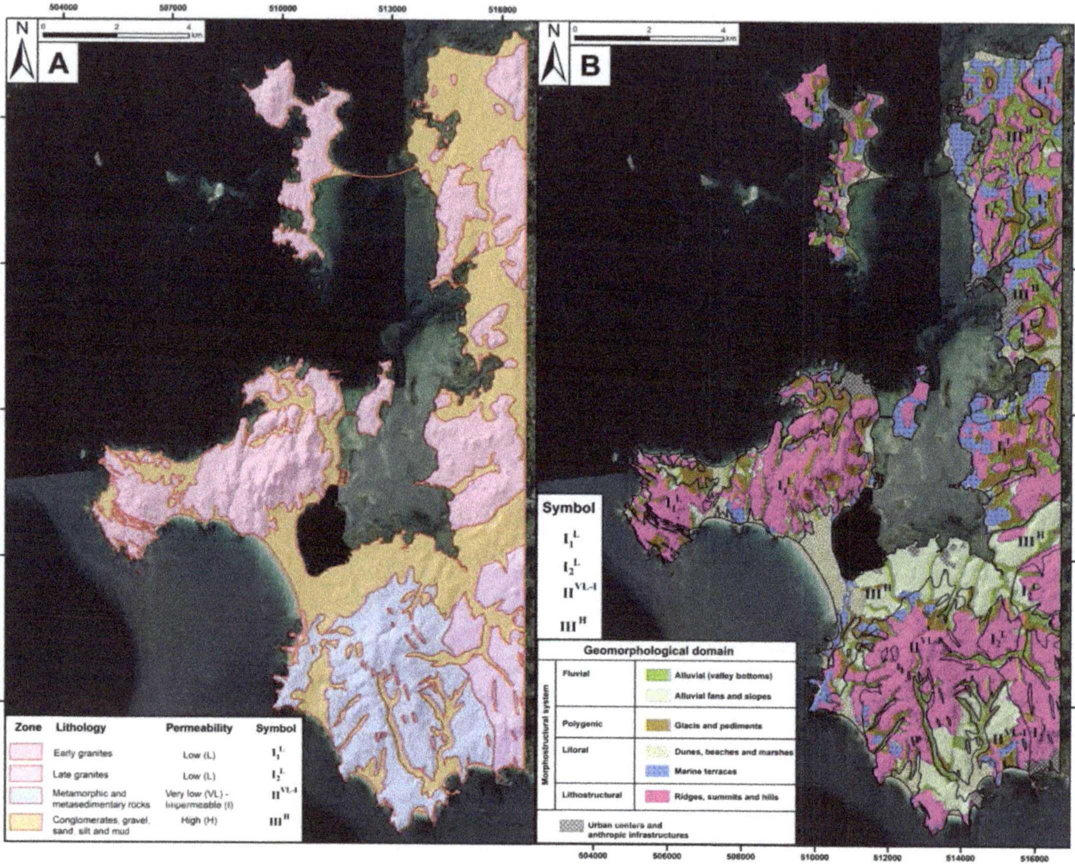

Figure 5. (**A**) Geotechnical zone mapping and (**B**) geotechnical characterization mapping of the SE margin of the Ría de Arosa.

3.1.4. Geotechnical Zones and Geotechnical Characterization Mapping

- Zone $I_1{}^L$: This zone exhibits a varied distribution and is prominently present in the El Grove Peninsula and the NE sector of the ría coastline. It occupies almost the entirety of the islands of Arosa and La Toja. Its presence around Punta Lanzada, on the NW margin of the Castrove Peninsula, is also notable. Lithologically, it corresponds to Caldas de Reyes granodiorite, which shows a slightly higher permeability than earlier facies (Figure 5A). This is particularly evident in tectonized areas, where fractures and weathering promote percolation processes and the development of minor aquifers. This area is well-defined by its residual litostructural relief, characterized by systems of summits, hills, and crests. In specific locations where the slopes are steeper, such as in the interior of the El Grove Peninsula or the northern sector of Arosa Island, dome-like morphologies or more mature rocky outcrops are visible (Figure 6A,B). Colluvium, along with pediments and glacis, is well represented in areas with more pronounced slope changes. Coastal environments make up the "marine rasas" (terraces) that show a smoothing of the relief, with almost nonexistent slopes (Figures 5B and 6C).
- Zone $I_2{}^L$: Located on the S–SE margin of the study area, this zone is entirely represented by early granite facies that are affected internally by tectonic deformation processes. This area shows very low permeability conditions that increase slightly where weakness planes (fracturing or jointing) are concentrated or where there is a

high degree of alteration (Figure 5A). The relief structures distinguished here are part of the regional modeling of granite, creating a residual relief where, in some cases, domed structures stand out (Figure 5B).

- Zone II^{VL-I}: This zone is found in the inner sector and SW margin of the Castrove Peninsula, where significant cliff areas are highlighted (Punta Fagilda, Punta de Cabicastro) (Figure 6D). Lithologically composed of metamorphic rocks (schists, gneisses) belonging to the "d'Home–La Lanzada Complex", these rocks show very low, almost impermeable, permeability values, with exceptions in sectors with a greater development of tectonic plates (schistosity) and higher concentrations of fractures and joint systems (Figure 5A). Geomorphologically, the inner areas where slopes are slightly steeper represent residual litostructural reliefs, primarily defined by hills and summits. In the southern zone, more gentle alluvial fans and cones have developed. The area closest to the coast contains some rasas (Figures 5B and 6D).
- Zone III^{H}: This zone has a heterogeneous distribution throughout the study area. The places where it is best recognized are in the Tombolo of La Lanzada and the N zone of the Castrove Peninsula, where the most significant alluvial fan systems are located. Composed of a substrate with a low or nonexistent consolidation of superficial formations, it has high permeability due to its high porosity (Figure 5A). Important coastal deposits such as dunes, beaches, or marsh areas are within this zone, as in the case of the Umia River mouth (Figure 6E,F). The poorly evolved drainage network of the territory also demonstrates the presence of alluvial bottoms. Significant rasas are identified in areas near the coast with very gentle slopes (Figure 5B).

3.2. Natural Hazards

By recognizing the active processes in the study area, potential risks of a lithological, geomorphological, geotechnical (Figure 7a), or hydrological (Figure 7b) nature can be identified. By overlapping these, we arrive at the natural hazard mapping (Figure 8).

3.2.1. Geotechnical Hazard Map

Based on the delineation of different geotechnical areas and supported by the construction conditions matrix (Table 1), the geotechnical hazard map is constructed (Figure 7A).

Table 1. Construction condition matrix.

Geomorphological Domains	Zone I_1^{L}	Zone I_2^{L}	Zone II^{VL-I}	Zone III^{H}
Alluvial (valley bottoms)	Unfavorable	Unfavorable	Unfavorable	Unfavorable
Alluvial fans and slopes	Unfavorable	Unfavorable	Unfavorable	Unfavorable
Glacis and pediments	Favorable	Favorable	Favorable	Acceptable
Dunes, beaches, and marshes	Unfavorable	Unfavorable	Unfavorable	Unfavorable
Marine terraces	Acceptable	Acceptable	Acceptable	Unfavorable
Ridges, summits, and hills	Unfavorable	Unfavorable	Acceptable	Acceptable

It is evident that most of the territory presents unfavorable construction conditions from a geotechnical point of view. Among all of these, the territories with poorly consolidated substrates, which exhibit low load-bearing capacity and are prone to potential settlements, stand out [39,45–48]. Critical areas include the dune systems of La Lanzada, the coastal environment influenced by tidal activity, and the alluvial deposits in valley bottoms. Other areas where the substrate is somewhat more consolidated and which display low slope gradients are represented by alluvial fan and cone systems. However, their active nature may favor the development of slope processes, such as landslides, in response to anthropogenic activities [39,49]. Despite being competent lithologies with high load-bearing capacities [39], the granite areas are generally steep regions with higher elevations, which complicate their stability in terms of potential construction, giving them low or no settlement potential. The instability of their slopes is often exacerbated by the presence of boulder fields, where rocks of various sizes appear. Metamorphic lithologies also exhibit high competence, granting them high load-bearing capacities and making settlement development impossible [39]. The areas with more pronounced reliefs, coinciding with the lithostructural system, offer acceptable construction conditions. These areas, unlike granite ones, are more stable due to their slightly gentler slopes and the absence of "boulder-like" morphologies. Additionally, acceptable construction conditions are found in coastal areas, while considering the retreat of cliffs due to marine and fluvial erosion processes. Across the entire metamorphic zone, the orientation and inclination of schistosity planes must be considered, as they may occasionally pose a problem when carrying out a project. This also prevents these outcrops from being classified as acceptable. Marine terraces or "rasas" are considered acceptable from a construction perspective, except for those with a significant thickness of poorly consolidated coarse-grained substrates (unfavorable). The pediments and glacis, which develop on highly competent rocks (granites and metamorphic rocks), have very gentle slopes and exhibit greater stability, making them favorable areas from a construction perspective.

Figure 6. (**A**) Granite dome on the south–west edge of the El Grove Peninsula. (**B**) Rocky outcrop with a large tor (Siradella Mount). (**C**) Coastal marine terraces on the south side of Isla de Arosa. (**D**) Cliffs and marine terraces at Punta Fagilda (Fagilda Cape). (**E**) Dune systems adjacent to Montalvo Cape. (**F**) Marshes at the mouth of the Umia River.

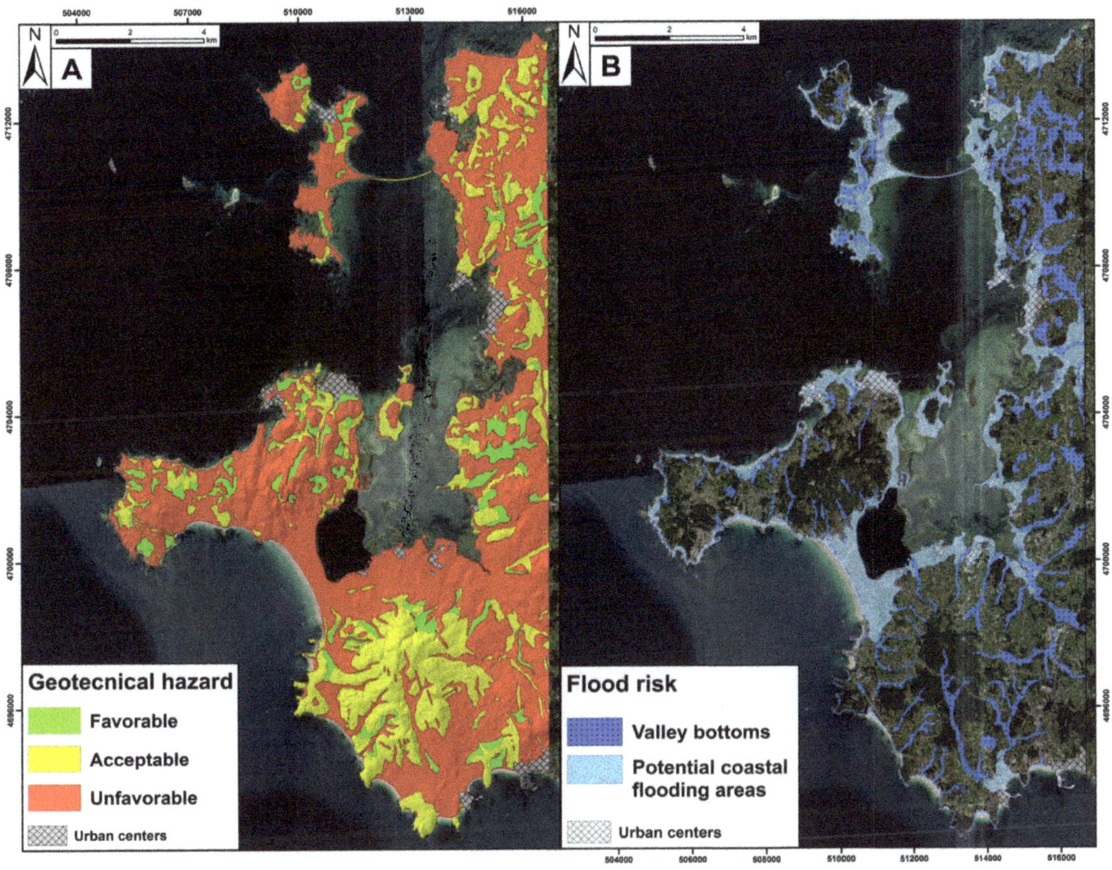

Figure 7. (**A**) Geotechnical hazard map. (**B**) Hydrological hazard map, modified from [28].

3.2.2. Hydrological Hazard Map

The coastal flood risk mapping proposed for this area was utilized (Figure 7b) [28]. In this case, areas at potential hydrological risk are those with the possibility of flooding within a return period of 25 to 100 years. Additionally, valley bottoms are added to this layer due to their potential risk of flooding. It is observed that the areas of greatest hazard are found in coastal environments with gentle slopes, which the water sheet can penetrate more easily, particularly in the area adjacent to the mouth of the Umia River.

3.2.3. Natural Hazards Mapping

The natural hazard mapping highlights the areas of this territory that are susceptible to various issues that could be triggered during the active phase of a construction project (Figure 7b). As such, it serves as an essential tool for the decision-making stage when selecting a project site. Consulting this map is recommended to enable effective risk management and thereby reduce the negative impacts that may arise in subsequent project phases. This, in turn, will directly reduce the economic resources needed to mitigate impacts from geomorphological, hydrological, geotechnical, and lithological issues, as well as contribute to sustainable land use. The main issues identified during the phases of a construction project are related to the rise in the water table and ground movement or settlement [51].

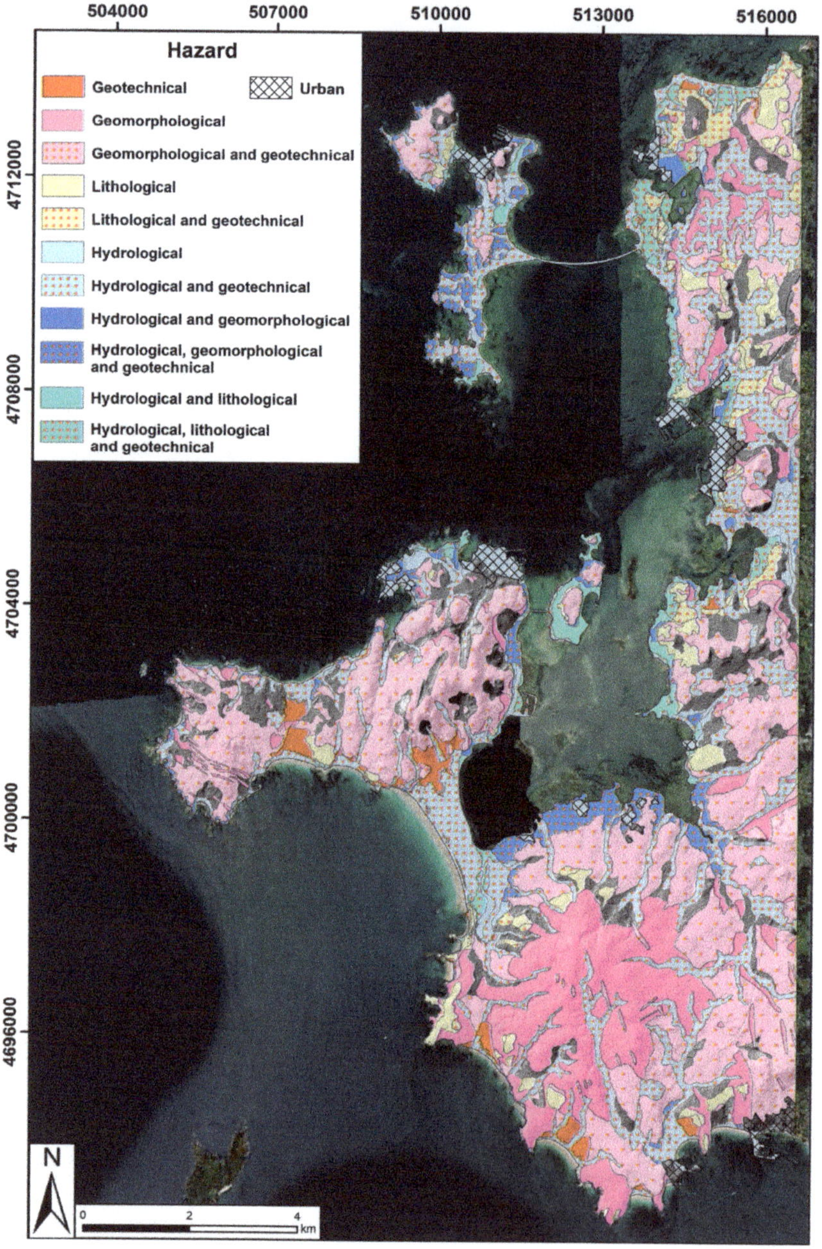

Figure 8. Natural hazard mapping of the SE margin of the Ría de Arosa.

The areas where hydrological issues are concentrated are those closest to the coastline or situated on alluvial valley floors. Under specific conditions, other natural issues may arise that increase the risk in these areas. Surface formations made up of weakly consolidated substrates are prone to problems related to rising water tables and waterlogging following heavy rainfall or flooding events. This is the case of the tombolo area of La Lanzada, where highly unconsolidated substrates (coastal formations such as beaches, dunes, or marshes) exhibit a low load-bearing capacity, high permeability, and susceptibility to flooding. These areas, which have combined hydrological and geotechnical issues, may also present geomorphological or lithological problems. Geomorphological issues are identified as being when active formations, such as alluvial fans, lose stability, potentially triggering mass movement processes. The northern flank of the Castrove Peninsula is an example of where these three issues converge. The poorly cemented conglomerate deposits found on some marine terraces cause lithological issues. They overlap with hydrological and geotechnical problems in many coastal areas around the mouth of the Umia River, north of the municipality of Cambados, or in the southern area near Villanueva de Arosa. In most cases, major urban or industrial projects are not recommended in these areas due to their combined issues.

Inland areas, beyond the valley floors, generally pose fewer hydrological issues. However, periodic monitoring of the water table is recommended in areas with high-permeability surface formations.

The most abundant zones in the study area are those with combined geomorphological and geotechnical issues. Generally, these are areas with a more pronounced relief where granite morphologies have been identified. There are no issues related to rising water tables or potential settlement due to load-bearing capacity loss thanks to the high competency and low permeability of the granitoids present here. The main issues that may arise in these locations are influenced by the terrain slopes, which can trigger mass movement processes. In many cases, the presence of granite boulders could lead to falls if stability is lost during the active phase of a project. The interior of the El Grove Peninsula (Mount Siradella) and the western coastal margin contain extensive areas with these characteristics. Similarly, the eastern margin of the Castrove Peninsula exhibits these same characteristics. Where surface formations associated with alluvial fan and cone systems are present, issues arise due to the mobile nature of these formations when stability is lost, potentially triggering mass movements, and, where the substrate thickness is greater, minor settlements. All alluvial fan and cone systems on the Castrove Peninsula present these issues.

Geomorphological issues appear in isolation in the rocky outcrops of the metamorphic facies found in inland areas and at various headlands along the western coast of the Castrove Peninsula. In these cases, the main issue stems from the steep slopes of the residual relief morphologies. In the peninsula's inland areas, mass movement processes may occur when instability is reached, which is sometimes promoted by tectonic planes of weakness in the metamorphic rocks. In coastal areas, where cliffs have significant escarpments, rockfall problems may arise. Due to the competence of metamorphic rocks, these locations are more suitable for civil works projects, provided that mass movement processes are considered. The continuous monitoring of stability would be advisable, with reinforcement in areas showing instability if necessary. Clifftop lands are not recommended for any projects, with the suggestion being to seek suitable sites further inland.

In areas with lithological issues, where poorly cemented conglomerates situated over marine terraces are found, geotechnical issues may be detected. It is recommended that this layer, which generally has a thin thickness, be removed in advance [39]. The removal of problematic lithology and the level nature of these erosion surfaces make these locations feasible for civil works projects.

4. Conclusions

Research in the southeast margin of the Ría de Arosa has allowed for a detailed characterization of the main lithological, geomorphological, hydrological, and geotechnical risks in the area. The integration of geotechnical characterization maps and natural hazard maps has facilitated the identification of areas with a greater susceptibility to one or several types of risks, a crucial aspect for planning and managing this territory in the context of increasing urban pressure and global climate change.

The results, based primarily on the geological and geomorphological categorization of the territory, allow for the identification of areas susceptible to natural hazards. Ground instability and susceptibility to flooding, rising water tables, or settlements are the main issues that may be triggered in this area. Zone III^H of our geotechnical characterization mapping, where poorly consolidated substrate surface formations are found, is especially vulnerable to geotechnical, geomorphological, and hydrological issues. The transitional areas containing dunes and marshes around La Lanzada, as well as the alluvial deposits in the northern part of the Peninsula of Castrove, stand out for their high environmental quality. Therefore, there arises a need to adopt sustainable strategies to protect wetland areas and enhance the environment's capacity to respond to natural and anthropogenic hazards, which is largely influenced by the large impact of tourism and infrastructure development.

On the other hand, Zone $I_1{}^L$, Zone $I_2{}^L$, and Zone II^{VL-I}, defined by a more competent lithic substrate, are primarily affected by geotechnical and geomorphological problems. This is due to their pronounced changes in slope, which increase ground instability for construction activities. In these areas, it would be advisable to prioritize locations with gentler slopes, a lower density of fractures or joints, and less alteration.

The preparation of these maps has created a valuable tool for decision-making in the pre-design phases of urban planning, facilitating the identification of critical areas that require attention and mitigation measures. This work not only contributes to our understanding of the interaction between natural processes and human activity but also lays the groundwork for future research and the implementation of risk management policies that protect both the population and coastal ecosystems. Additionally, it opens the door to a detailed analysis of high-problem areas, allowing for an examination of the issues causing the greatest challenges. This approach enables future risk analyses that can establish a hierarchy, revealing the relative importance of each issue in relation to others.

Author Contributions: Conceptualization, C.E.N., A.M.M.-G. and L.M.; methodology, C.E.N. and A.M.M.-G.; software, C.E.N.; validation, C.E.N., A.M.M.-G. and L.M.; formal analysis, C.E.N., A.M.M.-G. and L.M.; investigation, C.E.N. and A.M.M.-G.; resources, C.E.N., A.M.M.-G. and L.M.; data curation, C.E.N., A.M.M.-G. and L.M.; writing—original draft preparation, C.E.N. and A.M.M.-G.; writing—review and editing, A.M.M.-G.; visualization, C.E.N.; supervision, A.M.M.-G.; project administration, A.M.M.-G.; funding acquisition, A.M.M.-G. All authors have read and agreed to the published version of the manuscript.

Funding: Grant 131874B–I00 funded by MCIN/AEI/10.13039/501100011033. Ministry for the Ecological Transition and the Demographic Challenge.

Institutional Review Board Statement: Not applicable.

Informed Consent Statement: Not applicable.

Data Availability Statement: The data presented in this study are available within this article.

Acknowledgments: This research was assisted by the GEAPAGE research group (Environmental Geomorphology and Geological Heritage) of the University of Salamanca.

Conflicts of Interest: The authors declare no conflicts of interest.

References

1. Martínez, M.L.; Intralawan, A.; Vázquez, G.; Pérez–Maqueo, O.; Sutton, P.; Landgrave, R. The Coasts of Our World: Ecological, Economic and Social Importance. *Ecol. Econ.* **2007**, *63*, 254–272. [CrossRef]
2. Adger, W.N.; Hughes, T.P.; Folke, C.; Carpenter, S.R.; Rockstrom, J. Social–Ecological Resilience to Coastal Disasters. *Science* **2005**, *309*, 1036–1039. [CrossRef] [PubMed]
3. Jin, X.; Luan, W.; Yang, J.; Yue, W.; Wan, S.; Yang, D.; Xiao, X.; Xue, B.; Dou, Y.; Lyu, F.; et al. From the Coast to the Interior: Global Economic Evolution Patterns and Mechanisms. *Humanit. Soc. Sci. Commun.* **2023**, *10*, 723. [CrossRef]
4. Li, K.; Zhang, L.; Chen, B.; Zuo, J.; Yang, F.; Li, L. Analysis of China's Coastline Changes during 1990–2020. *Remote Sens.* **2023**, *15*, 981. [CrossRef]
5. Lee, H.; Calvin, K.; Dasgupta, D.; Krinner, G.; Mukherji, A.; Thorne, P.; Trisos, C.; Romero, J.; Aldunce, P.; Barret, K. IPCC 2023: Climate Change 2023: Synthesis Report, Summary for Policymakers. In *Contribution of Working Groups I, II and III to the Sixth Assessment Report of the Intergovernmental Panel on Climate Change*; CoreWriting Team, Lee, H., Romero, J., Eds.; IPCC: Geneva, Switzerland, 2023.

6. Brown, K.; Turner, R.K.; Hameed, H.; Bateman, I. Environmental Carrying Capacity and Tourism Development in the Maldives and Nepal. *Environ. Conserv.* **1997**, *24*, 316–325. [CrossRef]
7. O'Brien, K.; Leichenko, R.; Kelkar, U.; Venema, H.; Aandahl, G.; Tompkins, H.; Javed, A.; Bhadwal, S.; Barg, S.; Nygaard, L.; et al. Mapping Vulnerability to Multiple Stressors: Climate Change and Globalization in India. *Glob. Environ. Change* **2004**, *14*, 303–313. [CrossRef]
8. Bathrellos, G.D.; Gaki–Papanastassiou, K.; Skilodimou, H.D.; Papanastassiou, D.; Chousianitis, K.G. Potential Suitability for Urban Planning and Industry Development Using Natural Hazard Maps and Geological–Geomorphological Parameters. *Environ. Earth Sci.* **2012**, *66*, 537–548. [CrossRef]
9. Skilodimou, H.D.; Bathrellos, G.D.; Chousianitis, K.; Youssef, A.M.; Pradhan, B. Multi–Hazard Assessment Modeling via Multi–Criteria Analysis and GIS: A Case Study. *Environ. Earth Sci.* **2019**, *78*, 47. [CrossRef]
10. Bender, S. *Primer on Natural Hazard Management in Integrated Regional Development Planning*; Organization of American States, Department of Regional Development and Environment; Executive Secretariat for Economic and Social Affairs: Washington, DC, USA, 1991.
11. Alexander, D. *Natural Disasters*; Routledge: London, UK, 2018.
12. Mileti, D. *Disasters by Design: A Reassessment of Natural Hazards in the United States*; Joseph Henry Press: Washington, DC, USA, 1999.
13. Adger, W.N. Vulnerability. *Glob. Environ. Change* **2006**, *16*, 268–281. [CrossRef]
14. Cutter, S.L.; Barnes, L.; Berry, M.; Burton, C.; Evans, E.; Tate, E.; Webb, J. A Place–Based Model for Understanding Community Resilience to Natural Disasters. *Glob. Environ. Change* **2008**, *18*, 598–606. [CrossRef]
15. Burby, R.J.; May, P.J. *Making Governments Plan: State Experiments in Managing Land Use*; JHU Press: Baltimore, MD, USA, 1997.
16. Godschalk, D.R. Urban Hazard Mitigation: Creating Resilient Cities. *Nat. Hazards Rev.* **2003**, *4*, 136–143. [CrossRef]
17. UNISDR. *Hyogo Framework for Action 2005–2015*; United Nations International Strategy for Disaster Reduction: Geneva, Switzerland, 2007.
18. UNDRR. *Sendai Framework for Disaster Risk Reduction 2015–2030*; United Nations Office for Disaster Risk Reduction: Geneva, Switzerland, 2015.
19. Safani, J.; Matsuoka, T. Soft–Geotechnical Zone Determination Using Surface–Wave for Geotechnical Hazard Mitigation. *Procedia Environ. Sci.* **2013**, *17*, 354–360. [CrossRef]
20. Haeri, S.M. The Role of Geotechnical Engineering in Sustainable and Resilient Cities. *Sci. Iran.* **2016**, *23*, 1658–1674. [CrossRef]
21. EU. Council Directive 92/43/EEC on the conservation of natural habitats and of wild fauna and flora. *Off. J. Eur. Union* **1992**, *206*, 7–50.
22. Álvarez–Vázquez, M.Á.; González–Prieto, S.J.; Prego, R. Possible Impact of Environmental Policies in the Recovery of a Ramsar Wetland from Trace Metal Contamination. *Sci. Total Environ.* **2018**, *637–638*, 803–812. [CrossRef]
23. Austin, D.E. Coastal Exploitation, Land Loss, and Hurricanes: A Recipe for Disaster. *Am. Anthropol.* **2006**, *108*, 671–691. [CrossRef]
24. Mohammed, S.; Abdo, H.G.; Szabo, S.; Pham, Q.B.; Holb, I.J.; Linh, N.T.T.; Anh, D.T.; Alsafadi, K.; Mokhtar, A.; Kbibo, I.; et al. Estimating Human Impacts on Soil Erosion Considering Different Hillslope Inclinations and Land Uses in the Coastal Region of Syria. *Water* **2020**, *12*, 2786. [CrossRef]
25. Gill, J.C.; Malamud, B.D. Hazard Interactions and Interaction Networks (Cascades) within Multi–Hazardmethodologies. *Earth Syst. Dynam.* **2016**, *7*, 659–679. [CrossRef]
26. Gallina, V.; Torresan, S.; Critto, A.; Sperotto, A.; Glade, T.; Marcomini, A. A Review of Multi–Risk Methodologies for Natural Hazards: Consequences and Challenges for a Climate Change Impact Assessment. *J. Environ. Manag.* **2016**, *168*, 123–132. [CrossRef]
27. Zschau, J. *Where Are We with Multihazards, Multirisks Assessment Capacities?* European Union: Maastricht, The Netherlands, 2017.
28. Nieto, C.E.; Martínez–Graña, A.M.; Encinas, B. Analysis of the Risk of Coastal Flooding Due to Rising Sea Levels in Ría of Arosa (Pontevedra, Spain). *Appl. Sci.* **2023**, *13*, 12099. [CrossRef]
29. Nieto, C.E.; Martínez-Grana, A.M.; Merchán, L. Soil Erosion Risk Analysis in the Ría de Arosa (Pontevedra, Spain) Using the RUSLE and GIS Techniques. *Forests* **2024**, *15*, 1481. [CrossRef]
30. Martínez-Graña, A.M.; Goy, J.L.; Zazo, C. Peligrosidad Natural en el Espacio Protegido de las Batuecas–S Francia, Quilamas y su entorno (Salamanca). Análisis integrado de los factores potenciales de riesgo. *Geogaceta* **2004**, *36*, 71–74.
31. Merchán, L.; Martínez-Graña, A.M.; Nieto, C.E.; Criado, M. Natural Hazard Characterisation in the Arribes del Duero Natural Park (Spain). *Land* **2023**, *12*, 995. [CrossRef]
32. Gornitz, V. Global Coastal Hazards from Future Sea Level Rise. *Palaeogeogr. Palaeoclimatol. Palaeoecol.* **1991**, *89*, 379–398. [CrossRef]
33. Kunte, P.D.; Jauhari, N.; Mehrotra, U.; Kotha, M.; Hursthouse, A.S.; Gagnon, A.S. Multi–Hazards Coastal Vulnerability Assessment of Goa, India, Using Geospatial Techniques. *Ocean. Coast. Manag.* **2014**, *95*, 264–281. [CrossRef]
34. Gigović, L.; Pamučar, D.; Bajić, Z.; Milićević, M. The Combination of Expert Judgment and GIS–MAIRCA Analysis for the Selection of Sites for Ammunition Depots. *Sustainability* **2016**, *8*, 372. [CrossRef]
35. Stalhandske, Z.; Steinmann, C.B.; Meiler, S.; Sauer, I.J.; Vogt, T.; Bresch, D.N.; Kropf, C.M. Global Multi–Hazard Risk Assessment in a Changing Climate. *Sci. Rep.* **2024**, *14*, 5875. [CrossRef]
36. Wu, H.; Shi, A.; Ni, W.; Zhao, L.; Cheng, Z.; Zhong, Q. Numerical simulation on potential landslide–induced wave hazards by a novel hybrid method. *Eng. Geol.* **2024**, *331*, 107429. [CrossRef]
37. Cheng, K.; Ping, X.; Han, B.; Wu, H.; Liu, H. Study on particle loss-induced deformation of gap-graded soils: Role of particle stress. *Acta Geotech.* **2024**, 1–28. [CrossRef]

38. López Santiago, F. *Mapa Geotécnico y de Peligrosidad Natural de la Ciudad de León y Su Aglomeración Urbana;* ITGE: Madrid, Spain, 1991; p. 64.
39. IGME. *Mapa Geotécnico General E. 1:200.000, Pontevedra–La Guardia, Hoja 1–3/16, 1–4/26;* Instituto Geológico y Minero de España, Servicio de Publicaciones Ministerio de Industria y Energía: Madrid, Spain, 1972.
40. Chazarra Bernabé, A.; Lorenzo Marino, B.; Romero Fresneda, R.; Moreno García, J.V. *Evolución de Los Climas de Koppen en España En El Periodo 1951–2020;* Agencia Estatal de Meteorología: Madrid, Spain, 2022.
41. Martín–Serrano, A. Macizo Hespérico Septentrional. In *Geomorfología de España;* Gutiérrez, M., Ed.; Rueda: Madrid, Spain, 1994; pp. 25–62.
42. Llana–Fúnez, S.; Marcos, A. The Malpica–Lamego Line: A Major Crustal–Scale Shear Zone in the Variscan Belt of Iberia. *J. Struct. Geol.* **2001**, *23*, 1015–1030. [CrossRef]
43. Bisdom, E.B.A. Micromorphology of a weathered granite near the Ria de Arosa (NW Spain). *Leidse Geol. Med.* **1967**, *37*, 33–67.
44. Arps, C.E.S. Petrology of a part of the Western Galicia Basement between the río Jallas and the Ría de Arosa (NW Spain) with emphasis on zircon investigations. *Leidse. Geol. Med.* **1970**, *46*, 57–155.
45. IGME. *Mapa Geológico de Espana E. 1:50.000, Puebla de Caraminal, Hoja 151 (3–9), Segunda Serie—Primera Edición;* Instituto Geológico y Minero de Espana, Servicio de Publicaciones Ministerio de Industria y Energía: Madrid, Spain, 1981.
46. IGME. *Mapa Geológico de Espana E. 1:50.000, Villagarcía De Arosa, Hoja 152 (4–9), Segunda Serie—Primera Edición;* Instituto Geológico y Minero de Espana, Servicio de Publicaciones Ministerio de Industria y Energía: Madrid, Spain, 1982.
47. IGME. *Mapa Geológico de Espana E. 1:50.000, Grove, Hoja 184 (3–10), Segunda Serie—Primera Edición;* Instituto Geológico y Minero de Espana, Servicio de Publicaciones Ministerio de Industria y Energía: Madrid, Spain, 1981.
48. IGME. *Mapa Geológico de Espana E. 1:50.000, Pontevedra, Hoja 185 (4–10), Segunda Serie—Primera Edición;* Instituto Geológico y Minero de Espana, Servicio de Publicaciones Ministerio de Industria y Energía: Madrid, Spain, 1981.
49. Martinez–Graña, A.M.; Arias, L.; Goy, J.L.; Zazo, C.; Silva, P. Geomorphology of the mouth of the Arosa estuary (Coruña–Pontevedra, Spain). *J. Maps* **2017**, *13*, 554–562. [CrossRef]
50. Twidale, C.R. *Granite Landforms;* Elsevier: Amsterdam, The Netherlands, 1982.
51. Calik, I.; Gurgun, A.P.; Ozcoban, M. Methods and Tools Used for Management of Geotechnical Risks in Construction Projects. *Proc. Int. Struct. Eng. Constr.* **2023**, *10*, 1–7. [CrossRef]

sustainability

MDPI

Article

Analysis of Paddy Field Changes (1989–2021) Using Landsat Images and Flooding-Assisted MLC in an Urbanizing Tropical Watershed, Vientiane, Lao PDR

Iep Keovongsa [1], Atiqotun Fitriyah [2], Fumi Okura [3], Keigo Noda [4], Koshi Yoshida [5], Keoduangchai Keokhamphui [6] and Tasuku Kato [1,*]

1 Department of Environmental and Agricultural Engineering, United Graduate School of Agricultural Science, Tokyo University of Agriculture and Technology, 3-5-8 Saiwaicho, Fuchu, Tokyo 183-8509, Japan; s228618t@st.go.tuat.ac.jp

2 Research Center for Limnology and Water Resources, National Research and Innovation Agency, Cibinong, Bogor 16911, Indonesia; atiqotun.fitriyah@gmail.com

3 Japan International Research Center for Agricultural Sciences, 1-1 Owashi, Tsukuba 305-8686, Japan

4 Department of Biological and Environmental Engineering, Graduate School of Agricultural and Life Sciences, The University of Tokyo, 1-1-1 Yayoi, Bunkyo-ku, Tokyo 113-8657, Japan

5 Department of International Studies, Graduate School of Frontier Sciences, The University of Tokyo, 5-1-5 Kashiwanoha, Kashiwa Shi 277-8561, Japan

6 Faculty of Water Resources, National University of Laos, Vientiane P.O. Box 7322, Laos

* Correspondence: taskkato@go.tuat.ac.jp; Tel.: +81-42-367-5757

Abstract: Paddy fields are essential for food security and sustaining global dietary needs, yet urban expansion often encroaches on agricultural lands. Analyzing paddy fields and land use/land cover changes over time using satellite images provides critical insights for sustainable food production and balanced urban growth. However, mapping the paddy fields in tropical monsoon areas presents challenges due to persistent weather interference, monsoon-submerged fields, and a lack of training data. To address these challenges, this study proposed a flooding-assisted maximum likelihood classification (F-MLC) method. This approach utilizes accurate training datasets from intersecting flooded paddy field maps from the rainy and dry seasons, combined with the Automated Water Extraction Index (AWEI) to distinguish natural water bodies. The F-MLC method offers a robust solution for accurately mapping paddy fields and land use changes in challenging tropical monsoon climates. The classified images for 1989, 2000, 2013, and 2021 were produced and categorized into the following five major classes: urban areas, vegetation, paddy fields, water bodies, and other lands. The paddy field class derived for each year was validated using samples from various sources, contributing to the overall accuracies ranging from 83.6% to 90.4%, with a Kappa coefficient of between 0.80 and 0.88. The study highlights a significant decrease in paddy fields, while urban areas rapidly increased, replacing 23% of paddy fields between 1989 and 2021 in the watershed. This study demonstrates the potential of the F-MLC method for analyzing paddy fields and other land use changes over time in the tropical watershed. These findings underscore the urgent need for robust policy measures to protect paddy fields by clearly defining urban expansion boundaries, prioritizing paddy field preservation, and integrating these green spaces into urban development plans. Such measures are vital for ensuring a sustainable local food supply, promoting balanced urban growth, and maintaining ecological balance within the watershed.

Keywords: LULC change; urbanization; AWEI; maximum likelihood algorithm

1. Introduction

Paddy fields are crucial for ensuring food security for over half of the global population [1], accounting for approximately 12% of global agricultural land [2]. Changes in land surface, particularly rapid urbanization, driven by demographic growth and expanding economies, increasingly threatens these vital areas [3,4], disrupts hydrological systems, degrades soil, and causes ecosystem imbalances [5,6], leading to significant challenges for sustainable food production and environmental conservation. Understanding how paddy fields are affected by urbanization over time is essential not only for ensuring food security and balanced urban growth but also for promoting environmental sustainability [7,8].

Rice is the main crop in the Lao People's Democratic Republic where urban growth is expected, covering more than 60% of the total agricultural land [9]. Since achieving national rice self-sufficiency in 1999 [10], its production has mostly been on a family scale for self-consumption [9]. Meanwhile, urbanization has been ongoing since 1986, with a recent rapid expansion driven by new economic reforms and urban infrastructure development [11]. This growth attracted more people to urban areas, resulting in a 9.4% population increase over the past decade [12,13], particularly in Vientiane, the capital of Laos, leading to a significant 22% decrease in agricultural land from 2016 to 2020 [14]. Although some studies have examined land use/land cover (LULC) changes in the Vientiane, focusing on major land categories and short periods [14,15] may not provide sufficient information for sustainable urban planning. In particular, the impact of urbanization on paddy fields remains poorly understood, making detailed data on paddy field changes in this urbanizing watershed urgently needed to inform future development strategies.

Several studies, globally, have demonstrated that urban expansion significantly encroaches on land suitable for rice cultivation [14,16,17]. For instance, in the Hang–Jia–Hu plains of China, urban development replaced about 88% of paddy fields between 1980 and 2010 [18]. Similarly, urbanization in Japan and Indonesia has claimed 23.0% and 22.7% (5.68% annually) of the paddy fields from 1985 to 2019 and from 2015 to 2019, respectively [19,20]. This conversion has not only threatened global food security but also diminished the ability of paddy fields to provide vital ecosystem services, such as offering habitats [21–23], mitigating floods [24,25], and sequestering carbon [26].

Remote sensing (RS) technology has proven its effectiveness and is extensively used for mapping and quantifying changes in paddy fields and land use over time [27,28], offering crucial insights into the reduction of these fields due to urbanization and its environmental impacts. However, accurately tracking paddy field dynamics remains challenging due to their spectral similarity to other vegetation, variations in climate, and diverse cultivation practices [20], making algorithms highly region- and stage-dependent [29]. In contrast, urban areas are typically characterized by spectral diversity of surface materials [30], which differentiate them from natural surfaces like vegetation or water. This diversity enables supervised classification methods to leverage statistical techniques, such as the mean and covariance of the spectral values, for accurate mapping [31]. Meanwhile, in tropical monsoon Southeast Asia, where rice cultivation heavily relies on monsoon rains, the availability of images during the growing season is often limited due to frequent cloud coverage. Currently, research efforts have been concentrated on efficiently mapping paddy fields, with particular emphasis on phenology-based and supervised approaches.

The phenology-based methods leverage the unique phenological characteristics of paddy fields, enabling the extraction of initial paddy fields as training samples. The commonly used phenology-based method was designed to detect flooding signals during the flooding/transplanting stage. The land surface water index (LSWI) and enhanced vegetation index (EVI) are used as water signals and vegetation signals, respectively [32]. However, this assumption remains challenging because natural water bodies show similar spectral characteristics to paddy fields [33]. Consequently, several strategies in different regions have been proposed. For example, in temperate zones in China, land surface temperature data were incorporated to define natural wetlands [34], while surface moisture data were used to differentiate water bodies from initial flooding/transplanting events in paddy fields [35]. Other than considering the flooding/transplanting stage, Maiti et al. proposed the maximum greenness time approach (50–60 days after the transplanting date) to identify paddy fields in a monsoon area where images are available during the growing stage [36]. Most existing phenology-based methods are primarily based on detecting water signals during flooding/transplanting stage, and these methods rely heavily on the time-series data of vegetation indices (VIs) to distinguish natural water bodies derived from a moderate-resolution imaging spectroradiometer (MODIS) or Landsat. The difficulties of these methods are that VIs do not show rapid increases in values after the transplanting date [32,37–39]. Additionally, in a tropical monsoon climate, aside from the limited availability of images due to frequent cloud coverage, expansion of the submerged fields resulting from the monsoon rains during the flooding/transplanting stage present another challenge. The AWEI enhances the differentiation between water-covered areas and other land surfaces in complex landscapes and urban environments [40]. It effectively reduces false positives from shadows and dense vegetation common in tropical monsoon regions, providing a reliable method for water extraction in these dynamic environments [41,42].

In comparison to phenology-based methods, supervised methods typically do not require a high number of images and impose no restrictions on any stage of rice growth for classification [43]. One commonly used supervised approach is the maximum likelihood classification (MLC) algorithm, which requires a cloud-free optical band reflectance image. This algorithm assumes a normal distribution of spectral data for each class [44], making it particularly suitable for classifying homogeneous paddy fields at harvest time, where their spectral characteristics align with this assumption. It has been applied successfully in various regions to classify paddy fields, including the rainiest area in Iran [45], the tropical monsoon region in Myanmar [46], and a complex land surface area in Turkey [47]. Meanwhile, the algorithm leverages multiple normal distributions to represent different surface materials in urban areas, facilitating the classification of urbanization and effectively distinguishing urban areas from natural surfaces. This approach has been successfully used in various regions [44,48]. However, the MLC algorithm requires training data to perform the classification, and in some regions, such as paddy fields, statistical samples may not be available. Therefore, the issue of sample scarcity remains a significant challenge for classification using supervised approaches.

Analyzing long-term changes in paddy fields and other land uses is crucial for understanding the impact of urban expansion on rice cultivation over time. A few studies have focused on long-term paddy field mapping by using phenology-based or supervised approaches. Carrasco et al. extended the Landsat–RICE algorithm developed by Dong et al. [49] to map and quantify paddy field changes in Japan from 1985 to 2019 [20]. Their algorithms used a large number of images and applied specific thresholds for the normalized difference vegetation index (NDVI) to differentiate flooded paddy fields from natural water bodies. However, threshold values vary greatly across regions, potentially leading to an inaccurate mapping of active paddy fields. Jiang et al. mapped changes in paddy fields from 1990 to 2015 by using the NDVI difference between specific stages to identify the paddy fields [28]. In some regions, particularly tropical monsoon areas, monsoon rains can cause low NDVI values even during the growing stage. More recently, Zhang et al. proposed a phenology-assisted supervised method to map paddy fields in Heilongjiang Province, China, from 1990 to 2020 [43]. Their initial paddy field map was created by combining water-based and vegetation-based algorithms. However, both algorithms may overestimate paddy fields due to spectral similarities during the flooding/transplanting and growing stages. To avoid overestimating, we proposed a new approach to distinguish between monsoon-submerged fields and paddy fields by intersecting flooded paddy field maps from the rainy and dry seasons while using AWEI to exclude natural water bodies. This map is subsequently used to prepare training sample points for the MLC classifier. Based on the paddy fields' classification, we consider the factors on urban expansion and paddy fields to have decreased. Also, we would like to discuss rice production for food security in the study area.

The specific objectives of this study include the following: (1) map the paddy fields in an urbanizing tropical monsoon area using a proposed approach with Landsat images; (2) analyze spatiotemporal variations to understand the pace, magnitude, and conversion of LULC classes, with a particular focus on paddy fields; (3) examine how factors such as distance from roads and population density contribute to the reduction of paddy fields and rapid urbanization; and (4) estimate rice production based on spatiotemporal data of paddy fields.

2. Materials and Methods

2.1. Study Area

The Mak Hieo River (MHR) watershed, located in Vientiane (17°53′30″ N–18°5′30″ N latitude and 102°31′0″ E–102°56′0″ E longitude), the capital of Lao People's Democratic Republic, is an urbanizing watershed situated in the southern part of Vientiane. Figure 1 shows the geography and transportation infrastructure of this watershed, which covers an area of 454 km^2 and is drained by the river, a sub-tributary of the Mekong River. The region experiences a tropical monsoon climate with two dominant seasons—the wet season (May–November) and the dry season (December–April). Rice cultivation, a major crop in this study area, is divided into rainfed and irrigated types, with rainfed rice being the dominant form. Transplanting takes place from June to July and from mid-December to mid-February, while harvesting occurs from mid-October to mid-December and from mid-April to May in the rainy and dry seasons, respectively [50]. Vientiane, especially within this watershed, is undergoing rapid urbanization [14], which poses challenges to rice cultivation, such as urban infrastructural development, rural–urban migration [13], aging irrigation pumps [51], and poor water retention due to sandy soil [52,53], all of which may contribute to the conversion of paddy fields within the watershed.

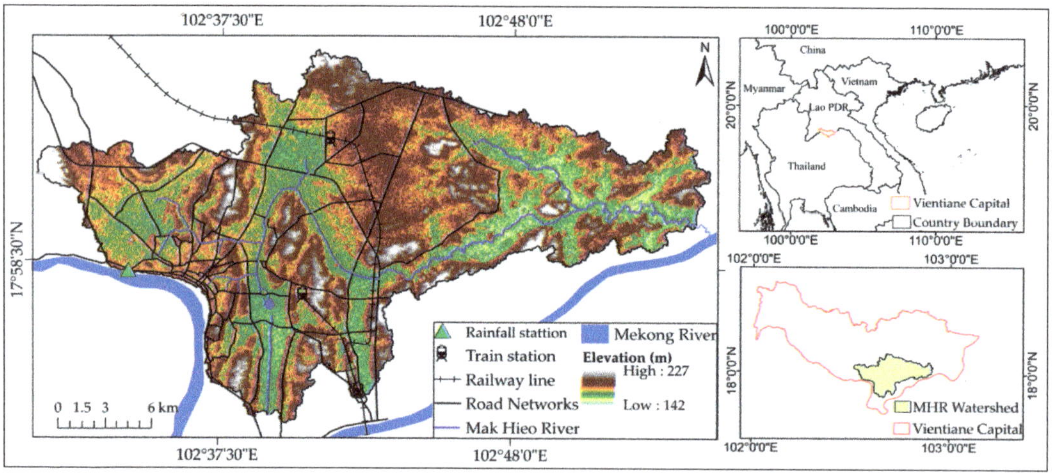

Figure 1. Location map of the study area.

2.2. Landsat Data and Preprocessing

2.2.1. Data Acquisition

In this study, the collection 2 Level-1 (C2 L1) top-of-atmosphere reflectance and Level-2 (C2 L2) surface reflectance products of Landsat 5 TM (1989, 2000) and Landsat 8 OLI (2013, 2021) were obtained at path/row (128/48) from the United States Geology Survey (USGS) [54], with a spatial and single temporal resolution of 30 m and 16 days, respectively (Table 1). Figure 2 shows the alignment of the rice growing stages with rainfall data, which guided the selection of multitemporal images with low rainfall accumulation and low cloud cover during the flooding/transplanting stage (day of the year (DOY): 152–210 [1 June–31 July]), (DOY: 349–46 [15 December–14 February]), and harvesting stage (DOY: 288–348 [15 October–14 December]). Due to frequent cloud cover during the late transplanting to early harvesting stages in the rainy season, some images from the early transplanting stage in the closest year were also used, ensuring no rainfall on the acquisition date.

Table 1. Usable Landsat images in the study.

Year	Satellite/Sensor	Spatial Resolution	Flooding/Transplanting Stage				Harvesting Stage			
			Acquisition Date	CC	P	DOY	Acquisition Date	CC	P	DOY
1989	Landsat 5 TM	30 m × 30 m	30 June 1989	<1	0	181	21 November 1989	0	0	325
			16 July 1989	<1	0	197	23 December 1989	0	0	357
			21 January 1989	0	0	21				
			06 February 1989	<1	0	37				
2000	Landsat 5 TM	30 m × 30 m	27 May 2000	<10	0	148	03 November 2000	0	0	308
			15 June 2001	<20	3	166	21 December 2000	0		356
			01 July 2001	<10	0	182				
			04 January 2000	0	0	4				
			06 January 2001	0	0	6				
			22 January 2001	0	0	22				
2013	Landsat 8 OLI	30 m × 30 m	16 June 2013	<5	0	167	06 October 2013	0	0	279
			10 January 2014	0	0	10	23 November 2013	0	0	327
			26 January 2014	0	0	26	09 December 2013	0	0	343
			27 February 2014	0	0	58				
2021	Landsat 8 OLI	30 m × 30 m	22 June 2021	<1	14.3	173	13 November 2021	0	0	317
			27 July 2022	<10	0	203	29 November 2021	0	0	333
			31 December 2021	0	0	365	15 December 2021	0	0	349
			08 January 2022	0	0	8				
			16 January 2022	<10	0	16				

CC = cloud cover (%); P = precipitation (mm).

In addition, we obtained daily rainfall data from the Department of Meteorology and Hydrology (DMH) of Laos, while the available statistical LULC maps for the years 2000, 2010, and 2021 were sourced from the Ministry of Agriculture and Forestry (MAF) of Laos, in which paddy fields are included as a sub-category of cropland. For further analysis of the factors influencing LULC changes, two explanatory factors—road networks (gathered from the official OpenStreetMap website [55]) and population density (derived from Lao Statistics Bureau [13])—were used to examine the significant decrease in paddy fields due to urbanization. These two socioeconomic factors are valuable for explaining LULC changes, especially given the limited accessibility of spatial data. We further analyzed the impacts of the regulated zones for urban expansion and industrial development outlined in the 2030 Vientiane Plan, obtained from the Ministry of Public Works and Transport (MPWT) [56]. Additionally, a field investigation was conducted to collect ground data and discuss rice-growing conditions with farmers in the study area.

2.2.2. Pre-Processing

All Landsat 5 TM and Landsat 8 OLI images used in this study were atmospherically corrected to surface reflectance for minimizing atmospheric effects, particularly the scenes used for vegetation index calculation [57,58]. This includes the terrain-corrected (L1T) and geometric-corrected images obtained using the Landsat ecosystem disturbance adaptive processing system (LEDAPS, Landsat TM) [59], and land surface reflectance code (LaSRC, Landsat OLI) [60]. Subsequently, relative radiometric correction or normalization was performed using metadata parameters to remove atmospheric effects via the SCP-plugin in the QGIS 3.28.3 software [61]. The pixels dominated by clouds and cloud shadows were removed using the Fmask algorithm, which has been extensively applied to Landsat imagery [62,63]. This process was carried out using the Cloud Masking plugin (version 23.3.30) in QGIS (version 3.28.3), which facilitates the implementation of Fmask by allowing users to define key parameters. In this study, a cloud probability threshold of 0.2 (20%)

was set, meaning that pixels with a probability score of greater than 20% were classified as clouds. Additionally, a 1-pixel buffer was applied to the cloud and shadow masks to enhance classification accuracy at the boundaries. The band combination (5/4/3 for Landsat 5) was utilized to effectively identify cloud and cloud shadow features. These boundaries were categorized as missing value or "no data" pixels [64] and masked in the corresponding bands to perform the calculation of vegetation indices.

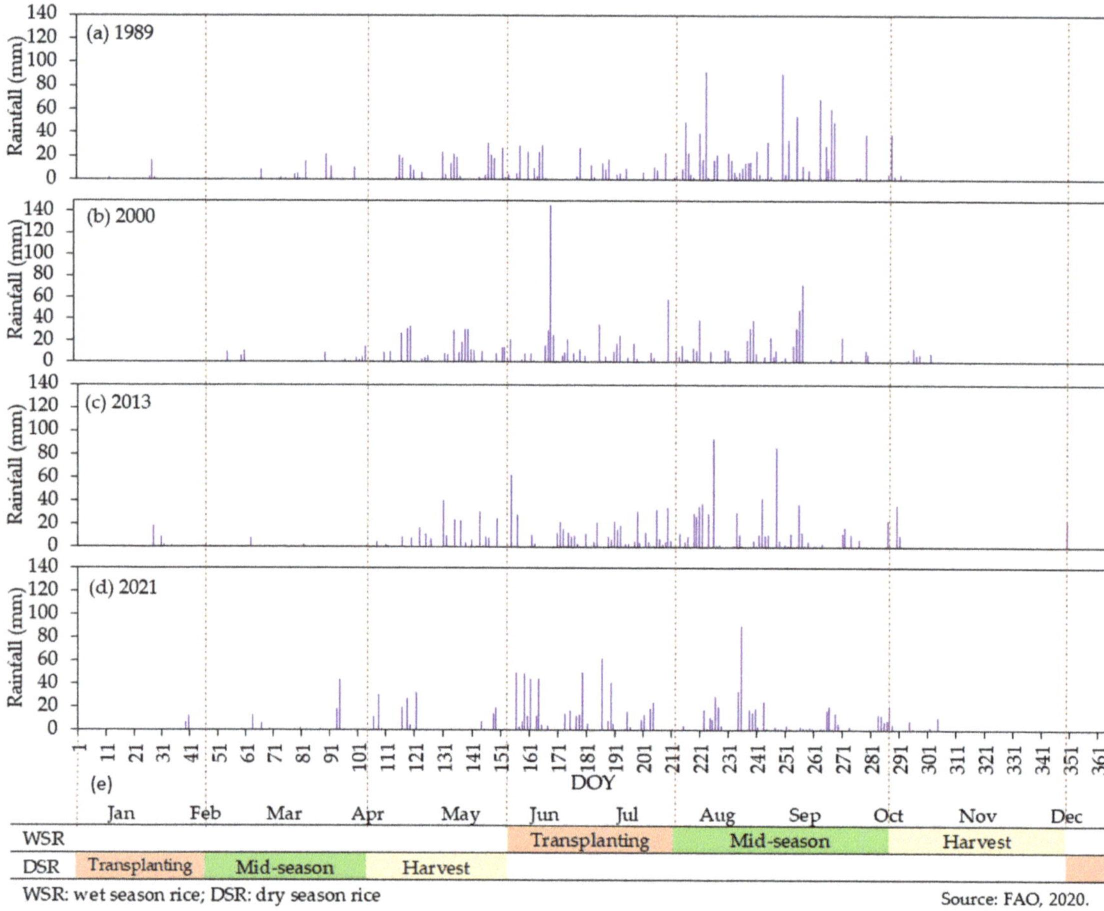

Figure 2. Referenced daily rainfall: 1989 (**a**), 2000 (**b**), 2013 (**c**), 2021 (**d**), and rice crop calendar in Laos (**e**) modified from FAO [65].

The socioeconomic factors (independent variables) were prepared for two different periods (2013 and 2021) as raster data with equal resolution (30 m × 30 m). The distance to roads was generated using the Euclidean Distance Tool in ArcMap 10.8.1 [66,67], while the population density was calculated from the ratio of population to the corresponding district's area and then converted from vector polygons to raster maps. To ensure comparability, these raster datasets were standardized before extracting the pixel values [67,68]. The dependent variables represented LULC classes, and each class was expressed by a binary value of "0" or "1". A value of "1" would indicate that a change had occurred, while a value of "0" would indicate that no change had occurred within a period [69,70]. The Extract Multi-Value to Points Tool in ArcMap 10.8.1 was employed to extract to raster values of both variable types at the same coordinate system (X, Y). Finally, the attribute table was exported in a dBASE format for statistical analysis in SPSS.

2.3. Methods

Figure 3 illustrates the conceptual framework of this study. To specifically detect the paddy field areas, we developed a new approach using a flooding-assisted supervised algorithm that focuses on the flooding/transplanting and harvesting stages. The approach is divided into four main steps. First, the multitemporal satellite data were collected and selected during flooding/transplanting and harvesting stages, based on daily observed rainfall accumulation. Various sources of the training sample data were also gathered. Second, in pre- and processing, cloud pixels were removed through masking, and spectral indices were calculated to generate an initial paddy field map by masking the monsoon water boundaries. This process provided training data points for the paddy field class in the MLC algorithm, while training data for the other classes were also included. Third, the output LULC maps were created and processed to assess classification accuracy. Finally, a post-classification analysis on change detection was conducted. Additionally, the collected socioeconomic factors and rice yield data were used for influencing factor analysis and for estimating rice production, respectively.

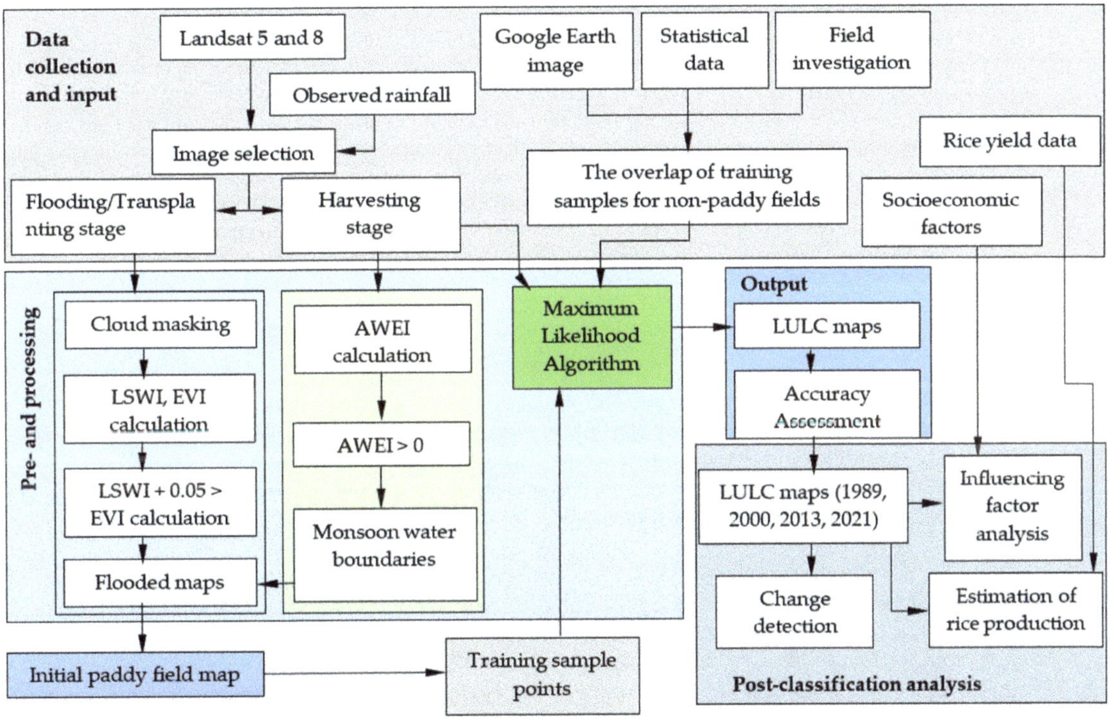

Figure 3. Methodological flowchart for detecting paddy fields and other classes.

2.3.1. Calculation of Vegetation Indices

We selected and calculated three vegetation indices—EVI, LSWI, and AWEI—using Equations (1)–(3). EVI is more sensitive to crop vegetation and reduces atmospheric interference compared to NDVI [36,71], making it particularly suitable for tropical monsoon climates characterized by high humidity and dense canopies due to seasonal rainfall [38]. LSWI is specifically designed to assess moisture content in vegetation and soil, enhancing its effectiveness in these climates, particularly in areas with dense vegetation and high moisture levels [29,32]. Importantly, AWEI addresses water extraction challenges in complex landscapes, as proposed by Feyisa et al. [40]. It effectively reduces false positives and demonstrates superior performance in distinguishing water bodies from non-water features compared to other water extraction indices, particularly in tropical monsoon climates, where a simple threshold of greater than 0 was used [41].

$$EVI = 2.5 \frac{(\rho_{nir} - \rho_{red})}{\rho_{nir} + (6 \times \rho_{red}) - (7.5 \times \rho_{blue}) + 1} \quad (1)$$

$$LSWI = \frac{(\rho_{nir} - \rho_{swir1})}{(\rho_{nir} + \rho_{swir1})} \quad (2)$$

$$AWEI_{sh} = \rho_{blue} + 2.5 \times \rho_{green} - 1.5 \times (\rho_{nir} + \rho_{swir1}) - 0.25 \times \rho_{swir2} \quad (3)$$

where ρ_{blue}, ρ_{green}, ρ_{red}, ρ_{nir}, ρ_{swir1}, and ρ_{swir2} are the values of wavelength in micrometer (μm). These ranges were (0.45–0.51) blue, (0.53–0.59) green, (0.64–0.67) red, (0.85–0.88) near-infrared, (1.57–1.65) shortwave infrared 1, and (2.11–2.29) shortwave infrared 2 in the Landsat OLI sensor, respectively. The subscript "sh" in the equation serves as an effective elimination of nonaqueous pixels.

2.3.2. Flooded Paddy Field Detection

In this study, the flooded/transplanting algorithm developed by Xiao et al. is employed (Equation (4)) [32]. This algorithm is designed to detect flooding signals of paddy fields during the flooding/transplanting stage, using a threshold of 0.05. However, natural water bodies may also be detected as flooded paddy fields at this stage. To address this concern, we focused on the harvesting stage, when the extent of natural water surfaces, including wetlands, increases due to accumulated monsoon rains and the paddy fields are dried up for harvesting work. We employed the AWEI to obtain these water surfaces [41]. Then, the actual flooded paddy field maps for the rainy and dry seasons were extracted by individually intersecting the natural water surfaces with the flooded paddy fields for each season.

$$Flooded/Transplanting\ paddy\ fields = \begin{cases} 1, & if\ LSWI + 0.05 > EVI \\ 0, & otherwize \end{cases} \quad (4)$$

Finally, the initial paddy field map was generated by intersecting the actual flooded paddy field maps from the rainy and dry seasons (Figure 4), which helped eliminate submerged fields caused by monsoon rains. The combined use of these maps reduces uncertainties and, fortunately, dry season rice in the study area is not affected by clouds or rainfall, making it a more reliable data source.

2.3.3. Derivation of Training Samples

a. Paddy fields

The obtained initial paddy field map above was utilized to create the training data. The original map still contained uncertainties, then a spatial filter technique with a 3 × 3 window size was applied to generate the training data [43]. We converted the raster map into point data using ArcMap for each year independently to ensure unbiased and consistent sampling across the time periods. The sample points were aligned with the "paddy rice" category in the statistical LULC maps and matched the shape of paddy plots as visualized in high-resolution Google Earth imagery. Unfortunately, the statistical LULC map for 1989 is not available, so the training sample points were randomly collected using the initial paddy field map, and they were referenced based on the nearest existing abandoned irrigation channels and information obtained from local people during field investigation.

b. Non-paddy fields

The non-paddy field class was categorized into the following four groups: urban areas, vegetation, water bodies, and other lands. Since these land cover classes are sensitive to spectral band reflectance, band combinations were used to digitize training samples. To enhance the accuracy of digitization, sample points for the non-paddy field classes were randomly collected from the statistical LULC maps (2000, 2010, and 2021) independently for each class and year. These points were then cross-referenced with high-resolution Google Earth imagery for refinement. In 1989, due to the unavailability of a statistical LULC map, training samples were primarily determined based on band combinations and geographical features rather than pre-selected points. According to the USGS, the optimal band combinations for identifying urban areas and vegetation using Landsat TM are 7/5/3 and 5/4/3, respectively [72], while training sample points for water bodies class were randomly selected from the 1989 monsoon water map processed with AWEI > 0.

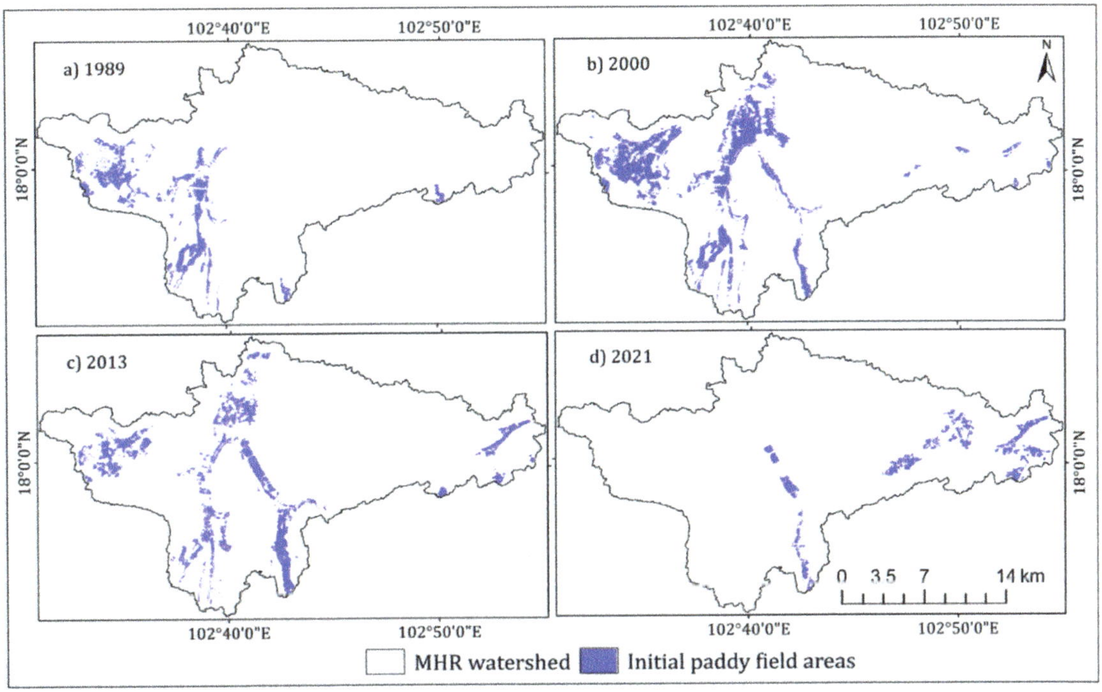

Figure 4. The initial paddy field maps for 1989 (**a**); 2000 (**b**); 2013 (**c**), and 2021 (**d**).

2.3.4. Image Classification

In this study, the MLC algorithm was selected to classify the paddy field class and other LULC classes. The algorithm uses probability theory to classify each pixel into the class it most likely belongs to, based on the highest probability from its spectral reflectance pattern [73]. This method is particularly effective for rice plants, which exhibit unique reflectance characteristics throughout their growth stages, especially with homogeneous biomass at harvest, aligning well with the normal distribution assumed by the algorithm. A cloud-free image with multispectral bands from the harvesting stage was used to precisely classify the paddy fields and other LULC classes for each year. The prepared training samples for the corresponding years and classes were input into the ArcGIS 10.8.1 software to carefully digitize the areas of interest (AOIs) surrounding the training sample points with similar spectral patterns for each class in two categories as follows: (a) Paddy fields: We collected 442 AOIs (10,347 pixels), 244 AOIs (8291 pixels), 204 AOIs (7583 pixels), and 162 AOIs (3044 pixels) over band combination to represent residual rice biomass training samples for 1989, 2000, 2013, and 2021, respectively; (b) Non-paddy fields: We collected 466 AOIs (16,252 pixels), 542 AOIs (9715 pixels), 513 AOIs (11,167 pixels), and 489 AOIs (10,158 pixels) for the year 1989, 2000, 2013, and 2021, respectively, covering all four classes. Given the numerous spectral reflectance analyses, we divided the training datasets into 27, 28, 35, and 30 subclasses for the years 1989, 2000, 2013, and 2021, respectively. These subclasses were then grouped into five classes for the corresponding years, as shown in Table 2.

Table 2. LCLU classes description in the MHR watershed.

LULC Classes	Description
Urban area	Built-up areas and roads, all constructions included golf courses, airport, industrial areas, and others.
Vegetation	Deciduous forest, unstocked forest, gallery forest, plantation, bamboo, savanna, and shrubs.
Paddy fields	Active rice paddy.
Water bodies	Rivers, channels, ponds, lakes, and marshlands.
Other lands	Residues of major LULC classes: fallow paddy fields, developing lands, other crops, etc.

2.4. Accuracy Assessment

Assessing the accuracy of classified images is critical for determining the reliability of the classification process [74,75]. In this study, validation data were collected from multiple sources for each year of classified images. Following standard practice, 250 validation points were generated using a stratified random sampling method for the years 1989, 2000, and 2013. Additionally, 50 points were allocated for each class across all years, including 2021. Field investigations were conducted in December 2022 and July 2023 to collect ground-truth data using a smartphone GPS camera (Figure 5). These data were then cross-referenced with high-resolution Google Earth imagery to ensure accurate coordinates for 2021. For the years 2000 and 2013, validation relied on statistical LULC maps for the corresponding years, along with checks against the closest-date available high-resolution Google Earth images. For 1989, validation was divided into paddy fields and non-paddy fields as follows: (a) Paddy fields: We adopted the method from Dong et al. using the best optical Landsat TM images from 16 July 1989, performing a 5/4/3 band combination that highlighted paddy fields in a distinctive dark blue tone (Appendix A Figure A1) [49], enabling the verification of points based on flooding pixels. Importantly, a monsoon water layer was added to exclude non-transplanted pixels, and there was no rainfall (0.0 mm) recorded on this date. (b) Non-paddy fields: The same procedure was applied using images from 21 November 1989, to identify corresponding LULC class territories.

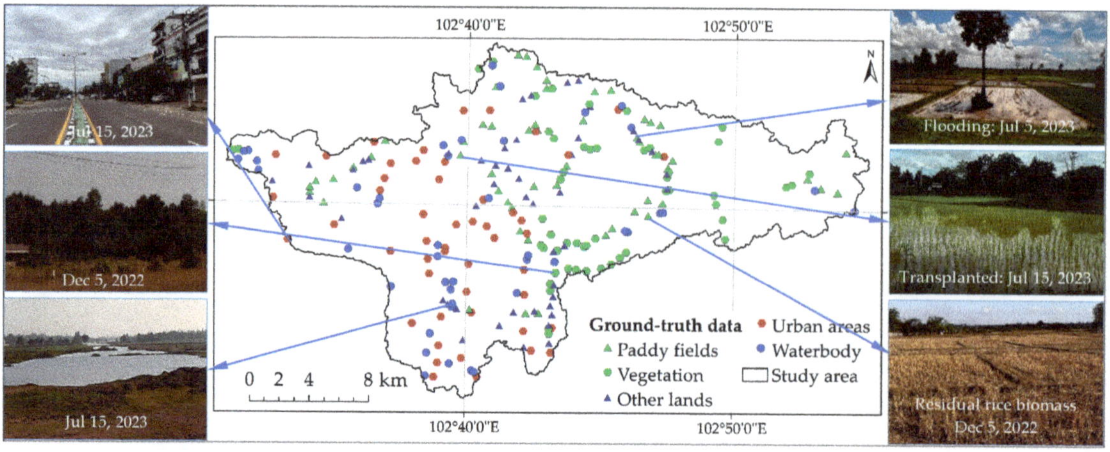

Figure 5. The ground-truth data samples in the MHR watershed from field survey in 2022 and 2023.

Furthermore, an accuracy assessment was performed using well-known measures as defined by Seyam et al. [44] and Ouedraogo et al. [76], as outlined in Equations (5)–(8). The user's accuracy (UA), producer's accuracy (PA), overall accuracy (OA), and kappa coefficient (K) are presented in the error matrix tables for the years 1989, 2000, 2013, and 2021. UA measures how many of the classified pixels, in fact, belong to that class, while PA measures how well a certain class has been correctly classified by the algorithm. OA represents the ratio of correctly categorized pixels to all validated pixels across all classes. Additionally, K is a statistical measure used to assess the agreement between classified and validated data, with values close to 0 indicating slight agreement and values close to 1 indicating an almost perfect agreement [74,77].

$$UA\ (\%) = \left(\frac{x_{ii}}{x_{+i}}\right) \times 100 \tag{5}$$

$$PA\ (\%) = \left(\frac{x_{ii}}{x_{i+}}\right) \times 100 \tag{6}$$

$$OA\ (\%) = \frac{\sum_{i=1}^{r} x_{ii}}{N} \times 100 \tag{7}$$

$$K = \frac{N\sum_{i=1}^{k} x_{ii} - \sum_{i=1}^{k}(x_{+i}.x_{i+})}{N^2 - \sum_{i=1}^{k}(x_{+i}.x_{i+})} \tag{8}$$

where r represents the number of classes, N is the total number of reference samples, x_{ii} represents the count of correctly classified samples in column i and row i, x_{i+} represents the sum of samples in column i, and x_{+i} is the sum of samples in row i in the confusion matrix.

To further compare results, the paddy field maps were analyzed alongside statistical LULC maps from the MAF, which were developed through a national-level classification system for Lao PDR to monitor LULC changes in Laos, particularly in forest and agricultural lands. The MAF maps categorize LULC into seven primary categories and twenty subcategories, with paddy fields included as a subcategory (i.e., rice paddy) under cropland. The vector maps were converted to a 30 m × 30 m resolution, allowing for a spatial comparison of the "rice paddy" category for 2000, 2010, and 2021 with the paddy fields identified in this study for 2000, 2013, and 2021. The year 1989 was excluded from the comparison due to the unavailability of MAF maps prior to 2000. These comparisons are expected to enhance the reliability of paddy field detection in the current study.

2.5. Analysis of the Explanatory Factors

Roads and population density are closely linked to urbanization, with roads enabling access and growing populations driving the demand for land. These two factors were analyzed for 2013 and 2021 (Figure 6) using a binary logistic regression model to illustrate spatial changes in the paddy fields and urban expansion. This approach is commonly used to verify the relationship between LULC classes and explanatory factors in terms of spatial changes [67,78], as shown in Equation (9). The analysis was conducted using SPSS 27 software by inputting dependent variables (LULC classes) and independent variables (socioeconomic factors). The model's performance was measured by the area under the curve (AUC) analysis. AUC values range from 0 to 1; values between 0.5 and 0.7 denote slight accuracy, values of 0.7 and 0.9 show acceptable accuracy, and values above 0.9 indicate exceptional precision [79].

$$\mathrm{Log}\left(\frac{\mathrm{P}}{1-\mathrm{P}}\right) = \beta_0 + \beta_1 X_1 + \beta_2 X_2 \tag{9}$$

where P is the grid cell probability of an absolute LULC class; β_0 is the intercept; β_1 and β_2 are the logistic regression coefficients for road distance and population density, respectively, with respect to each LULC class; X_1 and X_2 represent the road distance and population density factors, respectively.

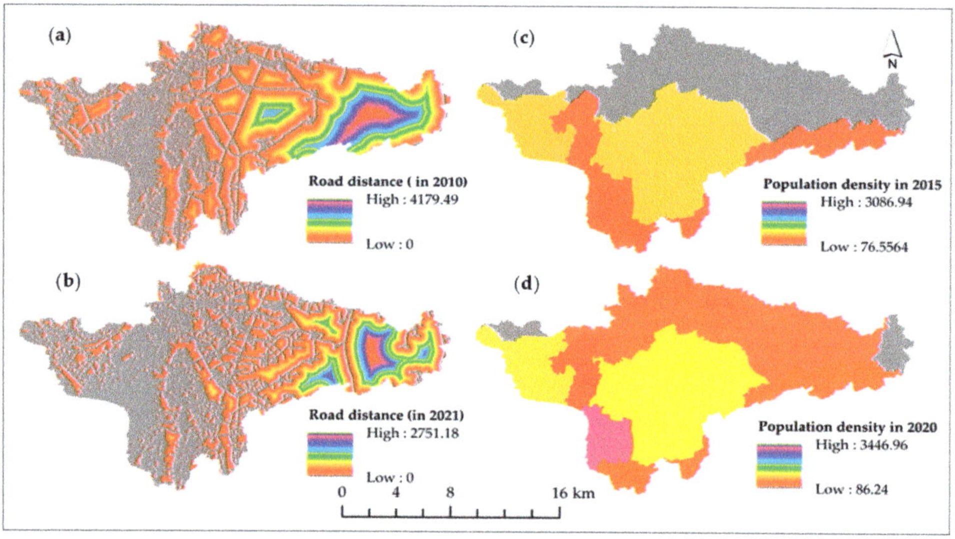

Figure 6. Spatial explanatory factors, road distance (m) for 2010 (**a**) and 2021 (**b**), and population density (person/km^2) for 2015 (**c**) and the projection for 2020 (**d**).

Further analysis using the regulated/permitted boundaries for urban expansion and industrial development outlined in the 2030 Vientiane plan published by the Ministry of Public Works and Transport [56] overlaid these boundaries onto the classified maps of 2013 and 2021 to identify how they impacted the reduction in paddy fields within the watershed.

2.6. Change Detection and Estimating Rice Production in the Study Area

Detecting LULC changes, particularly in paddy fields, is crucial for sustainable agricultural practices and environmental conservation [80]. Analyzing the conversion matrix and trends in paddy fields reveals how these areas have shifted over time, providing essential insights for promoting sustainable land use, enhancing food security, and optimizing urban growth. In this study, we examined the periods of 1989–2000, 2000–2013, 2013–2021, and 1989–2021 by converting classified raster images into vector layers using ArcMap 10.8.1 software. We then intersected the initial and final layers to generate conversion information for each LULC class, including the paddy fields. This approach allowed us to construct conversion matrices that track transitions between land use categories over time. The conversion matrices provided valuable insights into changes in the paddy fields, enabling us to identify spatial patterns and trends. By analyzing these trends from 1989 to 2021, we provide critical information for stakeholders to anticipate future land use changes and take proactive measures to mitigate negative impacts, supporting sustainable development goals in the watershed.

Additionally, to understand the critical influence of an increasing population on rice production, we estimated the quantity of rice demand and harvested rice based on population growth and the temporal paddy field areas derived from the LULC maps. This analysis is essential for ensuring food security and guiding resource management in response to the growing needs of the population. Rice demand was estimated based on the population within the watershed in a specific year and the per capita rice consumption rate. As is well known, rice is the primary grain consumed in Laos, with the annual consumption averaging about 350 kg of un-milled rice per capita [81,82]. Meanwhile, the amount of harvested rice was calculated by multiplying the rice yield by the total area of rice production during wet and dry seasons within the watershed. Statistical data on rice yield (for 1989, 2000, 2013, and 2021) and population census data (for 1995, 2005, 2015, as well as projected for 2021) were obtained from the Lao Statistics Bureau [13]. Temporal data of paddy field areas were based on the current study's results.

3. Results

3.1. Characteristics of LULC Changes in the MHR Watershed

In the MHR watershed, five LULC classes, namely urban areas, vegetation, paddy fields, water bodies, and other lands, were classified using Landsat images, as shown in Figure 7. Table 3 presents the statistical area coverage and percentage of each LULC class for four periods (1989, 2000, 2013, and 2021). In 1989, the major coverage classes were paddy fields and vegetation, accounting for 165.67 km^2 (36.48%) and 176.61 km^2 (38.89%), respectively. By 2021, urban areas predominantly increased to occupy 143.92 km^2 (31.69%). Over the four time periods, paddy fields showed a consistent decline (Figure 8). Significant declines occurred between 2000–2013 and 2013–2021, with 40.59 km^2 (−24.87%) and 31.27 km^2 (−25.50%) in paddy fields, respectively (Table 4). Conversely, urban areas saw a significant expansion—a 116.43 km^2 (423.62%) increase from 1989 to 2021. While urban areas are still growing, the amount of new urban land added in each successive period is less than in the previous one. Nevertheless, the highest annual growth rate occurred during 2013–2021, with an increase of 7.14 km^2 (8.23%) per year (Table 5). Vegetation area experienced a dramatic decline of 71.40% (Table 4), most notably between 1989 and 2013, where it decreased from 176.61 km^2 (38.89%) to 70.86 km^2 (15.61%) (Table 3). The most significant rate of reduction, 5.19 km^2 (3.75%) per year, occurred between 2000 and 2013 (Table 5), coinciding with the period just before urban expansion accelerated. Meanwhile, the other lands class expanded during the same period of declining vegetation, increasing from 67.49 km^2 (14.86%) in 1989 to 86.03 km^2 (18.95%) in 2000, and then doubling to 160.01 km^2 (35.51%) by 2013 before a slight decrease to 156.51 km^2 (34.47%) in 2021 (Table 3). Water bodies covered 16.84 km^2 (3.71%) in 1989 but declined to 11.81 km^2 (2.60%) by 2021 (Table 3), reflecting a slight decrease of 29.88% between 1989 and 2021 (Table 4). The most notable decreases occurred between 2000–2013 and 2013–2021, with annual rates of 1.18% and 1.82%, respectively (Table 5).

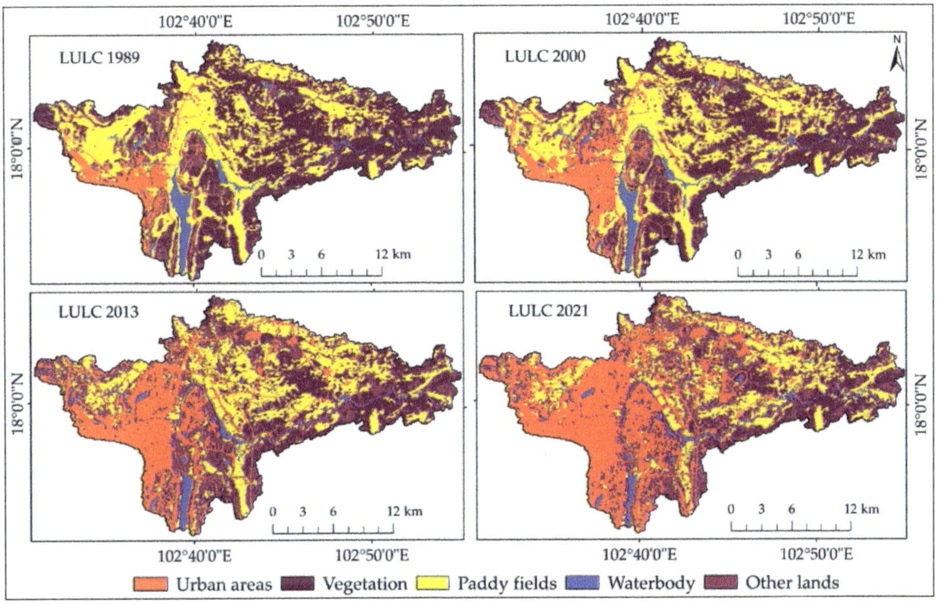

Figure 7. Classified LULC maps of the MHR watershed in 1989, 2000, 2013, and 2021.

Table 3. LULC patterns of the MHR watershed in 1989, 2000, 2013, and 2021.

LULC Class	1989		2000		2013		2021	
	Area (km²)	%	Area (km²)	%	Area (km²)	%	Area (km²)	%
Urban area	27.49	6.05	50.16	11.05	86.78	19.11	143.92	31.69
Vegetation	176.61	38.89	138.38	30.47	70.86	15.61	50.51	11.12
Paddy fields	165.67	36.48	163.20	35.94	122.61	27.00	91.34	20.12
Water bodies	16.84	3.71	16.32	3.59	13.82	3.04	11.81	2.60
Other lands	67.49	14.86	86.03	18.95	160.01	35.24	156.51	34.47
Total area	454.10	100	454.09	100	454.08	100	454.09	100

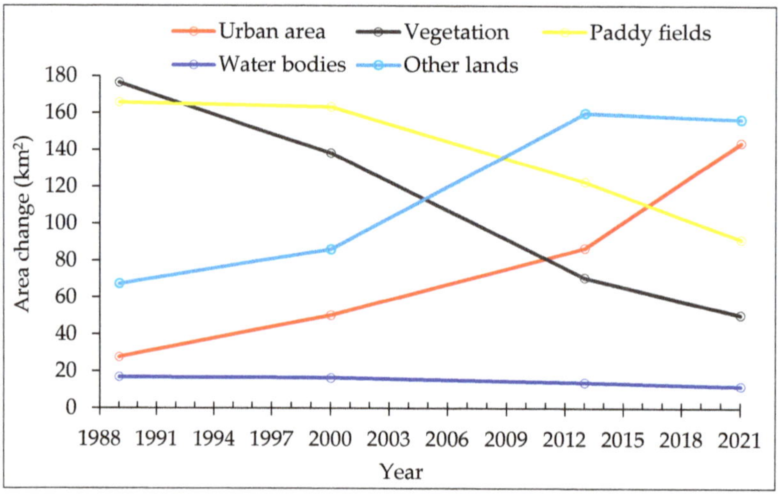

Figure 8. Change in Areas of LULC Classes in the MHR Watershed from 1989 to 2021.

Table 4. Area change in each LULC class over the MHR watershed.

LULC Class	1989–2000		2000–2013		2013–2021		1989–2021	
	Area (km²)	%	Area (km²)	%	Area (km²)	%	Area (km²)	%
Urban area	22.68	82.51	36.62	73.00	57.14	65.84	116.43	423.62
Vegetation	−38.23	−21.65	−67.52	−48.79	−20.35	−28.72	−126.10	−71.40
Paddy fields	−2.48	−1.49	−40.59	−24.87	−31.27	−25.50	−74.33	−44.87
Water bodies	−0.52	−3.08	−2.50	−15.32	−2.01	−14.56	−5.03	−29.88
Other lands	18.54	27.48	73.98	85.99	−3.50	−2.19	89.02	131.91

Table 5. Annual Rate of LULC change in the MHR watershed.

LULC Class	1989–2000		2000–2013		2013–2021		1989–2021	
	Area (km²)	%	Area (km²)	%	Area (km²)	%	Area (km²)	%
Urban area	2.06	7.50	2.82	5.62	7.14	8.23	3.64	13.24
Vegetation	−3.48	−1.97	−5.19	−3.75	−2.54	−3.59	−3.94	−2.23
Paddy fields	−0.23	−0.14	−3.12	−1.91	−3.91	−3.19	−2.32	−1.40
Water bodies	−0.05	−0.28	−0.19	−1.18	−0.25	−1.82	−0.16	−0.93
Other lands	1.69	2.50	5.69	6.61	−0.44	−0.27	2.78	4.12

3.2. Classification Accuracy

The accuracy assessments of the classified images for 1989, 2000, 2013, and 2021 were performed using the validation data described in Section 2.4, and the results showed that the classified images had reasonable accuracies (Table 6). The overall accuracies were 83.6%, 89.6%, 87.6%, and 90.4% in the years 1989, 2000, 2013, and 2021, respectively, while the Kappa coefficients were 0.80, 0.87, 0.85, and 0.88, respectively. The paddy field class achieved UA values of between 72% and 88% and PA values of between 86% and 92%, indicating some misclassification issues. The PA consistently being higher than the UA suggests that while a high proportion of actual paddy fields were correctly classified (high PA), some non-paddy field areas were mistakenly classified as paddy fields (lower UA). For example, the UA for paddy fields in 2021 is 84%, indicating the F-MLC method correctly classified paddy fields according to the validation data. However, 16% of the classified paddy fields were overestimated, coming from other lands and water bodies. Meanwhile, PA achieved 91%, reflecting that the actual paddy fields were correctly classified by the F-MLC method, although 9% of actual paddy fields were misclassified as other lands. Moreover, the water bodies class achieved UA values ranging from 94% to 96% and PA values from 92% to 96%, demonstrating successful water extraction. This accurate water classification contributed to the enhanced classification of paddy fields by reducing potential confusion between these classes.

Table 6. Error matrix of the classification accuracy for the year 1989, 2000, 2013, and 2021.

Year	LULC	UB	VE	PF	WB	OL	Total	UA (CE) %	K
1989	Urban area	43	2	1	0	4	50	86 (14)	0
	Vegetation	2	44	0	0	4	50	88 (12)	0
	Paddy field	2	1	36	2	9	50	72 (28)	0
	Water bodies	0	0	2	47	1	50	94 (6)	0
	Other land	1	7	3	0	39	50	78 (22)	0
	Total	48	54	42	49	57	250	0	0
	PA (OE) %	89 (11)	81 (19)	86 (14)	96 (4)	68 (32)	0	83.6	0
	K	0	0	0	0	0	0	0	0.80
2000	Urban area	45	0	0	1	4	50	90 (10)	0
	Vegetation	0	48	0	0	2	50	96 (4)	0
	Paddy field	0	0	44	1	5	50	88 (12)	0
	Water bodies	0	0	2	48	0	50	96 (4)	0
	Other land	3	5	2	1	39	50	78 (22)	0
	Total	48	53	48	51	50	250	0	0
	PA (OE) %	94 (6)	91 (9)	92 (8)	94 (6)	78 (22)	0	89.6	0
	K	0	0	0	0	0	0	0	0.87
2013	Urban area	46	0	0	0	4	50	92 (8)	0
	Vegetation	0	48	0	0	2	50	96 (4)	0
	Paddy field	1	1	39	1	8	50	78 (22)	0
	Water bodies	0	0	1	48	1	50	96 (4)	0
	Other land	2	3	5	2	38	50	76 (24)	0
	Total	49	52	45	51	53	250	0	0
	PA (OE) %	94 (6)	92 (8)	86 (14)	94 (6)	72 (28)	0	87.6	0
	K	0	0	0	0	0	0	0	0.85
2021	Urban area	46	1	0	0	3	50	92 (8)	0
	Vegetation	1	47	0	1	1	50	94 (6)	0
	Paddy field	0	0	42	2	6	50	84 (16)	0
	Water bodies	0	0	1	48	1	50	96 (4)	0
	Other land	2	1	3	1	43	50	86 (14)	0
	Total	49	49	46	52	54	250	0	0
	PA (OE) %	94 (6)	96 (4)	91 (9)	92 (8)	80 (20)	0	90.4	0
	K	0	0	0	0	0	0	0	0.88

UB—urban area; VE—vegetation; PF—paddy fields; WB—water bodies; OL—other lands; CE—commission errors; OE—omission errors.

These accuracy metrics reveal key sources of error in the classification when compared to the validation data. The lower UA suggests that some overestimations of paddy fields resulted from spectral similarities between the paddy fields and other land classes, such as fallow land or grasses. The slightly lower PA indicates that some actual paddy fields were misclassified as other land classes, reflecting the challenges of distinguishing paddy fields in complex landscapes. However, the algorithm has proven to be reliable and applicable for classifying paddy fields. Furthermore, the four LULC maps generated were found to be reasonable and suitable for further analysis, particularly for assessing the dynamics of paddy fields in the MHR watershed from 1989 to 2021.

3.3. Comparison of F-MLC's Paddy Fields with the Statistics

The comparison between the paddy field class derived from the current study and those from the MAF shows high agreement across all three periods (2000, 2013, and 2021), with a particularly strong alignment with 2000 (Figure 9). Overall, the trend indicates that paddy fields in this urbanizing watershed have been decreasing, while non-paddy field areas have expanded from 2000 to 2021 in both maps. The zoomed-in analysis of a peri-urban area revealed that the MAF map identified a larger area of paddy fields compared to the F-MLC map in all three periods (Figure 9a1–c1).

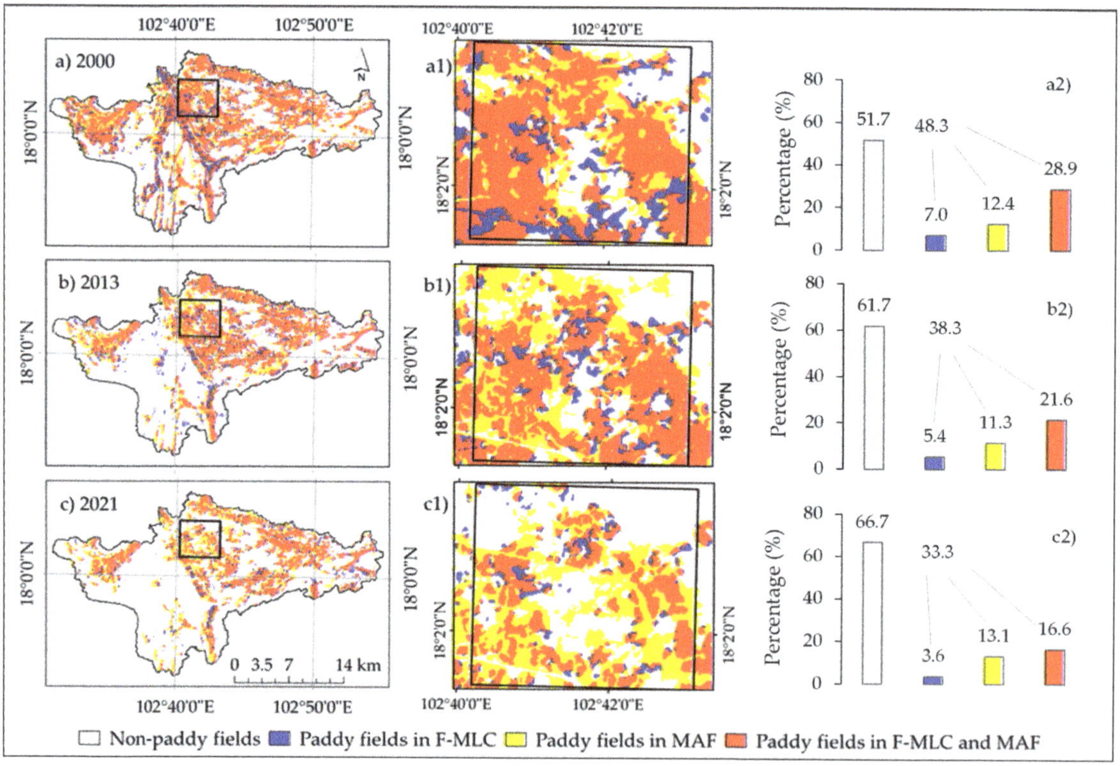

Figure 9. Spatial comparison and percent coverage: (**a**) 2000, (**b**) 2013, and (**c**) 2021 show spatial overlay of paddy fields between F-MLC and MAF's maps; (**a1–c1**) provide detailed zoom-ins of the spatial overlay; (**a2–c2**) indicate the percent coverage of both maps in the watershed.

To clarify this, we further validated the MAF map using the 2021 ground datasets for other land classes to determine the proportion of paddy fields mapped by both MAF and F-MLC that belong to other land classes. The results indicated that the MAF map overestimated paddy fields, with 38% of the areas classified as paddy fields actually belonging to other land classes, whereas the F-MLC map showed only a 13% overestimation.

3.4. Conversion Matrix and Trend of LULC Changes in the MHR Watershed

The conversion matrix of LULC changes illustrates the transitions among five LULC classes in the study area across the following four periods: 1989–2000, 2000–2013, 2013–2021, and 1989–2021 (Table 7). A significant transition of paddy fields and vegetation to urban areas and abandoned land was observed. Urban expansion had a notable impact on large areas of paddy fields and vegetation, especially during the periods 2000–2013 and 2013–2021, leading to the conversion of approximately 8.27 km^2 and 8.12 km^2 and 9.03 km^2 and 3.60 km^2 of these land classes, respectively. From 1989 to 2021, paddy fields and vegetation were predominantly converted into fallow and deforested lands, under the other land classes. This transition became more pronounced between 2000 and 2021, with about 51.70 km^2 of paddy fields and 54.34 km^2 of vegetation converted into other land classes during 2000–2013, and 39.01 km^2 of paddy fields and 24.54 km^2 of vegetation converted during 2013–2021. Additionally, the other land classes were transformed into urban areas, covering up to 48.64 km^2 between 2013 and 2021. This suggests that fallow and deforested lands served as preparatory stages for urban expansion rather than being left abandoned. Moreover, urban growth between 1989 and 2021 replaced approximately 2.51 km^2 of water bodies. Despite this loss, new artificial ponds as the water body class were created on former paddy fields for agricultural purposes, contributing to gains of 2.15 km^2 between 1989 and 2021. While a general decline in paddy fields was evident, the expansion of paddy fields to the middle and downstream parts of the watershed was observed during the periods 1989–2000 and 2000–2013, occurring at the expense of vegetation and accounting for 12.68 km^2 and 8.66 km^2, respectively.

Table 7. LULC conversion matrix of the MHR watershed for 1989–2000, 2000–2013, 2013–2021, and 1989–2021.

LULC Class		2000 Image (km^2), Gain					
		UB (%)	**VE (%)**	**PF (%)**	**WB (%)**	**OL (%)**	**Total**
1989 Image (km^2), Loss	Urban area	24.29 (88.35)	0.77 (2.81)	0.68 (2.47)	0.14 (0.51)	1.59 (5.85)	27.47
	Vegetation	5.54 (3.14)	126.96 (71.94)	12.68 (7.18)	0.87 (0.49)	30.42 (17.24)	176.48
	Paddy fields	4.84 (2.92)	2.19 (1.32)	137.66 (83.10)	2.08 (1.25)	18.88 (11.40)	165.65
	Water bodies	1.39 (8.25)	0.20 (1.18)	2.30 (13.66)	12.70 (75.43)	0.25 (1.48)	16.84
	Other lands	14.06 (20.86)	8.18 (12.14)	9.85 (14.61)	0.53 (0.78)	34.80 (51.62)	67.42
	Total	50.12	138.31	163.17	16.31	85.94	453.86
		2013 Image (km^2), Gain					
2000 Image (km^2), Loss	Urban area	47.14 (94.03)	0.52 (1.04)	0.62 (1.24)	0.36 (0.72)	1.49 (2.97)	50.13
	Vegetation	8.12 (5.88)	65.61 (47.47)	8.66 (6.26)	1.47 (1.07)	54.34 (39.32)	138.21
	Paddy fields	8.27 (5.07)	1.65 (1.01)	99.00 (60.68)	2.54 (1.55)	51.70 (31.69)	163.16
	Water bodies	0.83 (5.12)	0.18 (1.08)	0.83 (5.11)	8.39 (51.45)	6.08 (37.25)	16.31
	Other lands	22.35 (26.00)	2.82 (3.28)	13.46 (15.67)	1.05 (1.22)	46.27 (53.83)	85.95
	Total	86.71	70.79	122.57	13.81	159.88	453.77
		2021 Image (km^2), Gain					
2013 Image (km^2), Loss	Urban area	81.71 (94.21)	1.07 (1.23)	1.21 (1.40)	0.69 (0.80)	2.05 (2.37)	86.73
	Vegetation	3.60 (5.09)	39.99 (56.50)	1.99 (2.81)	0.65 (0.92)	24.54 (34.68)	70.78
	Paddy fields	9.03 (7.37)	0.31 (0.25)	73.41 (59.89)	0.83 (0.67)	39.01 (31.82)	122.57
	Water bodies	0.83 (5.99)	0.11 (0.82)	0.57 (4.12)	7.57 (54.78)	4.74 (34.28)	13.81
	Other lands	48.64 (30.42)	8.98 (5.62)	14.15 (8.85)	2.07 (1.29)	86.05 (53.82)	159.88
	Total	143.82	50.45	91.33	11.80	156.39	453.78
1989 Image (km^2), Loss	Urban area	24.33 (88.60)	0.54 (1.96)	0.91 (3.33)	0.45 (1.65)	1.26 (4.57)	27.49
	Vegetation	39.58 (22.44)	47.33 (26.83)	13.83 (7.84)	2.65 (1.50)	73.01 (41.39)	176.40
	Paddy fields	37.95 (22.91)	0.30 (0.18)	68.87 (41.58)	2.15 (1.30)	56.36 (34.03)	165.64
	Water bodies	2.51 (14.90)	0.16 (0.92)	1.08 (6.43)	5.88 (34.90)	7.21 (42.84)	16.83
	Other lands	39.45 (58.51)	2.13 (3.16)	6.62 (9.82)	0.67 (1.00)	18.55 (27.51)	67.41
	Total	143.80	50.46	91.32	11.80	156.38	453.75

UB—urban area; VE—vegetation; PF—paddy fields; WB—water bodies; OL—other lands.

The conversion trends of paddy fields over three distinct periods, 1989–2000, 2000–2013, and 2013–2021, are illustrated in Figure 10 (left), which details the extent of transformation from paddy fields to various LULC classes. The paddy fields showed a gradual increase in their transformation into urban areas, rising from 2.92% in 1989–2000 to 7.37% by 2013–2021. Conversely, natural conversions of paddy fields to vegetation and water bodies exhibited a slight decline, dropping 1.32% to 0.25% and 1.25% to 0.67%, respectively, between 1989 and 2021. Furthermore, the expansion of new paddy fields from vegetation demonstrated a downward trend, decreasing from 7.18% in 1989–2000 to 2.81% in 2013–2021. Conversely, there was a significant increase in the conversion of paddy fields to the other lands class, with a sharp rise from 11.40% to 31.69% between 1989–2000 and 2000–2013, followed by a slight increase to 31.82% from 2000–2013 to 2013–2021. This indicates an effort to expand paddy fields away from urban development zones. However, arable land under vegetation remains limited within the watershed.

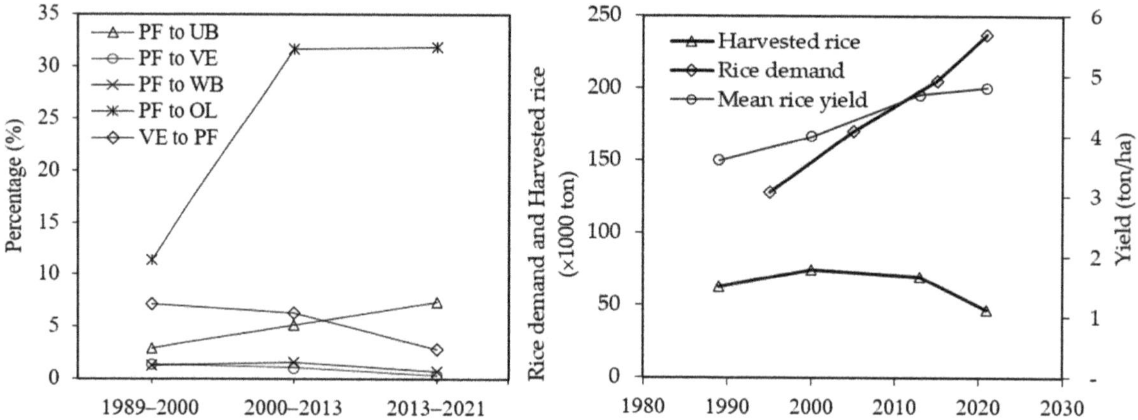

Figure 10. Conversion trends of paddy fields in the MHR watershed (PF—paddy fields, UB—urban area, VE—vegetation, WB—water bodies, OL—other land (**left**). Demand and harvested rice (**right**).

3.5. Results of Explanatory Factors Analysis

The logistic regression analysis indicated that the distance to major roads and population density significantly influenced the LULC changes in the study area. As indicated by the β coefficient values for the years 2013 and 2021 (Table 8). A positive β value represents a positive correlation between the explanatory factors and LULC classes.

Table 8. Logistic regression coefficients of explanatory factors in LULC changes.

Year	Factors	Urban Area		Vegetation		Paddy Fields		Water Bodies		Other Lands	
		β	exp(β)	β	exp(β)	β	exp(β)	β	exp(β)	β	exp(β)
2021	DR	−9.624	6.61×10^{-5}	0.637	1.891	0.256	1.292	0.238	1.269	−0.083	0.920
	PD	0.415	1.514	−0.76	0.468	−0.311	0.733	0.293	1.340	−0.352	0.703
	Intercept	−5.049		−2.464		−1.418		−3.768		−0.661	
	AUC value	0.900		0.780		0.700		0.660		0.540	
2013	DR	−11.888	6.87×10^{-6}	0.550	1.733	0.090 *	1.009	0.119	1.126	−0.180	0.835
	PD	0.499	1.647	−0.627	0.534	−0.270	0.763	0.29	1.336	−0.218	0.804
	Intercept	−7.670		−1.935		−0.729		−3.478		−0.843	
	AUC value	0.900		0.740		0.610		0.630		0.570	

DR—distance to road (m); PD—population density (person/km^2). β values were significantly statistic with p-value < 0.001; * p-value < 0.01.

In both 2013 and 2021, the distance to major roads presented an inverse and positive relationship with the urban area and paddy field classes, respectively. Conversely, population density demonstrated positive and inverse relationships with the urban area and paddy field classes. This implies that being closer to roads and an increase in population density are likely to drive urban expansion while leading to a reduction in paddy fields. Areas with higher population densities led to an increase in urban areas by 0.499 and 0.415 times, with probabilistic exponential functions of 1.647 and 1.514, respectively. Conversely, paddy field areas decreased by −0.27 and −0.311 times, corresponding the exponential probability functions of 0.763 and 0.733. The goodness-of-fit was evaluated using the AUC analysis, which indicated a satisfactory performance in the urban area, vegetation, and paddy field classes. However, the model's performance was not as highly successful in water bodies and other land classes when the AUC value < 0.7, as emphasized by Mandrekar [83], Manel et al. [79], and Rossiter & Loza [84]. Notably, Gil Pontius Jr and Schneider recommended that an AUC value above 0.52 significantly outperforms random classification [85].

Urban expansion was largely concentrated within the regulated zones near developed transportation infrastructures (Figure 11), contributing to a 52.69% (43.14 km^2) decrease in paddy fields within the permitted boundary between 2013 and 2021. In contrast, outside the permitted zones, where population density is lower and road networks are less developed, paddy fields expanded by approximately 29.15% (11.87 km^2).

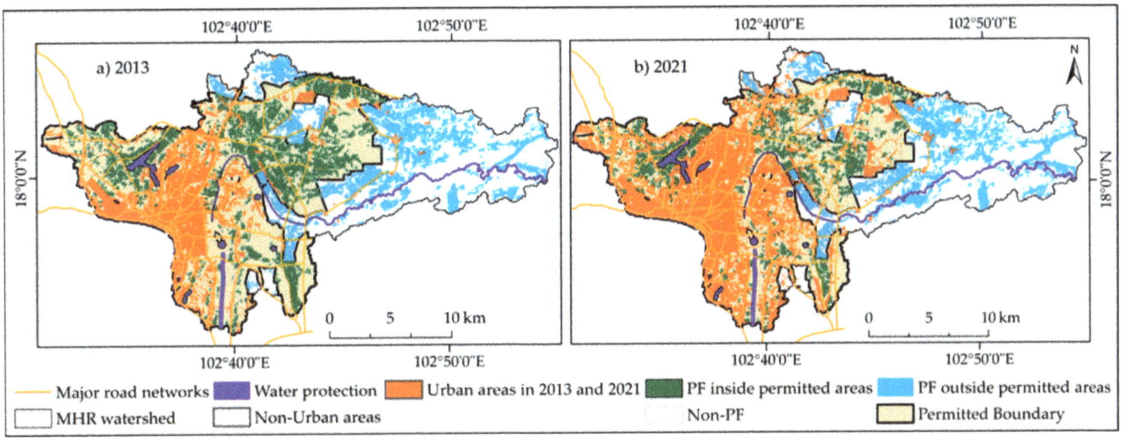

Figure 11. The boundary of the permitted areas for economic and urban development in Vientiane lies within the MHR watershed. "PF" refers to the paddy field areas.

3.6. Rice Production and Demand

The estimation of harvested rice against the demand for rice consumption within the watershed shows a concerning trend from 1989 to 2021 (Figure 10, right). Over these years, harvested rice quantities have generally decreased since 2000, while rice demand has sharply risen. Specifically, harvested rice production dropped from 74,500 tons in 2000 to 46,700 tons in 2021, whereas rice demand surged from 128,250 tons in 1989 to 237,000 tons in 2021. Notably, between 1989 and 2000, harvested rice production increased in parallel with rising demand, reflecting efforts to expand production areas and improve rice productivity during this period [82]. However, despite these efforts, rice production and demand continue to pose significant challenges to food security in the watershed, as well as in Vientiane.

4. Discussion

4.1. Advantages of the F-MLC Method

The method in this paper is straightforward and requires less data processing compared to supervised machine learning methods, which typically demand large datasets, while the F-MLC approach requires only a few Landsat images during the flooding/transplanting stage (rainy and dry seasons) and one during the harvesting stage, making it an efficient strategy to overcome the limitation of image availability in tropical monsoon climate areas.

Moreover, our flooding-assisted maximum likelihood classification (F-MLC) method is able to address the challenge posed by submerged fields due to monsoon rains, by intersecting flooded paddy field maps from the rainy and dry seasons. Although the F-MLC method was originally designed for tropical monsoon climates, it is adaptable to other regions with similar seasonal patterns, such as subtropical monsoon areas or regions with distinct wet and dry seasons. In these climates, seasonal rainfall can result in submerged fields, which the F-MLC method can discriminate by using seasonal imagery, distinguishing monsoon-induced flooding from actual flooded paddy fields. However, the F-MLC approach has limitations in regions with single-season rainfed rice cultivation, as it is difficult to distinguish monsoon-submerged fields. In large-scale paddy field mapping, this method can become cumbersome due to the extensive datasets required to cover wide geographic areas during training data collection. Additionally, in more heterogeneous landscapes or mixed cropping systems, the F-MLC may face challenges in distinguishing between land use classes with similar spectral characteristics, potentially leading to misclassification.

Based on the comparison and validation of the MAF and F-MLC maps using the same validation data from 2021, the F-MLC map showed a lower overestimation than the MAF map, indicating that the F-MLC method achieved a higher accuracy in detecting active paddy fields. Since the MAF utilized remote sensing data from periods outside the rainy season for forest and agricultural land monitoring, this likely contributed to the overestimation of paddy fields, as a similar reflectance of ground biomass (i.e., fallow paddy fields) may have been affected by rapid urbanization. The differences between the two datasets may also be partly attributed to their specific classification focuses and the different datasets utilized. The current study concentrated on a particular rice-growing stage, focusing mainly on the physical features of rice fields, whereas the MAF considered all land categories equally significant in its analysis. Through this comparison, the significant spatial correlation with statistical datasets suggests that our mapping results are reasonable.

4.2. Uncertainty and Improvement

In Laos, the rainy season is the main rice-growing period, but extensive cloud cover limits image availability, while the dry season is mostly cloudless. Therefore, the F-MLC method is proposed for monitoring or long-term paddy field mapping. However, this method is reliant on the training data generated from flooded paddy field maps. These maps are used as the baseline data to produce the resultant paddy field maps. In this study, flooded paddy field maps in rainy and dry seasons are intersected, so confusion pixels can be further minimized. Moreover, this combined map further reduced errors using monsoon water map. Despite various efforts to minimize errors in data processing, the remaining challenges are presented. First, the proposed F-MLC approach used the sample points to digitize the algorithm's training samples, which may result in overestimation. Thus, a statistical method should be developed for automated sample generation. Second, this method detected the residual rice biomass during the harvesting stage posed a challenge due to a similar reflectance between rice biomass and grasses (fallow paddy fields), leading to overestimation (Figure 12(a1,a2)). Increasing the training samples, possibly through automated sample generation or applying advanced machine learning algorithms [43,86], could help mitigate this heterogeneity. Additionally, integrating various VIs can further reduce uncertainty. For example, using multitemporal NDVI during the land preparation period can help discriminate between the fallow paddy fields and the land actively being prepared for rice cultivation [39]. Third, due to frequent droughts [87], farmers store rainwater in ponds, which is smaller than Landsat's spatial resolution. So, a minor underestimation in water surface detection using the AWEI was presented, even in the dry season rice, as shown in Figure 12(b1,b2,c1,c2). To address this issue, using satellite imagery with higher spatial resolution, such as SPOT 1-5 (launched on 21 February 1986), which offers three-day imagery with 10 m spatial resolution in the multispectral bands and 20 m in the SWIR band, could be beneficial. Unfortunately, SPOT primarily captures data from Europe and Africa, with limited coverage in Asia.

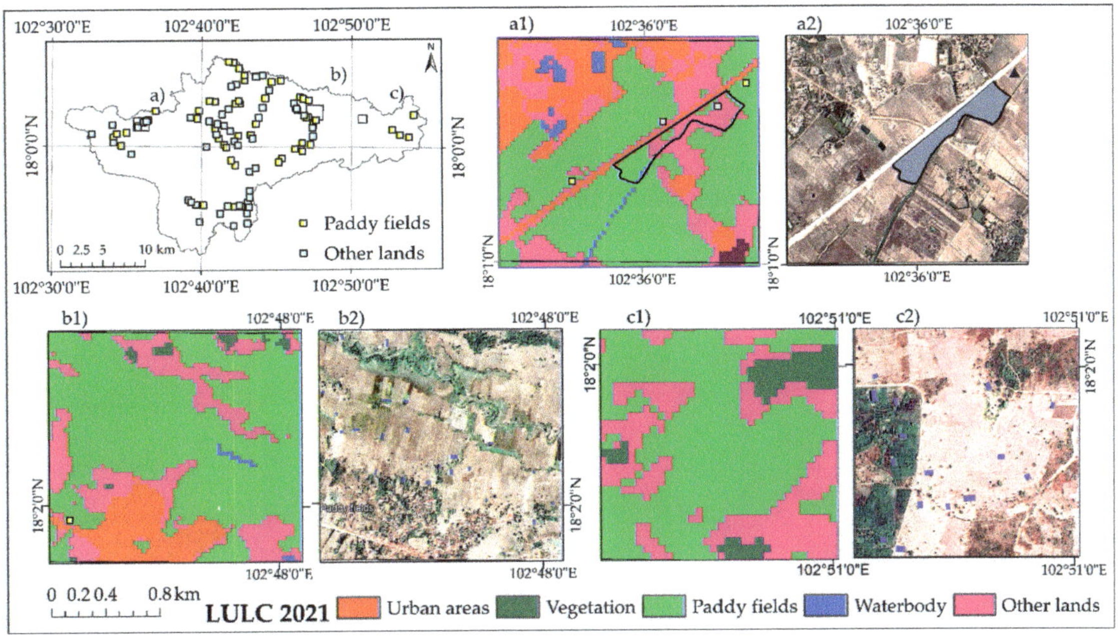

Figure 12. Observed points and Google Earth imagery: (**a1–c1**) show the zoomed-in sections of the classified image from 2021. (**a2**) shows that fallow paddy fields along the roadside were incorrectly classified as active paddy fields. (**b2,c2**) demonstrate the presence of rainwater storage ponds on paddy fields, as captured by Google Earth imagery.

4.3. Changes in LULC in the MHR Watershed

From 1989 to 2021, the MHR watershed experienced significant LULC changes, particularly a decrease in paddy fields, with a spatial change pattern shifting from the southwest to the northeast. Urban areas have expanded considerably, while paddy fields, vegetation, and water bodies have experienced a decline. A parallel study by Phompila et al. revealed that urban areas in Vientiane tripled from 2016 to 2020 [14]. Similarly, Faichia et al. also observed a rapid increase in urban areas from 2.8% in 1995 to 7.3% in 2018 [15].

The reason behind the LULC changes can be attributed to rapid urbanization driven by rural-to-urban migration, which in turn is a result of developing economic and urban infrastructure. This has led to an increasing demand for land for housing, services, and industrial purposes in areas near roads and in densely populated regions. Since the implementation of the New Economic Mechanism in 1986, aimed at enhancing domestic and foreign investment [11], the Lao government has particularly intensified economic reforms in the early 2000s to further attract investment [88]. According to the National Enterprise Database, the number of registered enterprises in Vientiane was only 19,481 in 2010, but this figure rapidly increased to 289,102 by 2021 [89]. These reforms have created numerous job opportunities, especially in the industrial and service sectors, which has resulted in substantial domestic migration to Vientiane. This shift offers peri-urban farmers the opportunity to transition from agricultural activities to off-farm employment [90], which aligns with the decreasing share of agriculture in the national gross domestic product

(GDP) from 57% in 1991 to around 16% by 2021 [91]. From 1995 to 2005, 72,800 in-migrants constituted approximately 12% of the local urban population. In the subsequent decade, the number of in-migrants more than doubled from 2005 to 2015 [13]. These migrants have greatly contributed to population growth and escalated the demand for housing areas. Notably, paddy field lands have emerged as one of the most feasible options for housing under land allocation schemes (Figure 13). This land use change contributed to a 4.8% and 45.2% reduction in rainfed and dry season rice areas, respectively, between 2012 and 2021 in Vientiane [13] and was accompanied by a sharp decline in rice production from 2013 to 2021 (Figure 10, right). A similar scenario unfolded in the Jakarta megacity, where urbanization replaced the paddy fields that once supplied rice to Western Java [16]. In the Hang–Jia–Hu plains of China, urban development displaced 88% of the paddy fields between 1980 and 2010 [18]. Likewise, in the Qinhuai River basin in Southern China, a 27% decrease in paddy fields was observed because of urbanization, coupled with extensive industrialization, over the past three decades [92]. Moreover, Zhang et al. revealed that population growth could be viewed as a major driving force for LULC changes in Collin County, Texas, as the population change accounted for 75.15% of the local-level built-up area change during 2010–2019 [93].

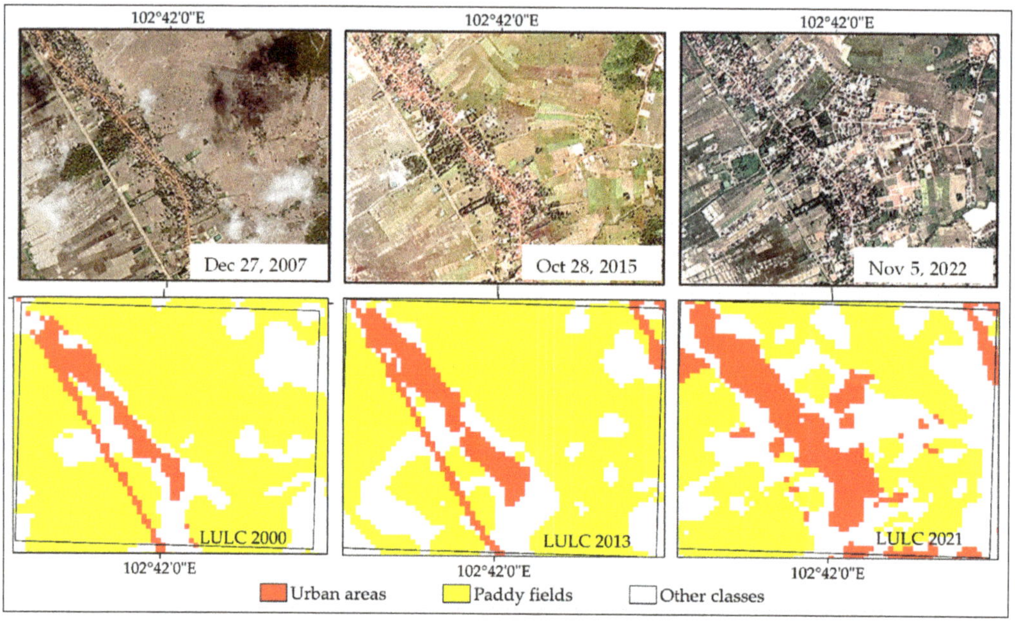

Figure 13. Detail of paddy field changes in peri-urban areas. The upper panels display imagery from Google Earth Pro for the clearest and most relevant time periods. The lower panels show the paddy fields affected by urban expansion from 2000 to 2021.

This kind of urbanization poses a significant threat to local food security and environmental sustainability. According to interviews with paddy farmers, they primarily grow rice for their own consumption and do not necessarily desire changes in land use. In reality, due to the inability by aging irrigation facilities, many farmers are forced to leave agriculture or move to other locations for farming. This trend increases the risk for food security. To mitigate these risks, it is essential to restrict urban expansion by excluding paddy fields from development areas. In the study area, new road networks were primarily constructed in the northwestern, northern, and northeastern parts of the urban center (see Figure 6a,b), indicating the spatial expansion of urban areas along these roads. This spatial pattern indicates that paddy fields near roads and in areas of higher population density have a higher probability of being converted to urban areas or becoming abandoned (Figure 14). This corresponds with the inverse relationships between the distance from roads, the population density, and the "other lands" class (see Table 8), indicating that the "other lands" class is more likely to be found near roads and less prevalent in areas of high population density due to a conversion to urban area class. These spatial changes are clear indicators of ongoing urbanization in the study area. As a result, there is an urgent need for robust urban land use planning to protect the existing paddy fields, in alignment with urban expansion strategies. For protection of paddy fields, probably, investment for irrigation facilities is a key indicator. This approach will help balance urban development with the preservation of paddy fields, ensuring a stable food supply and maintaining ecological integrity within the watershed.

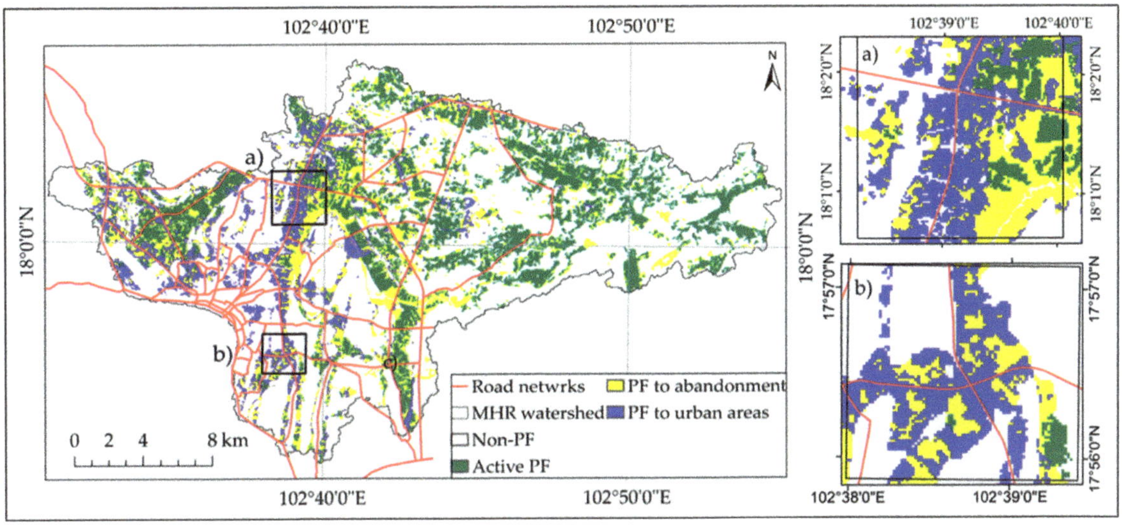

Figure 14. Spatial conversion of paddy fields (PF) from 1989 to 2021: (**a**,**b**) provide zoomed-in views of PF conversion into urban areas near major roads.

5. Conclusions

This study proposed a novel approach to understanding the decline of paddy fields in tropical monsoon regions, where image availability is often limited, and challenges arise due to monsoon-submerged fields during rainy season rice cultivation. These were addressed by developing the flooding-assisted maximum likelihood classification (F-MLC) method, which utilizes training samples from intersecting flooded paddy field maps from both the rainy and dry seasons, along with the AWEI to differentiate natural water bodies. This proposed approach has the potential to effectively map paddy fields and other land uses over time in such climates. The findings reveal a significant reduction in paddy fields by 44.87% over 32 years, while urban areas have expanded nearly fourfold, replacing 23% of the paddy fields in the MHR watershed. These results provide critical insights for policymakers and urban planners to revise urban land use strategies, balancing paddy field protection with sustainable food production and urban growth. Future studies will focus on refining the F-MLC method by automating training sample generation and integrating with VIs during the land preparation stage. These will enhance classification accuracy and scalability, enabling more comprehensive analyses of urbanization's impact on agriculture, contributing to sustainable urban planning, agricultural resilience, and ecological balance.

Author Contributions: Conceptualization, I.K., K.K. and T.K.; Methodology, I.K., A.F. and T.K.; Validation, I.K., K.K. and T.K.; Formal Analysis, I.K. and K.K.; Investigation, I.K., F.O., K.N., K.Y. and T.K.; Resources, F.O.; Data Curation, T.K.; Writing—Original Draft, I.K.; Writing—Review and Editing, A.F., F.O., K.N., K.Y. and T.K.; Visualization, I.K.; Supervision, T.K.; Project Administration, T.K.; Funding Acquisition, F.O. and T.K. All authors have read and agreed to the published version of the manuscript.

Funding: This work was partially supported by JSPS KAKENHI Grant No. JP20KK0242 and JP23K14039, and by JST e-Asia JRP 21338544.

Institutional Review Board Statement: Not applicable.

Informed Consent Statement: Not applicable.

Data Availability Statement: The datasets will be made available upon request from the authors.

Acknowledgments: The authors would like to appreciate the farmers in the study area, the Ministry of Agriculture and Forestry, and the Ministry of Public Works and Transport of Lao PDR for their valuable data support and cooperation during the field investigation. We also wish to acknowledge Enago for their invaluable assistance with English proofreading and the reviewers' and editors' constructive feedback and contributions to this paper.

Conflicts of Interest: The authors declare no conflicts of interest.

Appendix A

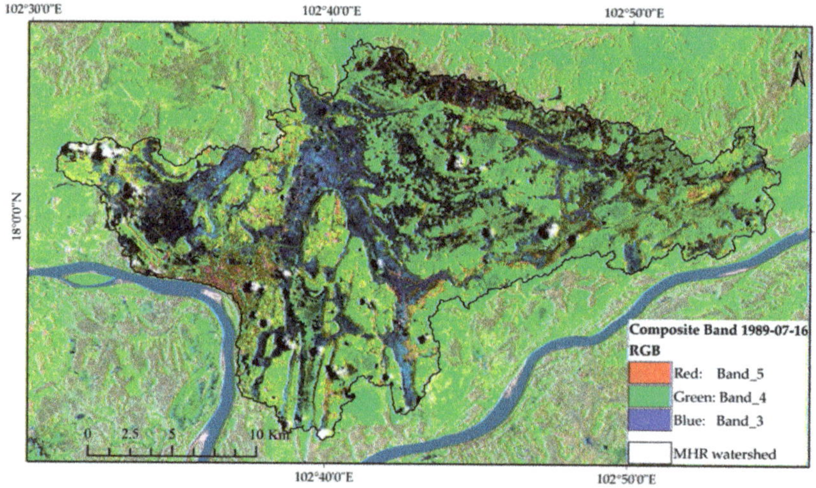

Figure A1. Band composition for identifying flooding/transplanting tone over the study area in July 1989.

References

1. Fukagawa, N.K.; Ziska, L.H. Rice: Importance for Global Nutrition. *J. Nutr. Sci. Vitaminol.* **2019**, *65*, S2–S3. [CrossRef] [PubMed]
2. FAOSTAT. Crops and Livestock Products. Available online: https://www.fao.org/faostat/en/#data/QCL (accessed on 26 September 2023).
3. Abebe, M.G. Impacts of Urbanization on Food Security in Ethiopia. A Review with Empirical Evidence. *J. Agric. Food Res.* **2024**, *15*, 100997. [CrossRef]
4. Döös, B.R. Population Growth and Loss of Arable Land. *Glob. Environ. Chang.* **2002**, *12*, 303–311. [CrossRef]
5. McGrane, S.J. Impacts of Urbanisation on Hydrological and Water Quality Dynamics, and Urban Water Management: A Review. *Hydro. Sci. J.* **2016**, *61*, 2295–2311. [CrossRef]
6. Cheng, Y.; Kang, Q.; Liu, K.; Cui, P.; Zhao, K.; Li, J.; Ma, X.; Ni, Q. Impact of Urbanization on Ecosystem Service Value from the Perspective of Spatio-Temporal Heterogeneity: A Case Study from the Yellow River Basin. *Land* **2023**, *12*, 1301. [CrossRef]
7. James, N. Urbanization and Its Impact on Environmental Sustainability. *J. Appl. Geogr. Stud.* **2023**, *3*, 54–66. [CrossRef]

8. Magigi, W. Urbanization and Its Impacts to Food Systems and Environmental Sustainability in Urban Space: Evidence from Urban Agriculture Livelihoods in Dar Es Salaam, Tanzania. *J. Environ. Prot.* **2013**, *04*, 1137–1148. [CrossRef]
9. Mullis, E.; Prasertsri, P. Laos Rice Report Annual. Available online: https://apps.fas.usda.gov/newgainapi/api/Report/DownloadReportByFileName?fileName=Laos%20Rice%20Report%20Annual_Bangkok_Laos_06-08-2020#:~:text=MY%202020/21%20rice%20consumption,according%20to%20the%20World%20Bank (accessed on 20 August 2024).
10. Bestari, N.G.; Shrestha, S.; Mongcopa, C.J. *LAO PDR: An Evaluation Synthesis on Rice*; Asian Development Bank: Manila, Philippines, 2006.
11. Phimphanthavong, H. Economic Reform and Regional Development of Laos. *Mod. Econ.* **2012**, *03*, 179–186. [CrossRef]
12. Lao Statistics Bureau. *The 3rd Lao Census of Agriculture 2019/2020*; Lao Statistics Bureau: Vientiane, Laos, 2021.
13. Lao Statistics Bureau. Laos Statistical Information Service. Available online: https://laosis.lsb.gov.la/ (accessed on 29 August 2023).
14. Phompila, C.; Sihapanya, V.; Chanthavong, B.; Thongmanivong, S.; Van Khoa, P.; Noda, K.; Vadrevu, K.P. Agricultural Land Use/Cover Changes in the Vientiane, Laos. In *Remote Sensing of Agriculture and Land Cover/Land Use Changes in South and Southeast Asian Countries*; Springer International Publishing: Berlin/Heidelberg, Germany, 2022; pp. 573–586. [CrossRef]
15. Faichia, C.; Tong, Z.; Zhang, J.; Liu, X.; Kazuva, E.; Ullah, K.; Al-Shaibah, B. Using Rs Data-Based CA–Markov Model for Dynamic Simulation of Historical and Future LUCC in Vientiane, Laos. *Sustainability* **2020**, *12*, 8410. [CrossRef]
16. Rustiadi, E.; Pravitasari, A.E.; Setiawan, Y.; Mulya, S.P.; Pribadi, D.O.; Tsutsumida, N. Impact of Continuous Jakarta Megacity Urban Expansion on the Formation of the Jakarta-Bandung Conurbation over the Rice Farm Regions. *Cities* **2021**, *111*, 103000. [CrossRef]
17. Li, S.; Nadolnyak, D.; Hartarska, V. Agricultural Land Conversion: Impacts of Economic and Natural Risk Factors in a Coastal Area. *Land. Use Policy* **2019**, *80*, 380–390. [CrossRef]
18. Song, J.; Cai, D.; Deng, J.; Wang, K.; Shen, Z. Dynamics of Paddy Field Patterns in Response to Urbanization: A Case Study of the Hang-Jia-Hu Plain. *Sustainability* **2015**, *7*, 13813–13835. [CrossRef]
19. Yafi, M.A.; Rondhi, M. Urbanisation, Paddy Field Conversion and Its Impact on Rice Production in Indonesia: A Synthesis of Panel Data 2015–2019. In Proceedings of the 2nd International Conference on Neural Networks and Machine Learning 2023 (ICNNML 2023), Jember, Indonesia, 7–8 November 2023; pp. 249–262. [CrossRef]
20. Carrasco, L.; Fujita, G.; Kito, K.; Miyashita, T. Historical Mapping of Rice Fields in Japan Using Phenology and Temporally Aggregated Landsat Images in Google Earth Engine. *ISPRS J. Photogramm. Remote Sens.* **2022**, *191*, 277–289. [CrossRef]
21. Choi, S.-H.; Choi, G.; Nam, H.-K. Impact of Rice Paddy Agriculture on Habitat Usage of Migratory Shorebirds at the Rice Paddy Scale in Korea. *Sci. Rep.* **2022**, *12*, 5762. [CrossRef]
22. Freed, S.; Kura, Y.; Sean, V.; Mith, S.; Cohen, P.; Kim, M.; Thay, S.; Chhy, S. Rice Field Fisheries: Wild Aquatic Species Diversity, Food Provision Services and Contribution to Inland Fisheries. *Fish. Res.* **2020**, *229*, 105615. [CrossRef]
23. Natuhara, Y. Ecosystem Services by Paddy Fields as Substitutes of Natural Wetlands in Japan. *Ecol. Eng.* **2013**, *56*, 97–106. [CrossRef]
24. Kobayashi, S.; Ide, H. Rice Crop Monitoring Using Sentinel-1 SAR Data: A Case Study in Saku, Japan. *Remote Sens.* **2022**, *14*, 3254. [CrossRef]
25. Osawa, T.; Nishida, T.; Oka, T. Potential of Mitigating Floodwater Damage to Residential Areas Using Paddy Fields in Water Storage Zones. *Int. J. Disaster Risk Reduct.* **2021**, *62*, 102410. [CrossRef]
26. Liu, Y.; Ge, T.; van Groenigen, K.J.; Yang, Y.; Wang, P.; Cheng, K.; Zhu, Z.; Wang, J.; Li, Y.; Guggenberger, G.; et al. Rice Paddy Soils Are a Quantitatively Important Carbon Store According to a Global Synthesis. *Commun. Earth Environ.* **2021**, *2*, 154. [CrossRef]
27. Zhang, G.; Xiao, X.; Dong, J.; Kou, W.; Jin, C.; Qin, Y.; Zhou, Y.; Wang, J.; Menarguez, M.A.; Biradar, C. Mapping Paddy Rice Planting Areas through Time Series Analysis of MODIS Land Surface Temperature and Vegetation Index Data. *ISPRS J. Photogramm. Remote Sens.* **2015**, *106*, 157–171. [CrossRef]
28. Jiang, M.; Xin, L.; Li, X.; Tan, M.; Wang, R. Decreasing Rice Cropping Intensity in Southern China from 1990 to 2015. *Remote Sens.* **2019**, *11*, 35. [CrossRef]
29. Dong, J.; Xiao, X.; Menarguez, M.A.; Zhang, G.; Qin, Y.; Thau, D.; Biradar, C.; Moore, B. Mapping Paddy Rice Planting Area in Northeastern Asia with Landsat 8 Images, Phenology-Based Algorithm and Google Earth Engine. *Remote Sens. Environ.* **2016**, *185*, 142–154. [CrossRef]
30. Thapa, R.B.; Murayama, Y. Urban Mapping, Accuracy, & Image Classification: A Comparison of Multiple Approaches in Tsukuba City, Japan. *Appl. Geogr.* **2009**, *29*, 135–144. [CrossRef]
31. Forget, Y.; Linard, C.; Gilbert, M. Supervised Classification of Built-Up Areas in Sub-Saharan African Cities Using Landsat Imagery and OpenStreetMap. *Remote Sens.* **2018**, *10*, 1145. [CrossRef]
32. Xiao, X.; Boles, S.; Liu, J.; Zhuang, D.; Frolking, S.; Li, C.; Salas, W.; Moore, B. Mapping Paddy Rice Agriculture in Southern China Using Multi-Temporal MODIS Images. *Remote Sens. Environ.* **2005**, *95*, 480–492. [CrossRef]
33. Dong, J.; Xiao, X. Evolution of Regional to Global Paddy Rice Mapping Methods: A Review. *ISPRS J. Photogramm. Remote Sens.* **2016**, *119*, 214–227. [CrossRef]
34. Yin, Q.; Liu, M.; Cheng, J.; Ke, Y.; Chen, X. Mapping Paddy Rice Planting Area in Northeastern China Using Spatiotemporal Data Fusion and Phenology-Based Method. *Remote Sens.* **2019**, *11*, 1699. [CrossRef]

35. Xiao, X.; Boles, S.; Frolking, S.; Li, C.; Babu, J.Y.; Salas, W.; Moore, B. Mapping Paddy Rice Agriculture in South and Southeast Asia Using Multi-Temporal MODIS Images. *Remote Sens. Environ.* **2006**, *100*, 95–113. [CrossRef]
36. Maiti, A.; Acharya, P.; Sannigrahi, S.; Zhang, Q.; Bar, S.; Chakraborti, S.; Gayen, B.K.; Barik, G.; Ghosh, S.; Punia, M. Mapping Active Paddy Rice Area over Monsoon Asia Using Time-Series Sentinel—2 Images in Google Earth Engine; a Case Study over Lower Gangetic Plain. *Geocarto Int.* **2022**, *37*, 10254–10277. [CrossRef]
37. Wang, J.; Xiao, X.; Qin, Y.; Dong, J.; Zhang, G.; Kou, W.; Jin, C.; Zhou, Y.; Zhang, Y. Mapping Paddy Rice Planting Area in Wheat-Rice Double-Cropped Areas through Integration of Landsat-8 OLI, MODIS, and PALSAR Images. *Sci. Rep.* **2015**, *5*, 10088. [CrossRef]
38. Sakamoto, T.; Yokozawa, M.; Toritani, H.; Shibayama, M.; Ishitsuka, N.; Ohno, H. A Crop Phenology Detection Method Using Time-Series MODIS Data. *Remote Sens. Environ.* **2005**, *96*, 366–374. [CrossRef]
39. Wibowo, A.; Ash Shidiq, I.P. Rice Productivity Estimation by Sentinel-2A Imagery in Karawang Regency, West Java, Indonesia. *Int. J. Geomate* **2020**, *19*, 49–53. [CrossRef]
40. Feyisa, G.L.; Meilby, H.; Fensholt, R.; Proud, S.R. Automated Water Extraction Index: A New Technique for Surface Water Mapping Using Landsat Imagery. *Remote Sens. Environ.* **2014**, *140*, 23–35. [CrossRef]
41. Laonamsai, J.; Julphunthong, P.; Saprathet, T.; Kimmany, B.; Ganchanasuragit, T.; Chomcheawchan, P.; Tomun, N. Utilizing NDWI, MNDWI, SAVI, WRI, and AWEI for Estimating Erosion and Deposition in Ping River in Thailand. *Hydrology* **2023**, *10*, 70. [CrossRef]
42. Yulianto, F.; Kushardono, D.; Budhiman, S.; Nugroho, G.; Chulafak, G.A.; Dewi, E.K.; Pambudi, A.I. Evaluation of the Threshold for an Improved Surface Water Extraction Index Using Optical Remote Sensing Data. *Sci. World J.* **2022**, *2022*, 4894929. [CrossRef]
43. Zhang, C.; Zhang, H.; Tian, S. Phenology-Assisted Supervised Paddy Rice Mapping with the Landsat Imagery on Google Earth Engine: Experiments in Heilongjiang Province of China from 1990 to 2020. *Comput. Electron. Agric.* **2023**, *212*, 108105. [CrossRef]
44. Seyam, M.M.H.; Haque, M.R.; Rahman, M.M. Identifying the Land Use Land Cover (LULC) Changes Using Remote Sensing and GIS Approach: A Case Study at Bhaluka in Mymensingh, Bangladesh. *Case Stud. Chem. Environ. Eng.* **2023**, *7*, 100293. [CrossRef]
45. Hedayati, A.; Vahidnia, M.H.; Behzadi, S. Paddy Lands Detection Using Landsat-8 Satellite Images and Object-Based Classification in Rasht City, Iran. *Egypt. J. Remote Sens. Space Sci.* **2022**, *25*, 73–84. [CrossRef]
46. Aung, T.O.; Sanga-Ngoie, K.; Kaoru, F. Land Cover Mapping for Agricultural Water Management of Rice-Based Irrigation Systems in Myanmar Using GIS and Remote Sensing. *J. Rainwater Catchment Syst.* **2007**, *12*, 1–15.
47. Inalpulat, M. Comparison of Different Supervised Classification Algorithms for Mapping Paddy Rice Areas Using Landsat 9 Imageries. *Türk Doğa Fen. Derg.* **2023**, *12*, 52–59. [CrossRef]
48. Jiménez, A.A.; Vilchez, F.F.; González, O.N.; Flores, S.M.L.M. Analysis of the Land Use and Cover Changes in the Metropolitan Area of Tepic-Xalisco (1973–2015) through Landsat Images. *Sustainability* **2018**, *10*, 1860. [CrossRef]
49. Dong, J.; Xiao, X.; Kou, W.; Qin, Y.; Zhang, G.; Li, L.; Jin, C.; Zhou, Y.; Wang, J.; Biradar, C.; et al. Tracking the Dynamics of Paddy Rice Planting Area in 1986-2010 through Time Series Landsat Images and Phenology-Based Algorithms. *Remote Sens. Environ.* **2015**, *160*, 99–113. [CrossRef]
50. United States Department of Agriculture. Sustainability of Future Rice Production Growth and Food Security Uncertain; United States Department of Agriculture. Available online: https://ipad.fas.usda.gov/highlights/2011/12/Laos_13Dec2011/ (accessed on 20 August 2024).
51. ADB. *Agriculture, Natural Resources, and Rural Development Sector Assessment, Strategy, and Road Map*; ADB: Manila, Philippines, 2018. [CrossRef]
52. Behnke, D.; Heft-Neal, S.; Roland-Holst, D. Early Warning Techniques for Local Climate Resilience: Smallholder Rice in Lao PDR. In *Natural Resource Management and Policy*; Springer: Berlin/Heidelberg, Germany, 2018; Volume 52, pp. 105–136. [CrossRef]
53. Sayavong, V. *Productivity and Technical Inefficiency of Paddy Rice Production in Laos: A Case Study of Farm Household Survey*; East Asian Development Network: Vientiane, Laos, 2017; Volume 118.
54. United States Geology Survey (USGS). EarthExplorer. Available online: https://earthexplorer.usgs.gov/ (accessed on 30 March 2023).
55. OpenStreetMap Foundation (OSMF). OpenStreetMap. Available online: https://openstreetmap.org/ (accessed on 13 June 2022).
56. The Ministry of Public Work and Transport. Vientiane Integrated Urban Information GIS-based Open data Platform. Available online: https://virgo.mpwt.gov.la (accessed on 8 August 2024).
57. Gao, B. NDWI—A Normalized Difference Water Index for Remote Sensing of Vegetation Liquid Water from Space. *Remote Sens. Environ.* **1996**, *58*, 257–266. [CrossRef]
58. Hadjimitsis, D.G.; Papadavid, G.; Agapiou, A.; Themistocleous, K.; Hadjimitsis, M.G.; Retalis, A.; Michaelides, S.; Chrysoulakis, N.; Toulios, L.; Clayton, C.R.I. Atmospheric Correction for Satellite Remotely Sensed Data Intended for Agricultural Applications: Impact on Vegetation Indices. *Nat. Hazards Earth Syst. Sci.* **2010**, *10*, 89–95. [CrossRef]
59. USGS. *Landsat Thematic Mapper (TM) Collection 2 (C2) Level 2 (L2) Data Format Control Book (DFCB), LSDS-1336*; USGS: Sioux Falls, SD, UAS, 2020; Volume 4.
60. USGS. *Landsat 8–9 Operational Land Imager (OLI)—Thermal Infrared Sensor (TIRS) Collection 2 (C2) Level 2 (L2) Data Format Control Book (DFCB), LSDS-1328*; USGS: Sioux Falls, SD, USA, 2022; Volume 7.
61. Young, N.E.; Anderson, R.S.; Chignell, S.M.; Vorster, A.G.; Lawrence, R.; Evangelista, P.H. A Survival Guide to Landsat Preprocessing. *Ecology* **2017**, *98*, 920–932. [CrossRef]

62. Zhu, Z.; Wang, S.; Woodcock, C.E. Improvement and Expansion of the Fmask Algorithm: Cloud, Cloud Shadow, and Snow Detection for Landsats 4-7, 8, and Sentinel 2 Images. *Remote Sens. Environ.* **2015**, *159*, 269–277. [CrossRef]
63. Qiu, S.; Zhu, Z.; He, B. Fmask 4.0: Improved Cloud and Cloud Shadow Detection in Landsats 4–8 and Sentinel-2 Imagery. *Remote Sens. Environ.* **2019**, *231*, 111205. [CrossRef]
64. Kussul, N.; Lavreniuk, M.; Skakun, S.; Shelestov, A. Deep Learning Classification of Land Cover and Crop Types Using Remote Sensing Data. *IEEE Geosci. Remote Sens. Lett.* **2017**, *14*, 778–782. [CrossRef]
65. FAO. *Special Report-2019 FAO/WFP Crop and Food Security Assessment Mission to the Lao People's Democratic Republic*; FAO: Rome, Italy, 2020. [CrossRef]
66. Jiang, W.; Chen, Z.; Lei, X.; Jia, K.; Wu, Y. Simulating Urban Land Use Change by Incorporating an Autologistic Regression Model into a CLUE-S Model. *J. Geogr. Sci.* **2015**, *25*, 836–850. [CrossRef]
67. Ren, Y.; Li, Z.; Li, J.; Ding, Y.; Miao, X. Analysis of Land Use/Cover Change and Driving Forces in the Selenga River Basin. *Sensors* **2022**, *22*, 1041. [CrossRef]
68. Azizi, P.; Soltani, A.; Bagheri, F.; Sharifi, S.; Mikaeili, M. An Integrated Modelling Approach to Urban Growth and Land Use/Cover Change. *Land* **2022**, *11*, 1715. [CrossRef]
69. Mahmoudzadeh, H.; Abedini, A.; Aram, F. Urban Growth Modeling and Land-Use/Land-Cover Change Analysis in a Metropolitan Area (Case Study: Tabriz). *Land* **2022**, *11*, 2162. [CrossRef]
70. Wang, M.; Cai, L.; Xu, H.; Zhao, S. Predicting Land Use Changes in Northern China Using Logistic Regression, Cellular Automata, and a Markov Model. *Arab. J. Geosci.* **2019**, *12*, 790. [CrossRef]
71. Sun, H.S.; Huang, J.F.; Huete, A.R.; Peng, D.L.; Zhang, F. Mapping Paddy Rice with Multi-Date Moderate-Resolution Imaging Spectroradiometer (MODIS) Data in China. *J. Zhejiang Univ. Sci. A* **2009**, *10*, 1509–1522. [CrossRef]
72. United States Geological Survey (USGS). Common Landsat Band Combinations. Available online: https://www.usgs.gov/media/images/common-landsat-band-combinations (accessed on 30 March 2024).
73. Suh, C.N.; Tsheko, R.; Kayombo, B.; Moroke, S.T. Analysis of Land Cover Land Use Change in the Greater Gaborone Area of South Eastern Botswana. *Acta Ecol. Sin.* **2023**, *43*, 1080–1089. [CrossRef]
74. Rwanga, S.S.; Ndambuki, J.M. Accuracy Assessment of Land Use/Land Cover Classification Using Remote Sensing and GIS. *Int. J. Geosci.* **2017**, *08*, 611–622. [CrossRef]
75. Sari, I.L.; Weston, C.J.; Newnham, G.J.; Volkova, L. Assessing Accuracy of Land Cover Change Maps Derived from Automated Digital Processing and Visual Interpretation in Tropical Forests in Indonesia. *Remote Sens.* **2021**, *13*, 1446. [CrossRef]
76. Ouedraogo, V.; Hackman, K.O.; Thiel, M.; Dukiya, J. Intensity Analysis for Urban Land Use/Land Cover Dynamics Characterization of Ouagadougou and Bobo-Dioulasso in Burkina Faso. *Land* **2023**, *12*, 1063. [CrossRef]
77. Loosvelt, L.; Peters, J.; Skriver, H.; Lievens, H.; Van Coillie, F.M.B.; De Baets, B.; Verhoest, N.E.C. Random Forests as a Tool for Estimating Uncertainty at Pixel-Level in SAR Image Classification. *Int. J. Appl. Earth Obs. Geoinf.* **2012**, *19*, 173–184. [CrossRef]
78. Luo, T.; Tan, R.; Kong, X.; Zhou, J. Analysis of the Driving Forces of Urban Expansion Based on a Modified Logistic Regression Model: A Case Study of Wuhan City, Central China. *Sustainability* **2019**, *11*, 2207. [CrossRef]
79. Manel, S.; Ceri Williams, H.; Ormerod, S.J. Evaluating Presence-Absence Models in Ecology: The Need to Account for Prevalence. *J. Appl. Ecol.* **2001**, *38*, 921–931. [CrossRef]
80. Malede, D.A.; Alamirew, T.; Kosgie, J.R.; Andualem, T.G. Analysis of Land Use/Land Cover Change Trends over Birr River Watershed, Abbay Basin, Ethiopia. *Environ. Sustain. Indic.* **2023**, *17*, 100222. [CrossRef]
81. Foppes, J.; Keonakone, T.; Chanthavong, N.; Chitpanya, S.; Phengkhammy, A. *Understanding Food Security in Northern Laos an Analysis of Household Food Security Strategies in Upland Production Systems*; National Agriculture and Forestry Research Institute: Vientiane, Laos, 2011.
82. Eliste, P.; Santos, N. *Lao People's Democratic Republic Rice Policy Study*; FAO: Rome, Italy, 2012.
83. Mandrekar, J.N. Receiver Operating Characteristic Curve in Diagnostic Test Assessment. *J. Thorac. Oncol.* **2010**, *5*, 1315–1316. [CrossRef] [PubMed]
84. Rossiter, D.G.; Loza, A. *Analyzing Land Cover Change with Logistic Regression in R*; Cornell University: Ithaca, NY, USA, 2016.
85. Gil Pontius Jr, R.; Schneider, L.C. Land-Cover Change Model Validation by an ROC Method for the Ipswich Watershed, Massachusetts, USA. *Agric. Ecosyst. Environ.* **2001**, *85*, 239–248. [CrossRef]
86. Zhang, H.; He, B.; Xing, J. Mapping Paddy Rice in Complex Landscapes with Landsat Time Series Data and Superpixel-Based Deep Learning Method. *Remote Sens.* **2022**, *14*, 3721. [CrossRef]
87. United Nations Development Programme. *National Adaptation Programme of Action to Climate Change*; United Nations Development Programme: Vientiane, Laos, 2009.
88. World Bank. *Linking Laos, Unlocking Policies Lao PDR Country Economic Memorandum*; World Bank: Washington, DC, USA, 2022.
89. Ministry of Industry and Commerce. National Enterprise Database. Available online: http://www.ned.moic.gov.la/ (accessed on 29 December 2023).
90. Noda, K.; Maki, M.; Miyaoka, K.; Homma, K.; Shirakawa, H.; Oki, K. A Decision-Making Model for Rice Paddy Cropping in an Urbanizing Area of the Lao PDR. *Paddy Water Environ.* **2015**, *13*, 487–493. [CrossRef]
91. Bank of the Lao PDR. Annual Report. Available online: https://www.bol.gov.la/en/annualreports (accessed on 22 August 2024).

92. Hao, L.; Sun, G.; Liu, Y.; Wan, J.; Qin, M.; Qian, H.; Liu, C.; Zheng, J.; John, R.; Fan, P.; et al. Urbanization Dramatically Altered the Water Balances of a Paddy Field-Dominated Basin in Southern China. *Hydrol. Earth Syst. Sci.* **2015**, *19*, 3319–3331. [CrossRef]
93. Zhang, B.; Li, W.; Zhang, C. Analyzing Land Use and Land Cover Change Patterns and Population Dynamics of Fast-Growing US Cities: Evidence from Collin County, Texas. *Remote Sens. Appl.* **2022**, *27*, 100804. [CrossRef]

sustainability

MDPI

Article

Urban Ecological Quality Assessment Based on Google Earth Engine and Driving Factors Analysis: A Case Study of Wuhan City, China

Weiwei Zhang [1,2,3], Wanqian Zhang [2,*], Jianwan Ji [1,3] and Chao Chen [1,3]

1 School of Geography Science and Geomatics Engineering, Suzhou University of Science and Technology, Suzhou 215000, China; zhangweiwei@usts.edu.cn (W.Z.); jijw@usts.edu.cn (J.J.); chenchao@usts.edu.cn (C.C.)
2 School of Environmental Science and Engineering, Suzhou University of Science and Technology, Suzhou 215000, China
3 Suzhou Key Laboratory of Spatial Information Intelligent Technology and Application, Suzhou 215000, China
* Correspondence: 2213021121@post.usts.edu.cn

Abstract: Ecological quality is a critical factor affecting the livability of urban areas. Remote sensing technology enables the rapid assessment of ecological quality (EQ), providing scientific theoretical support for the maintenance and management of urban ecology. This paper evaluates and analyzes the EQ and its driving factors in the city of Wuhan using remote sensing data from five periods: 2001, 2006, 2011, 2016, and 2021, supported by the Google Earth Engine (GEE) platform. By employing principal component analysis, a Remote Sensing Ecological Index (RSEI) was constructed to assess the spatiotemporal differences of EQ in Wuhan City. Furthermore, the study utilized the optimal parameter-based geographical detector model to analyze the influence of factors such as elevation, slope, aspect, population density, greenness, wetness, dryness, and heat on the RSEI value in 2021 and further explored the impact of changes in precipitation and temperature on the EQ in Wuhan. The results indicate that (1) principal component analysis shows that greenness and wetness positively affect Wuhan's EQ, while dryness and heat have negative impacts; (2) spatiotemporal analysis reveals that from 2001 to 2021, the EQ in Wuhan showed a trend of initial decline followed by improvement, with the classification grades evolving from poor and average to good and better; (3) the analysis of driving factors shows that all nine indicators have a certain impact on the EQ in Wuhan, with the influence ranking as NDVI > NDBSI > LST > WET > elevation > population density > GDP > slope > aspect; (4) the annual average temperature and precipitation in Wuhan have a non-significant impact on the EQ. The EQ in Wuhan has improved in recent years, but comprehensive management still requires enhancement.

Keywords: urban ecological quality; Google Earth Engine; remote sensing ecological index; optimal parameter-based geographical detector

1. Introduction

Human survival and development are dependent on the stability and health of natural ecosystems [1]. With the acceleration of global urbanization, urban expansion and human activities have profoundly impacted the ecological environment [2]. Rapid urbanization has led to dramatic changes in land use patterns, intensified habitat fragmentation, reduced biodiversity, and degraded ecosystem service functions [3,4]. The burgeoning population and rapid economic development in urban areas have placed tremendous pressure on limited environmental resources and fragile ecological foundations [5]. This problem is prevalent in both developed and developing countries. For example, the United States experienced drastic urban expansion in the second half of the 20th century, with large amounts of farmland, forests, and wetlands being converted into urban land, resulting in ecosystem degradation and decreased environmental quality [6]. The urbanization process in Europe has been relatively moderate, but urban sprawl has still led to ecosystem fragmentation and biodiversity loss [7]. In developing countries, urbanization is progressing rapidly, and ecological and environmental problems are more prominent [8,9]. Therefore, coordinating the relationship between urban development and ecological protection to achieve sustainable urban development has become a pressing global issue [10].

In response to this challenge, The United Nations' Sustainable Development Goals (SDGs) underscore the necessity of integrating ecological conservation with sustainable urban development, urging the adoption of advanced technological solutions in ecological assessments [11]. Particularly, Goal 11 [12] (to make cities and human settlements inclusive, safe, resilient, and sustainable) and Goal 15 [13] (to protect, restore, and promote the sustainable use of terrestrial ecosystems, sustainably manage forests, combat desertification, halt and reverse land degradation, and halt the loss of biodiversity) are directly related to the sustainable management and protection of urban ecological environments. To support the implementation of SDGs at the urban scale, it is imperative to carry out urban ecosystem monitoring and comprehensive assessment research. The long-term and dynamic monitoring of the spatiotemporal differentiation characteristics and driving mechanisms of urban ecological quality can provide a scientific basis for urban ecological protection, spatial planning, and management decisions [14].

Currently, there are various methods for evaluating the quality of urban ecology, such as the Technique for Order Preference by Similarity to Ideal Solution (TOPSIS), the entropy weight method, and the Analytic Hierarchy Process (AHP) [15]. However, these methods rely heavily on socio-economic statistical data and questionnaire survey data, which are often difficult to obtain and lack timeliness. Remote sensing technology, with its advantages of being macroscopic, rapid, dynamic, and economical, has been widely used in urban ecological environment monitoring since the 1970s [16,17]. Many scholars have used remote sensing data to assess the impact of urban expansion on the ecological environment in different regions of the world. For example, Weng [18] used Landsat imagery to evaluate the changes in landscape patterns and ecological processes caused by urban expansion in Indianapolis, USA. Rimal [19] utilized remote sensing and GIS techniques to assess land use/land cover changes and their impacts on the urban environment in the Haldia Municipality, India. Estoque [20] assessed the relationship between urban expansion and ecosystem service changes in the Manila metropolitan area, Philippines. Other scholars have focused on analyzing the spatiotemporal changes in urban surface parameters and their ecological effects, such as vegetation cover, surface temperature, and moisture [21,22]. However, these studies mostly focused on specific aspects of urban ecological processes or patterns and are difficult to comprehensively reflect the overall characteristics of complex urban ecosystems.

In recent years, constructing ecological composite indices by comprehensively utilizing multi-source remote sensing data and geo-models to conduct the overall evaluation and mapping of regional ecological environments has become a research hotspot in the field of ecological remote sensing [23]. Among them, the Remote Sensing Ecological Index (RSEI) proposed by Xu [24] comprehensively evaluates regional ecological conditions from four dimensions: greenness, wetness, heat, and dryness, and has strong comprehensiveness and regional comparability. The RSEI and its improved models have been applied to urban ecological quality assessments at multiple scales worldwide with good results [25–27]. For example, Zhou [28] and others used this index to assess the ecological environment changes in the Dongjiang source area over nearly 20 years (2000–2019), identifying urban construction land expansion driven by human activities as the main reason for the changes in the region's ecological environment quality. In addition, Paudel [29] used RSEI to evaluate the changes in the ecological environment in the Middle Hills of Western Nepal from 2000 to 2015. The results showed that the overall ecological environmental quality in the region showed a deteriorating trend, mainly due to deforestation and land degradation.

However, previous studies mostly used traditional remote sensing data processing and analysis methods, which have limitations in data acquisition and computational efficiency.

The Google Earth Engine (GEE) cloud computing platform that has emerged in recent years provides a new solution for remote sensing data acquisition, processing, analysis, and sharing [30]. It integrates a variety of commonly used remote sensing datasets and offers near-real-time data updates, with storage reaching petabyte levels. Users can develop and test algorithms and process and share data outcomes swiftly through the client, significantly boosting the efficiency of geographical information data processing and analysis [31,32]. Since its inception, GEE has been widely applied in diverse research fields such as ecology, environment, and agriculture [33,34].

Furthermore, when analyzing the driving factors of urban ecological quality, the geographical detector model is an effective tool for exploring the spatial heterogeneity and interaction of variables [35]. However, the traditional geographical detector often neglects the scale effect and zoning effect caused by the modifiable areal unit problem (MAUP), which may affect the reliability of the results [36,37]. Addressing this oversight, the optimal parameter-based geographical detector (OPGD) model proposed by Song [38] can identify the optimal spatial scale parameters and discretization scheme, providing a more robust framework for factor analysis.

In this context, we take Wuhan, a rapidly urbanizing city in central China, as a case study. As a major industrial base and transportation hub, urban ecology in Wuhan, Hubei Province, China is relatively fragile and unstable, more susceptible to the impacts of human economic activities [39]. Therefore, it is in urgent need of systematic and dynamic ecological quality monitoring and assessment. This study implements the RSEI model on the GEE platform to evaluate the spatiotemporal patterns of ecological quality in Wuhan from 2001 to 2021 based on Landsat imagery. The OPGD model is further applied to analyze the driving factors behind the ecological quality dynamics. The main innovations of this study include (1) integrating GEE and RSEI for efficient urban-scale ecological quality assessment, overcoming the limitations of data acquisition and processing in previous studies; (2) characterizing the long-term ecological quality dynamics under rapid urbanization using multi-temporal remote sensing data; and (3) quantifying the influence of natural and anthropogenic factors on ecological quality patterns with the OPGD model, providing decision support for urban ecological planning and management.

The findings of this study not only have important implications for ecological civilization construction and sustainable development in Wuhan but also contribute to global research on urban ecological assessment and management. Moreover, it provides a novel methodological framework that integrates cloud computing, remote sensing indices, and geographical detectors, which can be extended to other urban areas worldwide. The research outcomes are expected to deepen the understanding of urban ecological quality dynamics and support the realization of SDGs at the city level.

2. Materials and Methods

2.1. Study Area

Wuhan, the capital city of the Hubei Province, China, is geographically positioned between 29°58′ N to 31°22′ N latitude and 113°41′ E to 115°05′ E longitude, situated in the eastern part of the Jianghan Plain and the middle reaches of the Yangtze River. The city is comprised of six central urban districts and seven peripheral districts (as detailed in Figure 1), covering an administrative area of 57,943.92 km^2 and hosting a substantial permanent resident population of 31.9874 million [40]. Wuhan is characterized by its varied topography in all directions, particularly at the confluence of the Yangtze and Han rivers, which forms the unique geographical layout of "Two Rivers and Three Towns". The city experiences a subtropical humid monsoon climate, with abundant rainfall and sufficient heat throughout the year, boasting an average annual temperature between 15.8 °C and 17.5 °C. It possesses an exceptional ecological environment, serving as a confluence for a vast array of plant species from both northern and southern regions. Nearly half of the city's area is covered with green vegetation, providing more than the expected average of 10 square meters of green space per capita. These ecological assets are vital for Wuhan's endeavor to become an ecological civilization city and stand at the core of its environmental protection objectives [41].

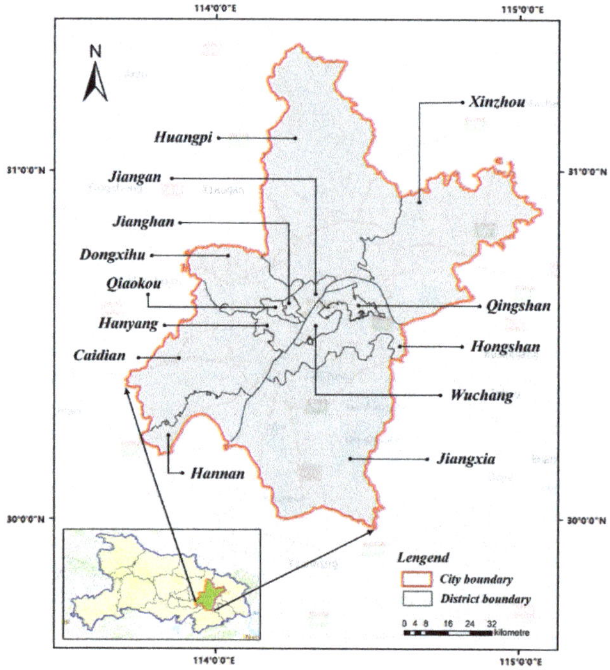

Figure 1. The location of Wuhan City, Hubei Province, China.

2.2. Data

The powerful data computation capabilities of GEE enable the batch and rapid processing of image datasets, making it an ideal platform for preprocessing and calculating various indices. This capability allows for the efficient and swift processing of image datasets, including the selection of images with minimal cloud coverage to ensure optimal image quality. To maximize the observation of vegetation greenness, this study primarily utilizes Landsat 8 OLI remote sensing imagery from the GEE platform, aiming to capture images during the peak vegetation growth periods to extract NDVI, WET, LST, and NDBSI. High-quality image data from Wuhan City, spanning from 2001 to 2021 with less than 10% cloud and a spatial resolution of 30 m, were selected for image preprocessing. The image data include cloud processing using the CFMASK (The C Function of Mask) algorithm to mask the quality assessment (QA) bands. Moreover, manual identification and processing were conducted to address areas within cloud regions that could not be labeled as clouds and anomalies due to sensor issues, with these being progressively eliminated through stepwise masking.

Additional data sources utilized in this study are detailed in Table 1. Digital Elevation Model (DEM) data were sourced from the Geospatial Data Cloud, providing essential topographical information. Population density and Gross Domestic Product (GDP) data [42], pivotal for analyzing human impact on the landscape, were obtained from the Resource Environment Science and Data Center. Furthermore, slope and aspect data, crucial for understanding terrain influences on vegetation patterns [43], were derived via detailed elevation data analysis. To ensure the reliability and accuracy of these data in experimental analyses, a comprehensive preprocessing regimen was employed. This regimen included cropping to the study area, reclassification to align with the research objectives, and geometric correction to ensure spatial accuracy. These meticulous preprocessing steps are vital for the integrity of the data and the validity of the subsequent analysis, underlining the study's commitment to methodological rigor and precision.

Table 1. Source information of the data.

Data Name	Resolution	Data Source	Data Preprocessing
WET	30 m	Google Earth Engine	Mosaicking, Reclassification
NDVI	30 m		Mosaicking, Reclassification
LST	30 m		Mosaicking, Reclassification
NDBSI	30 m		Mosaicking, Reclassification
DEM	30 m	Geospatial data cloud (www.gscloud.cn) (accessed on 30 September 2023)	Mosaicking, Clipping, Reclassification
Slope	1000 m		Extraction, Clipping
Aspect	1000 m		Extraction, Clipping
GDP	1000 m		Geometric correction
Population	1000 m	CAS Resource and Data Center (www.resdc.cn) (accessed on 30 September 2023)	Clipping, Geometric correction, Reclassification

2.3. Methods

2.3.1. Remote Sensing Ecological Index (RSEI)

This article utilized the RSEI to evaluate the EQ in Wuhan. Dryness, heat, wetness, and greenness indices are selected as the main research indicators, and the PCA method is used to construct the RSEI. Combined with the EQ grade difference calculation, it provides a detailed assessment and analysis of the spatiotemporal ecological changes in Wuhan's urban area. Figure 2 presents the framework of this study.

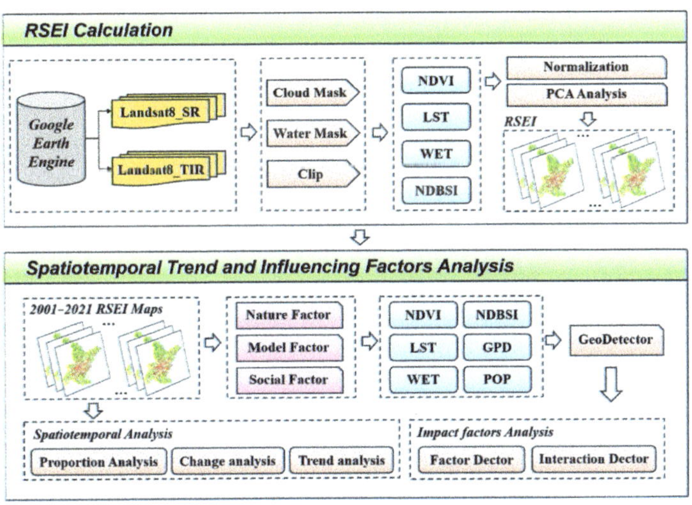

Figure 2. Technical Flowchart.

1. NDVI

The greenness index, which measures the biomass, leaf area index, and vegetation coverage of green plants, utilizes the normalized difference vegetation index (NDVI) by calculating the difference between the near-infrared (NIR) and red (RED) bands of remote sensing data [44]. The calculation formula is

$$NDVI = \frac{\rho_{nir} - \rho_{\mathrm{red}}}{\rho_{nir} + \rho_{red}} \tag{1}$$

In the formula, ρ_{nir} and ρ_{red} represent the reflectance in the near-infrared and red-light bands, respectively.

2. WET

The wetness index, which includes the moisture content of both soil and vegetation, is obtained from remote sensing data through the Tasseled Cap Transformation (K-T) [45]. Due to the differences in spectral resolution between the Landsat TM and OLI sensors, the formula for the wetness index is as follows:

$$\begin{aligned}\mathrm{Wet}_{TM} = 0.0315\rho_{\mathrm{blue}} + 0.2021\rho_{\mathrm{green}} + 0.3102\rho_{\mathrm{red}} \\ +0.1594\rho_{nir} - 0.6806\rho_{\mathrm{swir1}} - 0.6109\rho_{\mathrm{swir2}}\end{aligned} \tag{2}$$

$$\begin{aligned}\mathrm{Wet}_{OLI} = 0.01511\rho_{\mathrm{blue}} + 0.1973\rho_{\mathrm{green}} + 0.3283\rho_{\mathrm{red}} \\ +0.3407\rho_{nir} - 0.7117\rho_{\mathrm{swir1}} - 0.4559\rho_{\mathrm{swir2}}\end{aligned} \tag{3}$$

In the formula, ρ_{green}, ρ_{blue}, ρ_{swir1} and ρ_{swir2}, respectively, represent the reflectance of the green, blue, shortwave infrared 1 (SWIR1), and shortwave infrared 2 (SWIR2) bands.

3. LST

Land surface temperature represents the heat level. It reflects the level of radiant heat from the Earth's surface and plays a crucial role in the ecological environment, constituting an essential variable within it. Various methods for calculating land surface temperature include the radiative transfer equation, single-window algorithm, and single-channel algorithm [46]. This paper employs the single-channel algorithm formula:

$$T_s = \frac{\left(\frac{c_2}{\lambda}\right)}{\ln\left(\frac{c_1}{\lambda^1 S_{B(T_s)}} + 1\right)} \tag{4}$$

$$B(T_s) = a_0 + a_1 w + \left(a_2 + a_3 w + a_4 w^2\right) \cdot (1/\varepsilon) + \left(a_5 + a_6 w + a_7 w^2\right) \cdot \left(L_{\mathrm{Sen}}/\varepsilon\right) \tag{5}$$

In the formula, L_{sen} represents the radiance received by the sensor; ε denotes the emissivity of the land surface; w stands for the water vapor content in the atmosphere; $B(T_S)$ represents the Planck's radiance value at temperature; λ is the effective wavelength; and $a_{\mathrm{i}}(i = 1, 2, 3 \cdots 7)$ are the coefficients within the $B(T_S)$ model for Landsat series data.

4. NDBSI

The dryness index, known as the Normalized Difference Built-up and Soil Index (NDBSI), is composed of the average values of the Index-Based Built-up Index (IBI) and the Soil Index (SI) [47]. The calculation formula is

$$NDBSI = \frac{(SI + IBI)}{2} \tag{6}$$

$$\mathrm{SI} = \frac{\left(\rho_{\mathrm{swir1}} + \rho_{\mathrm{red}}\right) - \left(\rho_{\mathrm{blue}} - \rho_{nir}\right)}{\left(\rho_{\mathrm{swir1}} + \rho_{\mathrm{red}}\right) + \left(\rho_{\mathrm{blue}} + \rho_{nir}\right)} \tag{7}$$

$$IBI = \frac{2\frac{\rho_{swir2}}{(\rho_{swir1} + \rho_{nir})} - \frac{\rho_{nir}}{\rho_{nir} + \rho_{red}} + \frac{\rho_{green}}{\left(\rho_{swir1} + \rho_{green}\right)}}{2\frac{\rho_{swir2}}{(\rho_{swir1} + \rho_{nir})} + \frac{\rho_{nir}}{\rho_{nir} + \rho_{red}} + \frac{\rho_{green}}{\left(\rho_{swir1} + \rho_{green}\right)}} \tag{8}$$

In the formula, ρ_{green}, ρ_{blue}, ρ_{red}, ρ_{nir}, ρ_{swir1} and ρ_{swir2}, respectively, represent the reflectance of the green, blue, red, near-infrared, shortwave infrared 1 (SWIR1), and shortwave infrared 2 (SWIR2) bands.

5. Construction of the RSEI

The RSEI is an index that evaluates the ecological environment by integrating wetness, greenness, heat, and dryness [48]. Due to the differences in numerical units and magnitudes among these components, it is necessary to normalize the data before integration to remove unit discrepancies. The specific formula for the normalization process is as follows:

$$NI = \frac{NI - NI_{min}}{NI_{max} - NI_{min}} \text{ or } NI = \frac{NI_{max} - NI}{NI_{max} - NI_{min}} \quad (9)$$

In the formula, N represents the normalized value of the index, I is the value of the index itself, and I_{max} and I_{min} represent the peak and trough values among all the indices, respectively. After the normalization of all component indices, we employ PCA to determine the variance contribution of each principal component. These contributions are used as weights for the component indices, which are then further transformed into the four original component indices. The extraction formula is as follows:

$$RSEI = PCA[f(NDVI, WET, NDBSI, LST)] \quad (10)$$

The initial RSEI is normalized to fall within the [0, 1] range, where values closer to 1 indicate better and superior ecological environment quality [49]. Based on the ecological environment grading standards set forth in the Technical Specifications for Ecological Environment Evaluation, the RSEI is classified into five levels, as shown in Table 2:

Table 2. The ecological environment situation scale.

Level Index	Feature Description
Worse ($0 < RSEI \leq 0.2$)	Low vegetation cover, drought and low rainfall, rock exposure, soil drying, and obvious limitations on human life.
Poor ($0.2 < RSEI \leq 0.4$)	Relatively low vegetation coverage, dry weather, sparse rainfall, fewer species, are notable factors limiting human habitation.
Fair ($0.4 < RSEI \leq 0.6$)	Medium coverage, moderate rainfall, suitable for human habitation, and factors limiting human survival.
Good ($0.6 < RSEI \leq 0.8$)	High vegetation coverage, rich biodiversity, soil rich in organic matter, and favorable climate; suitable for human residence.
Better ($0.8 < RSEI \leq 1.0$)	High vegetation coverage, rich biodiversity, high organic matter content in the soil, moist and pleasant climate, and ecological stability

2.3.2. Optimal Parameter-Based Geographical Detector

1. Spatial Scale Optimization

Based on the scope of the research area, two scales of 2 km and 3 km were established, generating 2613 and 899 grids, respectively. By comparing the 90th percentile of all driving factors q at these two different spatial scales, the scale at which this percentile reaches its maximum value is identified as the optimal spatial scale.

2. Geographical Detector Model

The Geographical Detector is a tool used to study spatially varying geographical phenomena. It includes four detectors: differentiation and factor detection, interaction detection, risk area detection, and ecological detection. Its advantages lie in the lack of a need for a linear hypothesis and its clear physical significance [50].

Differentiation and factor detection evaluate the impact of the independent variable X (an evaluation index) on the dependent variable Y (EQ), measured with the q value:

$$q = 1 - \frac{\sum_{h=1}^{L} N_h \sigma_h^2}{N\sigma^2} = 1 - \frac{SSW}{SST} \tag{11}$$

$$SSW = \sum_{h=1}^{L} N_h \sigma_h^2,\ SST = N\sigma^2 \tag{12}$$

In the formula, $h = 1, 2 \ldots; L$ represents the stratification of variable Y or factor X; N and N_h denote the total number of units in the entire region and in the stratum h, respectively; σ_h^2 and σ^2 are the variances of Y values within stratum h and across the entire region, respectively; SSW and SST represent the sum of within-stratum variances and the total variance across the region. The q value serves as an indicator of the influence of factor X on Y, with larger values indicating a stronger influence.

Interaction detection is utilized to explore the interactions between different factors, assessing whether the interaction effect between every two independent variables enhances or reduces the explanatory power on the dependent variable. There are also instances where the influence of independent variables on the dependent variable is mutually independent. Refer to the Table 3 for the types of interactions between factors:

Table 3. Interaction types of the detection factors.

Interaction Type	Judgment Criteria
Non-linear Weakening	$q(X_1 \cap X_2) < Min[q(X_1), q(X_2)]$
Non-linear Attenuation	$Min[q(X_1), q(X_2)] < q(X_1 \cap X_2 < Max[q(X_1), q(X_2)]$
Bifactor Enhancement	$q(X_1 \cap X_2) > Max[q(X_1), q(X_2)]$
Mutually Independent	$q(X_1 \cap X_2) = q(X_1) + q(X_2)$
Non-linear Enhancement	$q(X_1 \cap X_2) > q(X_1) + q(X_2)$

3. Results

3.1. Factor Index Principal Component Analysis (PCA)

PCA was applied to four indicators of Wuhan City in 2001, 2006, 2011, 2016, and 2021, yielding the proportion, RSEI, and contribution rate of the first principal component. According to the data in Table A1, the average contribution of the first principal component exceeds 66%, indicating its dominant role in the overall framework. This demonstrates that PCA is an effective and impartial method to reflect the comprehensive environmental status of Wuhan City.

The PCA of Wuhan City's four indicators in 2001, 2006, 2011, 2016, and 2021 (see Table A1) reveals that the eigenvalue contribution of the first principal component consistently surpasses 50%, signifying its representation of the majority of information across the four indicators. Therefore, this principal component can substitute for the four components of greenness, wetness, dryness, and heat. To minimize the influence of subjective factors in the composite of multiple indicators, the variance contribution rate is selected as the weight for each component indicator. Utilizing these weights, a comprehensive evaluation model for the RSEI is constructed to conduct an in-depth assessment of EQ in Wuhan. The contribution rate of greenness has been increasing over the study period, indicating that the ecological protection has improved in Wuhan in recent years, resulting in a positive trend in vegetation cover and its increasing impact on the overall EQ compared to other indicators. Similarly, the contribution rate of the heat indicator is rising, reflecting the continuous increase in Wuhan's surface temperature during the study period, which is closely related to the urban heat island effect, thereby increasing the weight of the heat indicator in the RSEI evaluation. In the annual first principal component, the greenness (NDVI) and wetness (WET) indicators are positive, suggesting a beneficial impact on the ecological environment, while the dryness (NDBSI) and heat (LTS) indicators are negative, indicating potential adverse effects. High values of greenness and wetness suggest good vegetation cover and soil moisture content, reflecting a favorable EQ. Conversely, high dryness and heat values may indicate issues like sparse vegetation, exposed bedrock, soil desertification, and urbanization, signifying poor ecological conditions. Since other principal components do not show significant trends or clear reflections of EQ, this study only utilizes the contribution rate of the first principal component to construct the RSEI [51].

3.2. Spatiotemporal Distribution of EQ in Wuhan

From 2001 to 2021, the average value of RSEI in Wuhan and the area and proportion of the area corresponding to each grade are shown in Figure 3 and Table 4. Observing the data for individual years, in 2001, the area rated as poor in EQ reached a peak, accounting for 43.25% of the total area, while the area rated as good was the least, making up only 10% of the total area, approximately 823.62 km^2. In the same year, the areas rated as average and better accounted for 29.53% and 12%, respectively. By 2006, the proportion of areas rated as poor increased to 25.58%, covering an area of 2146.92 km^2, while the proportion of average areas decreased to 24.51%, covering 2056 km^2. In 2011, compared to 2006, there was little change; the proportion of poor areas decreased to 21.56%, covering 1809 km^2, while the proportion of better areas increased to 7.01%. In 2016, the proportion of areas rated as good rose to 31.04%, and the proportion of better areas also significantly increased to 36.43%, covering 3057.07 km^2. In 2021, the proportion of average areas increased to 27.70%, good areas continued to rise to 33.52%, and the proportion of better areas slightly decreased to 12.43%, covering 1043.21 km^2.

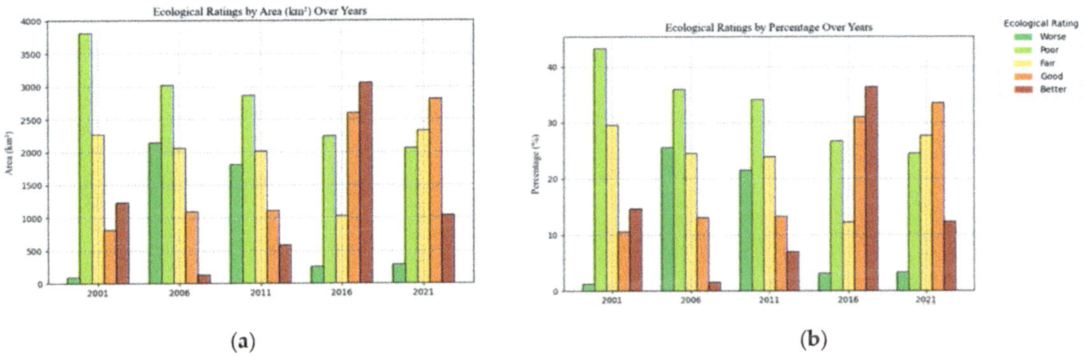

Figure 3. Area and proportion of Wuhan EQ classification areas from 2001 to 2021. (**a**) Area of different grades; (**b**) proportion of different grades.

Table 4. Area statistics of RSEI levels from 2001 to 2021 in Wuhan.

Ecological Rating	2001		2006		2011		2016		2021	
	Area /km^2	Area /%	Area /km^2	Area /%	Area /km^2	Area /%	Area /km^2	Area /%	Area /km^2	Area /%
Worse	93.23	1.21	2146.92	25.58	1809.03	21.56	262.37	3.13	283.63	3.38
Poor	3815.65	43.25	3021.72	36.01	2863.97	34.13	2240.60	26.70	2063.85	24.59
Fair	2274.25	29.53	2056.62	24.51	2012.68	23.98	1027.33	12.24	2324.56	27.70
Good	823.62	10.62	1099.56	13.1	1118.14	13.32	2604.64	31.04	2813.52	33.52
Better	1235.25	14.72	134.89	1.61	588.66	7.01	3057.07	36.43	1043.21	12.43
Total	8392.62	100.00	8392.25	100.00	8392.48	100.00	8392.01	100.00	8392.37	100.00

Looking at the overall trend, from 2001 to 2021, Wuhan City's RSEI mean value shows a trend of first decreasing and then increasing, indicating an improvement in the EQ in recent years, which may be related to the city's economic development policies.

According to statistical data, Wuhan's GDP grew from CNY 39.91 billion in 1978 to CNY 13,410.34 billion in 2017, and the permanent population increased from 8.58 million in 2004 to 10.33 million in 2014. This growth led to more human activities, causing ecological issues such as vegetation destruction and soil pollution. With the government's continuous efforts in environmental protection and governance, the implementation of relevant policies, and the promotion of ecological civilization construction over the past two decades, the environmental awareness of Wuhan's residents has significantly improved. Through the combined efforts in various aspects, the trend of ecological environment deterioration in Wuhan has been successfully curbed, shifting towards improvement.

To more directly illustrate the geographic distribution of EQ in Wuhan, as shown in Figure 4, the surrounding urban districts exhibit better ecological conditions. These areas, characterized by slower economic development, primarily utilize land for agriculture and forestry, boasting rich vegetation and high biodiversity. In contrast, regions such as Hongshan, Hanyang, Wuchang, and Qingshan exhibit relatively poorer ecological conditions. The Hongshan District, anchored by academic and educational institutions, hosts numerous higher education entities and has a dense population, whereas Qingshan, Hanyang, and Wuchang are predominantly industrial areas, housing major industrial enterprises like Wuhan Iron and Steel, Wuhan Petrochemical, Dongfeng Motor, and Wuhan Shipbuilding. Industrial production and human activities in these areas exert significant pressure on the ecological environment.

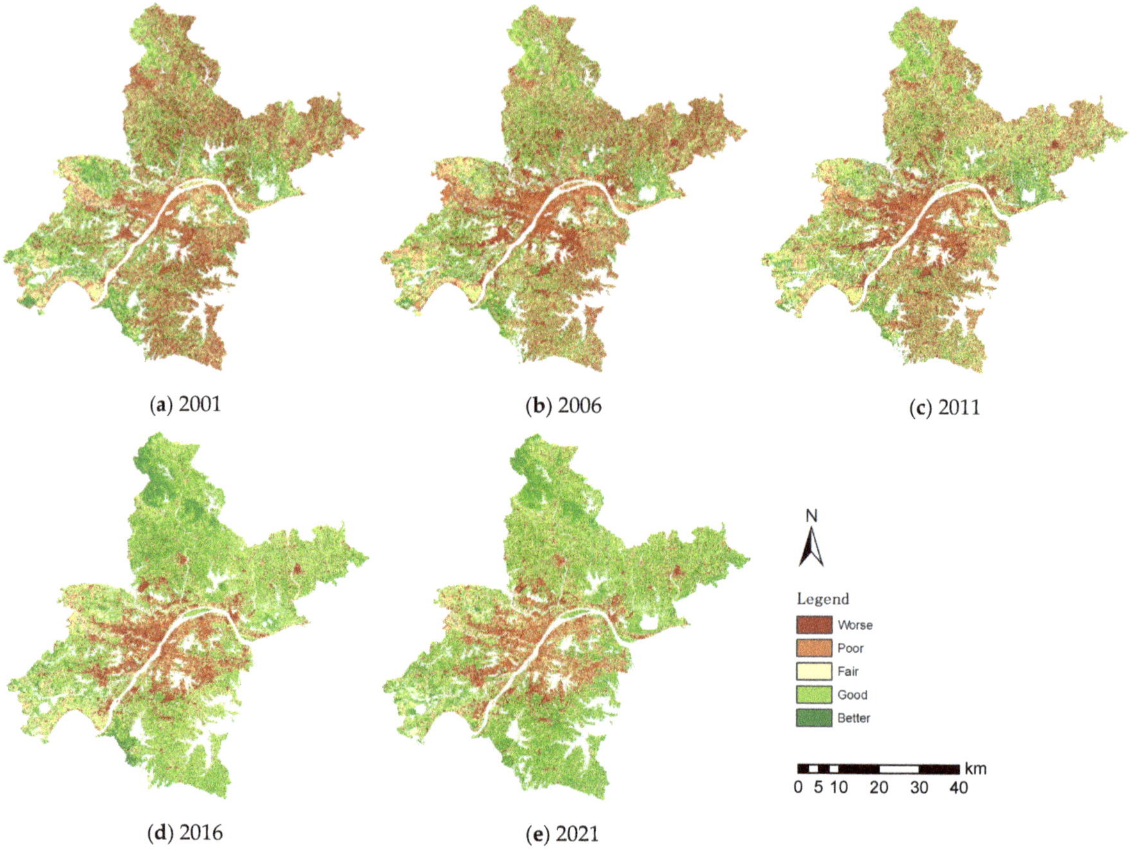

Figure 4. Classification map of EQ from 2001 to 2021 in Wuhan.

Observing the trend from 2001 to 2021, Wuhan's ecological environment displays a ring-shaped distribution pattern, with poorer conditions in the core urban areas and relatively better conditions in the suburbs. Over time, the scope of the central urban area has gradually expanded, especially towards the Yangtze and Han rivers. By 2021, areas with poorer ecological conditions have extended across the east–west axis, covering most of Wuhan and its core urban areas. This pattern indicates that the newly developed surrounding areas of Wuhan respond quickly to ecological changes, reflecting the city's urban expansion. Large tracts of farmland and forests have been transformed into urban land, altering land use and cover patterns on the surface, and leading to reduced vegetation, decreased biodiversity, and diminished soil retention capability, thereby degrading EQ. The rapid urban expansion has significantly impacted these changes. For instance, in 2006, ecological problem areas were relatively dispersed, and issues in the urban core were not particularly pronounced. By 2016, the most noticeable ecological changes were concentrated along the western bank of the Han River and along both sides of the Yangtze River. Recently, as Wuhan's urban development has approached saturation and with the advancement of ecological civilization, the central urban area has not shown a trend of continuous deterioration, and the overall EQ is improving.

3.3. Spatiotemporal Analysis of EQ Differences

Based on the RSEI, this study analyzes the spatiotemporal differences in the EQ in Wuhan from 2001 to 2021, with five-year intervals. Figure 5 illustrates that from 2001 to 2006, Wuhan's EQ exhibited a declining trend, with the area of deteriorated EQ accounting for 59.62%, while the areas of improvement and no change accounted for 10.64% and 29.75%, respectively. Specifically, the areas of mild and significant deterioration were 2910.42 km^2 and 2108.8 km^2, respectively. Between 2006 and 2011, the EQ in Wuhan remained largely unchanged, with stable areas accounting for 56.35%, areas of decline accounting for 23.45%, and areas of improvement representing 10.10%, including a mild deterioration area of 1936.36 km^2 and a mild improvement area of 1639.9 km^2. From 2011 to 2016, the EQ in Wuhan showed a slight downward trend, with 46.9% of the area worsening, 16.48% remaining unchanged, and 36.63% improving. The area of mild improvement was 2162.23 km^2, while significant deterioration covered 2681.44 km^2. Between 2016 and 2021, Wuhan's EQ showed an improving trend, with 43.33% of the area improving, and 27.6% and 29.08% remaining unchanged or worsening, respectively. The areas of significant and mild improvement were 1613.02 km^2 and 2054.53 km^2, respectively. Overall, from 2001 to 2021, the area showing improvement in Wuhan's EQ accounted for 47.32%, approximately 3962.526 km^2, while the areas of no change and deterioration accounted for 30.43% and 21.81%, respectively.

In terms of the spatial distribution of EQ changes, Figure 6 shows that from 2001 to 2016, the areas of EQ decline in Wuhan were mainly concentrated on the city's outskirts and near water bodies. The ecological degradation of water bodies is associated with the illegal discharge of domestic sewage and industrial wastewater from urban residents and factories, as well as the burgeoning aquaculture industry in recent years. The construction land in Wuhan's center mainly originated from the encroachment on surrounding green spaces and lakes, as well as the expansion of new urban areas toward the two rivers. From 2016 to 2021, the urban EQ improved, particularly in the city center, where the EQ remained stable or even enhanced, except for the water bodies. This improvement reflects the Wuhan government's efforts in water environment management, increasing urban greening coverage, and ecological restoration projects, as well as the heightened environmental awareness among the citizens [52].

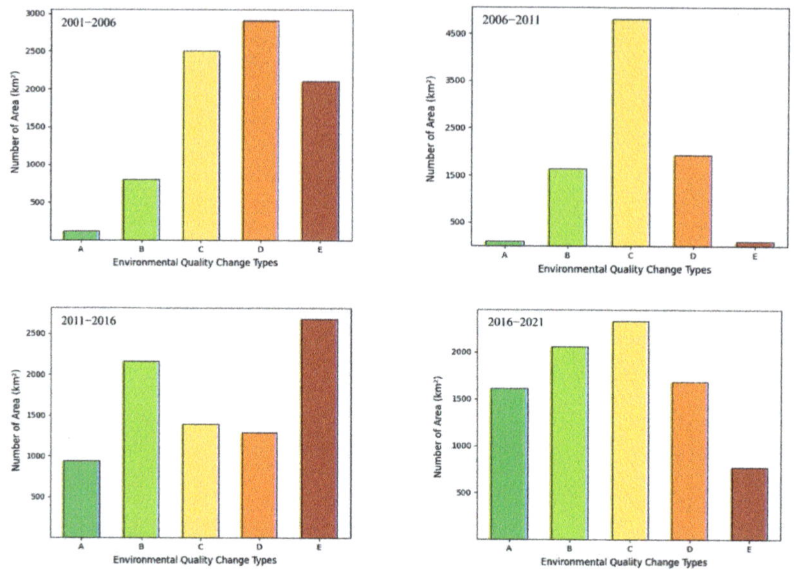

Figure 5. Changes in EQ type area in Wuhan City from 2001 to 2021. Note: A. Significant Improvement; B. Mild Improvement; C. Unchanged; D. Mild Deterioration; E. Significant Deterioration.

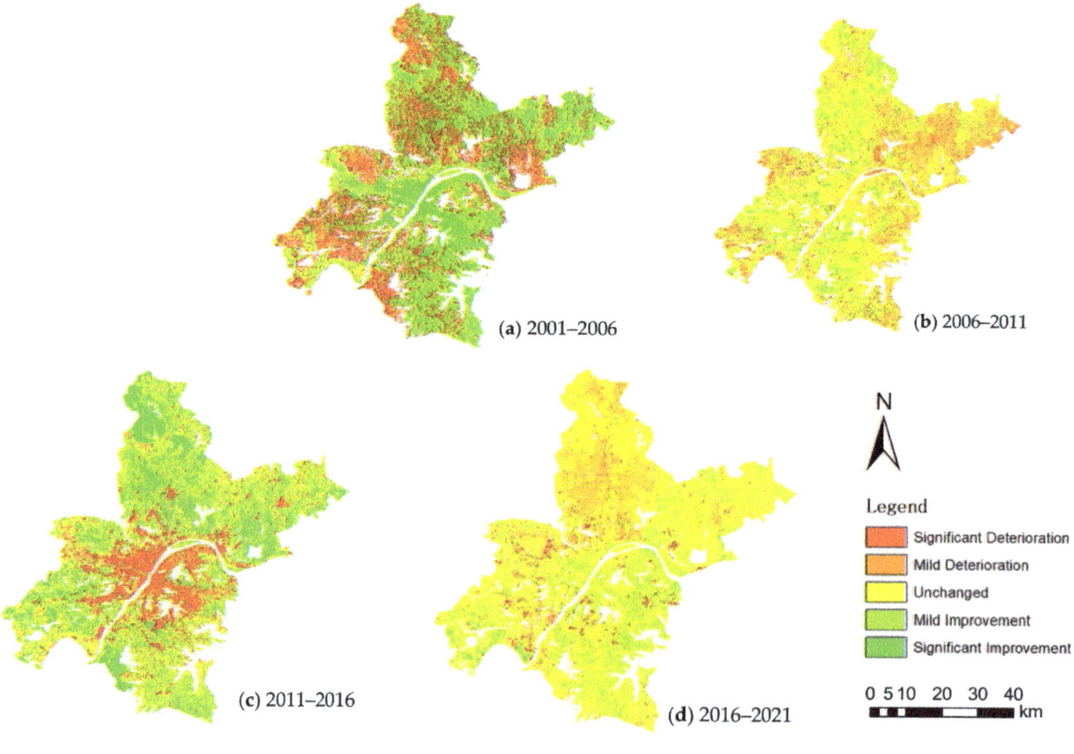

Figure 6. Spatial Distribution Map of EQ Changes in Wuhan from 2001 to 2021.

3.4. *Analysis of Driving Factors Affecting EQ in Wuhan*

3.4.1. Identification of Optimal Spatial Scale

At different spatial scales, there are variations in the degree of influence of various factors, as indicated in Table 5. With an increase in spatial grid size, several driving factors (q) tend to exhibit minor fluctuations. A common practice in existing research involves comparing the sizes of the 90th percentile of all driving factors (q) at different spatial scales, considering the scale at which this percentile reaches its maximum as the optimal spatial scale. The trend of decreasing 90th percentiles for all driving factors (q) reaches its peak at a spatial grid of 2 km, with a maximum decrease of 0.756. Consequently, among the two grid sizes, the 2 km grid is better suited to reflect the impact of latent variables on the changes in ecological and environmental quality.

Table 5. Comparative Spatial Scale Effects of Driving Factor q and 90th Percentile.

Grid Size \ Factor	LST	NDVI	WET	NDBSI	POP	GDP	DEM	PD	PX	The 90th Percentile of q
2 km	0.353	0.869	0.145	0.743	0.028	0.027	0.068	0.025	0.024	0.756
3 km	0.287	0.843	0.175	0.721	0.026	0.029	0.071	0.021	0.019	0.733

3.4.2. Analysis of Differentiation Factor Detection Results

This article employs the geographical detector model to uncover the natural factors influencing the changes in EQ in Wuhan City. The study utilized the fishnet tool in ArcGIS to establish research grid points of 2 km × 2 km within the study area. It classified elevation, slope, aspect, population density, GDP, and four remote sensing ecological indicators into five levels. Through the fishnet points, the RSEI values were spatially associated with the values of nine driving factors. Subsequently, the spatially matched results were imported into the geographical detector model for factor detection analysis, which determined the influence values (q values, with higher q values indicating a greater impact of a specific index factor on RSEI) and explanatory power values (p values, with higher p values indicating a lesser explanatory power of an index factor on RSEI) of the nine driving factors on RSEI.

As shown in Table 6, the p values for LST, NDVI, WET, NDBSI, population density, GDP, and elevation are all zero, indicating that these seven driving factor indices have sufficient explanatory power for the EQ in Wuhan City. From the perspective of q values, NDVI and NDBSI have the highest values, indicating that these two driving factors had the most significant impact on Wuhan City's RSEI value in 2021. The influence of LST and WET ranks third and fourth, respectively. The q values for slope and aspect are smaller, and their p values are larger; hence, their impact on RSEI values can be considered negligible. Finally, the study concludes that the impact strength of the nine factors on RSEI values within this research area, from strongest to weakest, is as follows: NDVI, NDBSI, LST, WET, elevation, population density, GDP, slope, and aspect.

Table 6. Spatial Heterogeneity Response of RSEI Value to Nine Driving Factors.

Factor	LST	NDVI	WET	NDBSI	POP	GDP	DEM	PD	PX
q	0.353	0.869	0.145	0.743	0.028	0.027	0.068	0.025	0.024
p	0.000	0.000	0.000	0.000	0.000	0.000	0.000	0.015	0.017
Ranking	3	1	4	2	6	7	5	8	9

3.4.3. Analysis of the Results from the Detection of Factor Interactions

To detect the interactions between various factors, this study evaluated whether the interaction between every two independent variables enhances or weakens the explanatory power on the dependent variable. The interaction detection analysis, conducted using a geographical detector model, categorized the results into bifactor enhancement and non-linear enhancement, as shown in Figure 7. The analysis indicates that, compared to individual influencing factors, all interaction factors significantly amplify the impact on the spatial heterogeneity of the dependent variable RSEI. Key interaction factors with a relatively high impact on RSEI's spatial heterogeneity include NDVI ∩ NDBSI, WET ∩ NDBSI, LST ∩ NDBSI, NDBSI ∩ GDP, NDBSI ∩ DEM, and NDBSI ∩ POP, demonstrating that interaction factors have a more substantial effect on spatial heterogeneity. The detection results of differentiation factors revealed that slope and aspect have minimal impact on RSEI values; hence, their interactions were not included in the result analysis.

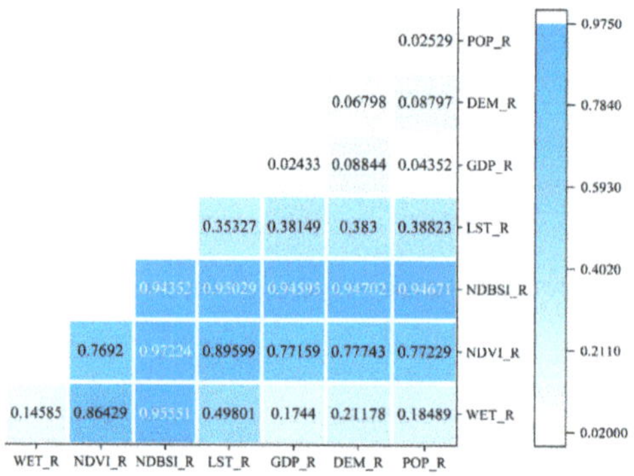

Figure 7. Heat map of factor interaction detection results.

3.4.4. Analysis of the Causative Factors of Natural Elements

From 2001 to 2021, the annual average temperature and annual precipitation in Wuhan City both exhibited a slight upward trend. As illustrated in Figure 8, the fluctuation range of the annual average temperature is relatively significant, especially in 2006 and 2016, when the decline in Wuhan's EQ may be closely related to the rise in temperature. Although the annual precipitation also shows an increasing trend, its significance is not prominent. Therefore, this indicator can be relatively overlooked when analyzing the changes in the EQ in Wuhan.

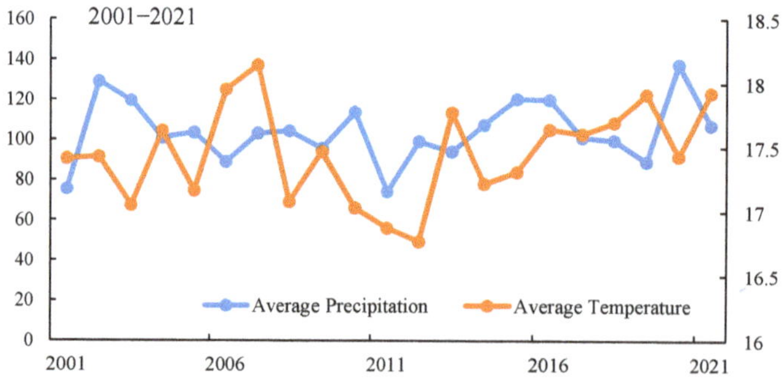

Figure 8. Changes in average annual temperature and precipitation in Wuhan from 2001 to 2021.

4. Discussion

4.1. The Advantages of the GEE Platform for Constructing the RSEI Model

Compared to conventional RSEI modeling approaches, the GEE platform empowers researchers to focus on the core objectives of their studies rather than on repetitive technical tasks [53]. GEE provides a plethora of built-in codes and functions that are highly accessible, including cloud masking, image compositing, principal component analysis, and ridge and linear regression functions [54]. These embedded resources guarantee that researchers can accurately and promptly detect changes in regional RSEI and forecast future EQ in specific areas. This study demonstrates that, through direct coding on the GEE platform, principal component analysis no longer necessitates external software such as MATLAB 2022 or SPSS 2022, thereby markedly improving research efficiency.

The use of GEE for RSEI modeling is a relatively novel approach in the field of urban ecological quality assessment [55]. Previous studies have primarily relied on traditional software and methods, such as using ENVI 5.3 for image preprocessing, ArcGIS for spatial analysis, and MATLAB or SPSS for statistical analysis. While these tools are powerful, they often require significant time and effort to process large datasets and integrate multiple data sources. In contrast, GEE's cloud-based platform and extensive library of datasets and functions streamline the entire process, enabling researchers to analyze vast amounts of data more efficiently [56].

Moreover, GEE's ability to handle multi-temporal and multi-source data is particularly advantageous for assessing urban ecological quality, which often involves analyzing changes over time and integrating various environmental and socio-economic factors [57]. By leveraging GEE's capabilities, this study demonstrates the potential for more comprehensive and efficient assessments of urban ecological quality, which can inform sustainable urban planning and management strategies [58].

However, it is important to acknowledge the limitations and challenges of using GEE for RSEI modeling. One potential issue is the need for reliable internet connectivity and sufficient computing power to process large datasets on the cloud platform. Additionally, while GEE offers a wide range of datasets and functions, it may not include all the specific data or analysis tools required for a particular study, necessitating the integration of external data or custom code development.

Despite these limitations, the advantages of using GEE for RSEI modeling and urban ecological quality assessment are significant. As demonstrated in this study, GEE enables researchers to efficiently process and analyze large volumes of multi-source data, providing a more comprehensive understanding of the spatiotemporal patterns and drivers of urban ecological quality. This innovative approach has the potential to advance the field of urban ecological research and inform more effective strategies for sustainable urban development.

4.2. Temporal Changes of EQ and Causal Analysis

The temporal analysis of Wuhan's ecological quality (EQ) from 2001 to 2021 reveals significant changes over time, with notable improvements in recent years. In 2001, the proportion of ecologically impoverished areas in Wuhan reached a staggering 43.25%, with the average EQ value at a relative low point. This poor ecological state can be attributed to several factors, including rapid industrialization, accelerated urbanization, population growth, and intensified human activities. During this period, Wuhan experienced a surge in economic development, which exerted tremendous pressure on natural resources and ecological systems, leading to environmental degradation.

From 2006 to 2011, the city's EQ remained relatively low but fluctuated less, indicating that environmental governance policies had begun to curb the deteriorating trend to some extent. This period marked the beginning of a shift in Wuhan's development strategy, with increasing emphasis on environmental protection and sustainable development [59]. The implementation of stricter environmental regulations, such as the "Environmental Protection Law of the People's Republic of China" in 2008, likely contributed to the stabilization of Wuhan's EQ during this time.

After 2016, Wuhan's EQ began to improve significantly, with the proportion of areas experiencing improvement reaching 43.33% between 2016 and 2021. This positive trend can be attributed to several factors, including the government's increased efforts in environmental management, the transformation of urban development patterns, industrial restructuring, and the heightened public awareness of environmental protection [60]. Wuhan's government implemented a series of ecological restoration projects, such as the "Green Wuhan" initiative, which aimed to increase green space and improve the city's ecological infrastructure [52]. Moreover, the promotion of low-carbon and green industries, coupled with stricter pollution control measures, helped to reduce the environmental impact of economic activities. These targeted actions, driven by a shift in government priorities and growing public concern for the environment, have played a crucial role in enhancing Wuhan's EQ in recent years.

The temporal patterns of Wuhan's EQ are consistent with findings from other studies on urban ecological quality in China. For example, a study by Wang [61] on the spatiotemporal evolution of urban ecological quality in Beijing found a similar trend, with a decline in ecological quality during the early 2000s, followed by a gradual improvement in recent years due to government-led environmental protection efforts. Similarly, research by Zhang [62] on the ecological quality of Shenzhen revealed a significant improvement in EQ from 2010 to 2020, attributed to the city's sustainable development policies and ecological restoration projects.

However, it is important to note that the improvement in Wuhan's EQ is a relatively recent phenomenon, and the city still faces significant ecological challenges. As indicated by the relatively high proportion of ecologically impoverished areas in 2021 (25.58%), there is a need for continued efforts to promote sustainable urban development and address the underlying drivers of ecological degradation. Future research could explore the long-term effectiveness of Wuhan's ecological protection measures and identify potential strategies for further improving the city's EQ.

4.3. Spatial Patterns of EQ and Causal Analysis

The spatial analysis of Wuhan's ecological quality (EQ) reveals a distinct urban–rural disparity and a concentric distribution pattern [63]. The core urban areas, such as Wuchang, Hanyang, and Hongshan, which are densely populated and industrially developed, generally exhibit poorer EQ due to the intensity of human activities. In contrast, the relatively less economically active urban–rural fringes and suburban areas maintain a favorable ecological state [64]. This spatial pattern is consistent with findings from other studies on urban ecological quality in China and globally.

Although both the urban core and the urban–rural fringe areas have experienced ecological degradation, the degree of deterioration varies, resulting in the concentric distribution pattern of EQ. The urban core areas have suffered the most severe ecological degradation due to the intense urbanization pressures, such as the massive conversion of natural land to built-up areas and the high concentration of population and industrial activities [65]. In comparison, the urban–rural fringe areas, while also affected by urban expansion, have managed to retain a higher proportion of natural and semi-natural land covers, such as forests, grasslands, and agricultural lands, owing to the relatively lower development intensity [66]. These areas provide important ecosystem services, such as carbon sequestration, water regulation, and biodiversity conservation, which contribute to their higher EQ. The gradient change in EQ from the urban core to the periphery reflects the spatially heterogeneous impact of urban expansion on the ecological environment.

The formation of this spatial pattern is closely related to the process of urban expansion and land use change. As Wuhan has grown and developed, the central urban areas have experienced significant land cover changes, with the conversion of green spaces, wetlands, and agricultural lands into built-up areas [67]. This process has led to the degradation of natural habitats, reduction in biodiversity, and increased environmental pollution [68]. Moreover, the high concentration of population and industrial activities in the urban core has placed additional pressure on the local ecosystem, contributing to the lower EQ in these areas [69].

The spatial pattern of Wuhan's EQ highlights the need for a more balanced and sustainable approach to urban development. While the concentration of population and economic activities in the urban core is a common feature of many cities, it is crucial to ensure that this growth does not come at the expense of the local ecosystem. Strategies such as urban greening, ecological restoration, and the protection of natural habitats in both urban and peri-urban areas can help mitigate the negative impacts of urbanization and improve the overall EQ of the city. Moreover, the spatial analysis of Wuhan's EQ underscores the importance of considering the spatial heterogeneity of urban ecological quality in urban planning and management. A one-size-fits-all approach to ecological conservation and restoration may not be effective, given the distinct characteristics and challenges of different urban areas. Instead, a more targeted and context-specific approach, which takes into account the local socio-economic, environmental, and institutional factors, may be necessary.

4.4. Analysis of Influencing Factors and Recommendations

The analysis of the influencing factors on Wuhan's ecological quality (EQ) using the geographic detection model reveals that vegetation cover (NDVI), built-up land (NDBI), land surface temperature (LST), and wetness (WET) are the dominant drivers of EQ in the city. These findings are consistent with previous studies that have identified land cover composition, urban heat island effect, and moisture conditions as key determinants of urban ecological quality [70–72]. The strong influence of NDVI on Wuhan's EQ highlights the crucial role of vegetation in maintaining and improving the urban ecosystem. Vegetation provides numerous ecological benefits, such as reducing air and noise pollution, regulating microclimate, and supporting biodiversity. In Wuhan, the expansion of built-up areas and the consequent loss of vegetation cover have been major contributors to the decline in EQ, particularly in the urban core. Therefore, protecting and restoring green spaces, such as parks, forests, and wetlands, should be a key priority in Wuhan's ecological management strategy.

The significant impact of NDBI on Wuhan's EQ underscores the negative ecological consequences of rapid urbanization and the expansion of impervious surfaces. Built-up areas not only lead to the direct loss of natural habitats but also contribute to a range of environmental problems, such as increased surface runoff, reduced groundwater recharge, and the urban heat island effect. To mitigate these impacts, Wuhan should promote sustainable urban design and planning practices, such as compact development, mixed land use, and the integration of green infrastructure.

The influence of LST on Wuhan's EQ highlights the need to address the urban heat island effect, which is a common problem in many cities worldwide. The high concentration of impervious surfaces and the lack of vegetation in urban areas contribute to higher surface and air temperatures, which can have negative impacts on human health, energy consumption, and ecosystem functioning. In Wuhan, the urban heat island effect has been intensifying in recent years, particularly in the summer months. To mitigate this problem, the city should implement strategies such as increasing the coverage of green spaces, promoting the use of cool materials and green roofs, and improving the efficiency of energy systems.

The role of WET in influencing Wuhan's EQ emphasizes the importance of maintaining and restoring the city's water resources and wetland ecosystems. Wuhan is known as the "city of a hundred lakes" due to its abundant water resources, including the Yangtze and Han rivers and numerous lakes and wetlands. These water bodies not only provide important ecosystem services, such as water purification, flood control, and biodiversity conservation but also contribute to the city's unique landscape and cultural identity. However, rapid urbanization and industrial development have led to the degradation and loss of many of Wuhan's water resources and wetland habitats. To address this issue, the city should strengthen the protection and management of its water resources, implement wetland restoration projects, and promote sustainable water use practices.

In addition to these key influencing factors, the analysis also reveals the significant impact of socio-economic factors, such as population density and GDP, on Wuhan's EQ. This finding highlights the complex interplay between urban development, human activities, and ecological quality. To achieve sustainable urban development, Wuhan needs to find a balance between economic growth, social well-being, and environmental protection.

This requires a multi-faceted approach that involves not only technical solutions, such as ecological restoration and green infrastructure, but also policy and institutional reforms, such as strengthening environmental regulations, promoting public participation, and fostering cross-sectoral collaboration.

Furthermore, the study highlights the potential impact of climate change on Wuhan's EQ, as indicated by the negative influence of temperature increases on EQ in certain years. Wuhan, like many other cities in China and worldwide, is facing the challenges of climate change, such as rising temperatures, changing precipitation patterns, and more frequent extreme weather events [65]. These changes can have significant impacts on urban ecosystems, such as altering species distributions, disrupting ecological processes, and exacerbating environmental problems, such as air and water pollution [73]. To build resilience to climate change, Wuhan needs to mainstream climate considerations into its urban planning and management practices, such as developing climate adaptation plans, promoting low-carbon development, and strengthening early warning and disaster response systems.

Finally, it is important to recognize the limitations and uncertainties of this study and the need for further research. While the RSEI model and the geographic detection method provide valuable insights into the spatiotemporal patterns and influencing factors of Wuhan's EQ, they are based on a limited set of indicators and data sources. Future studies could incorporate a wider range of ecological, social, and economic indicators, such as biodiversity, ecosystem services, and human well-being, to provide a more comprehensive assessment of urban ecological quality. Moreover, the study focuses on a single city, and the findings may not be directly applicable to other urban contexts. Comparative studies across different cities and regions could help to identify common patterns and context-specific factors influencing urban ecological quality. In summary, this study provides a valuable contribution to the understanding of urban ecological quality in Wuhan and offers important insights for sustainable urban development in China and beyond. The findings highlight the need for a multi-dimensional and integrative approach to urban ecological management, which takes into account the complex interactions between environmental, social, and economic factors. By adopting a more holistic and adaptive approach, cities like Wuhan can strive to achieve a balance between human well-being and ecological sustainability and contribute to the global goals of sustainable development.

5. Conclusions

In this study, in leveraging the GEE platform we conducted a rapid and detailed assessment of the changes in EQ in Wuhan from 2001 to 2021 and employed geographical detector model technology to uncover the key factors affecting the region's EQ. The conclusions of this research not only provide vital scientific evidence for environmental management and sustainable development in Wuhan City but also offer empirical support for achieving the SDGs. The main findings are summarized as follows:

(1) Trend of EQ: Between 2001 and 2021, Wuhan City experienced an initial decline followed by a subsequent increase in EQ. This trend is closely associated with the city's rapid economic development, reduction in vegetation due to human activities, shrinkage of lake areas, and urban expansion. This finding underscores the importance of implementing effective environmental governance measures while pursuing economic growth. It also highlights the need for a balanced approach to urban development that prioritizes ecological sustainability alongside economic progress.

(2) Spatial Distribution Differences: There is a significant difference in the ecological conditions between the core urban areas and the peripheral regions, which is related to the economic development level and strategic positioning of each district. Adjusting industrial layout and promoting industrial upgrading can provide new impetus for improving the urban ecology. This finding suggests that a spatially differentiated approach to ecological management, which takes into account the unique

characteristics and challenges of different urban areas, may be more effective than a one-size-fits-all strategy.

(3) Analysis of Driving Forces: Through the analysis of nine driving factors, it was found that greenness and dryness have the most significant impact on the EQ, while the effects of slope and aspect are relatively minor. This insight provides guidance for urban planning and ecological restoration, emphasizing the importance of maintaining and enhancing vegetation cover and managing built-up areas to improve urban ecological quality. Moreover, the significant influence of socio-economic factors, such as population density and GDP, highlights the need for an integrated approach that addresses the complex interplay between the environmental, social, and economic dimensions of urban sustainability.

(4) Areas of Focus: The EQ along the Yangtze and Han riversides and in the city center remains a concern. Urban planning and development strategies need to place greater emphasis on ecological protection to achieve harmonious coexistence between humans and nature. This finding underscores the importance of prioritizing the conservation and restoration of critical ecological assets, such as rivers, lakes, and wetlands, in urban development plans. It also calls for a more proactive and integrated approach to urban ecological management, which involves not only technical solutions but also policy and institutional reforms to promote sustainable land use and environmental stewardship.

(5) Methodological Contributions: This study demonstrates the effectiveness of the RSEI model and the geographical detector method for assessing and analyzing urban ecological quality. The integration of multi-source remote sensing data and socio-economic data within the GEE platform enables a comprehensive and efficient assessment of the spatiotemporal patterns and driving factors of urban EQ. This approach offers a promising tool for monitoring and evaluating urban ecological conditions, which can inform evidence-based decision-making for sustainable urban management. The methodology developed in this study can be applied to other cities and regions, providing a valuable reference for comparative studies and global assessments of urban ecological quality.

(6) Implications for Sustainable Urban Development: The findings of this study have important implications for sustainable urban development in Wuhan and beyond. They highlight the need for a multi-dimensional and integrative approach to urban ecological management, which takes into account the complex interactions between environmental, social, and economic factors. By adopting a more holistic and adaptive approach, cities can strive to achieve a balance between human well-being and ecological sustainability and contribute to the global goals of sustainable development. This requires not only technical solutions, such as ecological restoration and green infrastructure but also policy and institutional reforms, such as strengthening environmental regulations, promoting public participation, and fostering cross-sectoral collaboration.

In conclusion, this study provides a comprehensive assessment of the spatiotemporal patterns and driving factors of urban ecological quality in Wuhan, using advanced remote sensing and spatial analysis techniques. The findings offer valuable insights into the complex dynamics of urban ecosystems and the challenges and opportunities for sustainable urban development. By highlighting the key areas of concern and the critical factors influencing urban ecological quality, this study provides a scientific basis for informed decision-making and targeted interventions to promote urban sustainability. The methodology and conclusions of this study can be extended to other cities and regions, contributing to the growing body of knowledge on urban ecological assessment and management. As cities around the world face the pressing challenges of rapid urbanization, environmental degradation, and climate change, this study underscores the importance of developing and applying innovative tools and approaches to monitor, assess, and enhance urban ecological quality, as a key component of sustainable development.

Author Contributions: Conceptualization, W.Z. (Weiwei Zhang) and W.Z. (Wanqian Zhang); methodology, W.Z. (Wanqian Zhang) and W.Z. (Weiwei Zhang); validation, W.Z. (Wanqian Zhang); formal analysis, J.J. and W.Z. (Wanqian Zhang); investigation, C.C.; resources, W.Z. (Weiwei Zhang); data curation, W.Z. (Weiwei Zhang); writing—original draft preparation, W.Z. (Wanqian Zhang) and W.Z. (Weiwei Zhang); writing—review and editing, W.Z. (Weiwei Zhang), J.J. and C.C.; visualization, J.J.; supervision, W.Z. (Weiwei Zhang); project administration, W.Z. (Weiwei Zhang); funding acquisition, W.Z. (Weiwei Zhang). All authors have read and agreed to the published version of the manuscript.

Funding: This study was supported by the National Natural Science Foundation of China (42171311, 41701477).

Institutional Review Board Statement: Not applicable.

Informed Consent Statement: Not applicable.

Data Availability Statement: The data will be made available upon request.

Acknowledgments: We would like to thank the editor and anonymous reviewers for their constructive comments and suggestions for improving the manuscript.

Conflicts of Interest: The authors declare that they have no known competing financial interests or personal relationships that could have appeared to influence the work reported in this paper.

Appendix A

Table A1. Result of principal component analysis.

Year	Parameter	PC1	PC2	PC3	PC4
2001	NDVI	0.323	0.665	−0.269	−0.616
	WET	0.689	−0.133	−0.680	−0.207
	LTS	−0.586	0.189	0.678	−0.399
	NDBSI	−0.275	−0.709	0.056	0.646
	Eigenvalue	0.178	0.102	0.026	0.003
	Contribution Rate/%	58	33	8	1
2006	NDVI	0.327	0.712	0.161	0.598
	WET	0.688	0.098	0.642	0.321
	LTS	−0.515	0.060	0.748	0.412
	NDBSI	−0.391	−0.691	−0.008	−0.607
	Eigenvalue	0.154	0.118	0.020	0.003
	Contribution Rate/%	53	40	6	1
2011	NDVI	0.306	−0.692	0.222	0.613
	WET	0.704	0.116	0.655	0.247
	LTS	−0.563	0.148	0.720	−0.374
	NDBSI	−0.303	−0.695	0.043	−0.649
	Eigenvalue	0.153	0.116	0.021	0.003
	Contribution Rate/%	52	39	8	1
2016	NDVI	0.361	0.468	−0.485	−0.644
	WET	0.532	0.678	478	−0.165
	LTS	−0.668	−0.028	0.719	−0.188
	NDBSI	−0.373	0.565	−135	722
	Eigenvalue	0.120	0.019	0.008	0.002
	Contribution Rate/%	80	13	6	1
2021	NDVI	0.370	0.472	−0.453	−0.658
	WET	0.537	0.691	0.464	−0.126
	LTS	−0.641	−0.033	0.745	−0.176
	NDBSI	−0.401	0.544	−0.150	0.720
	Eigenvalue	0.125	0.019	0.011	0.002
	Contribution Rate/%	80	12	7	1

References

1. Xie, H.L.; Zhang, Y.W.; Choi, Y.; Li, F.Q. A scientometrics review on land ecosystem service research. *Sustainability* **2020**, *12*, 2959. [CrossRef]
2. Grimm, N.B.; Faeth, S.H.; Golubiewski, N.E.; Redman, C.L.; Wu, J.; Bai, X.; Briggs, J.M. Global change and the ecology of cities. *Science* **2008**, *319*, 756–760. [CrossRef]
3. Seto, K.C.; Güneralp, B.; Hutyra, L.R. Global forecasts of urban expansion to 2030 and direct impacts on biodiversity and carbon pools. *Proc. Natl. Acad. Sci. USA* **2012**, *109*, 16083–16088. [CrossRef]
4. Erlwein, A. Exploring Ecosystems Health: Effects of Increments of Biodiversity and Trophic Complexity on the Stability of a Simple Gaian Ecosystem Model. *Agro. Sur.* **2022**, *50*, 192–204. [CrossRef]
5. Gong, P.; Li, X.; Zhang, W. 40-Year (1978–2017) human settlement changes in China reflected by impervious surfaces from satellite remote sensing. *Sci. Bull.* **2019**, *64*, 756–763. [CrossRef] [PubMed]
6. Lopez, R. Urban sprawl and risk for being overweight or obese. *Am. J. Public Health* **2004**, *94*, 1574–1579. [CrossRef]
7. European Environment Agency. *Urban Sprawl in Europe—The Ignored Challenge*; EEA Report No. 10/2006; European Environment Agency: Copenhagen, Denmark, 2006.
8. United Nations. *World Urbanization Prospects: The 2018 Revision*; United Nations: New York, NY, USA, 2019.
9. Hao, L.; Sun, G.; Liu, Y.; Wan, J.; Qin, M.; Qian, H.; Liu, C.; Zheng, J.; John, R.; Fan, P.; et al. Urbanization dramatically altered the water balances of a paddy field-dominated basin in southern China. *Hydrol. Earth Syst. Sci.* **2015**, *19*, 3319–3331. [CrossRef]
10. Angel, S.; Parent, J.; Civco, D.L.; Blei, A.; Potere, D. The dimensions of global urban expansion: Estimates and projections for all countries, 2000–2050. *Prog. Plann.* **2011**, *75*, 53–107. [CrossRef]
11. Long, H. Analysis of the Key Factors of Ecological Environment Protection in the National Economic Sustainable Development Goals. *J. Environ. Public Health* **2022**, *2022*, 3593587. [CrossRef]
12. Blasi, S.; Ganzaroli, A.; De Noni, I. Smartening sustainable development in cities: Strengthening the theoretical linkage between smart cities and SDGs. *Sustain. Cities Soc.* **2022**, *80*, 103793. [CrossRef]
13. Dickens, C.; McCartney, M.; Tickner, D.; Harrison, I.J.; Pacheco, P.; Ndhlovu, B. Evaluating the global state of ecosystems and natural resources: Within and beyond the SDGs. *Sustainability* **2020**, *12*, 7381. [CrossRef]
14. Haas, J.; Ban, Y. Urban growth and environmental impacts in Jing-Jin-Ji, the Yangtze, River Delta and the Pearl River Delta. *Int. J. Appl. Earth Obs. Geoinf.* **2014**, *30*, 42–55. [CrossRef]
15. Sopandi, A.S.; Gustian, D.; Sembiring, F.; Muslih, M.; Arianti, N.D.; Setiawatif, A.; Kurniawan. Sistem pendukung keputusan penerima bantuan sosial tunai dengan metode technique for order preference by similarity to ideal solution. *J. Rekayasa Nusa Putra* **2022**, *8*, 268.
16. Lillesand, T.; Kiefer, R.W.; Chipman, J. *Remote Sensing and Image Interpretation*; John Wiley & Sons: Hoboken, NJ, USA, 2015.
17. Jensen, J.R. *Remote Sensing of the Environment: An Earth Resource Perspective*; Pearson Prentice Hall: Upper Saddle River, NJ, USA, 2007.
18. Weng, Q. Remote sensing of impervious surfaces in the urban areas: Requirements, methods, and trends. *Remote Sens. Environ.* **2012**, *117*, 34–49. [CrossRef]
19. Rimal, B.; Baral, H.; Stork, N.E.; Paudyal, K.; Rijal, S. Growing city and rapid land use transition: Assessing multiple hazards in the Kathmandu Valley, Nepal. *Land* **2018**, *7*, 10.
20. Estoque, R.C.; Murayama, Y. Monitoring surface urban heat island formation in a tropical mountain city using Landsat data (1987–2015). *ISPRS J. Photogramm. Remote Sens.* **2017**, *133*, 18–29. [CrossRef]
21. Ranagalage, M.; Estoque, R.C.; Murayama, Y. An urban heat island study of the Colombo metropolitan area, Sri Lanka, based on Landsat data (1997–2017). *ISPRS Int. J. Geo-Inf.* **2017**, *6*, 189. [CrossRef]
22. Du, S.; Xiong, Z.; Wang, Y.C.; Guo, L. Quantifying the multilevel effects of landscape composition and configuration on land surface temperature. *Remote Sens. Environ.* **2016**, *178*, 84–92. [CrossRef]
23. Turner, M.G.; Gardner, R.H. *Landscape Ecology in Theory and Practice*; Springer: New York, NY, USA, 2015.
24. Xu, H.Q. A remote sensing urban ecological index and its application. *Acta Ecol. Sin.* **2013**, *33*, 7853–7862.
25. Li, X.; Meng, Q.; Gu, X.; Jancso, T.; Yu, T.; Wang, K.; Mavromatis, S. A hybrid method combining pixel-based and object-oriented methods and its application in Hungary using Chinese HJ-1 satellite images. *Int. J. Remote Sens.* **2013**, *34*, 4655–4668. [CrossRef]
26. Gu, H.; Singh, A.; Townsend, P.A. Detection of gradients of forest composition in an urban area using imaging spectroscopy. *Remote Sens. Environ.* **2015**, *167*, 168–180. [CrossRef]
27. Zhao, S.; Zhu, W.; Shen, W.; Zhang, J.; He, B.; Li, Q. Gauging the ecological and environmental influences of the guangdong-hong kong-macao greater bay area on the pearl river estuary using the remote sensing ecological index. *Water* **2021**, *13*, 2256.
28. Zhou, M. Evaluation of ecological environment quality of Dongjiang River headwaters based on remote sensing ecological index during 2000–2019. *Bull Soil Water Conserv.* **2021**, *4*, 231.
29. Paudel, B.; Andersen, P. Monitoring ecological conditions at multiple scales using the Remote Sensing Ecological Index (RSEI) in the Middle Hills of Western Nepal. *Remote Sens.* **2020**, *12*, 1543.
30. Gorelick, N.; Hancher, M.; Dixon, M.; Ilyushchenko, S.; Thau, D.; Moore, R. Google Earth Engine: Planetary-scale geospatial analysis for everyone. *Remote Sens. Environ.* **2017**, *202*, 18–27. [CrossRef]
31. Zhou, J.; Menenti, M.; Jia, L.; Gao, B.; Zhao, F.; Cui, Y.; Xiong, X.; Liu, X.; Li, D. A scalable software package for time series reconstruction of remote sensing datasets on the Google Earth Engine platform. *Int. J. Digit. Earth* **2023**, *16*, 988–1007. [CrossRef]

32. de Raus Maure, E.; Ilyushchenko, S.; Terauchi, G. A Simple Procedure to Preprocess and Ingest Level-2 Ocean Color Data into Google Earth Engine. *Remote Sens.* **2022**, *14*, 4906. [CrossRef]
33. Habibie, M.I. The application of machine learning using google earth engine for remote sensing analysis. *J. Teknoinfo* **2022**, *16*, 1872. [CrossRef]
34. Zhai, H.M.; Xie, W.Q.; Li, S.Q.; Zhang, Q. Evaluation of urban ecological environment based on remote sensing based ecological index model. *Fresen. Environ. Bull.* **2021**, *30*, 2527–2535.
35. Wang, J.F.; Xu, C.D. Geodetector: Principle and prospective. *Acta Geogr. Sinica.* **2017**, *72*, 116–134.
36. Wang, J.F.; Zhang, T.L.; Fu, B.J. A measure of spatial stratified heterogeneity. *Ecol. Indic.* **2016**, *67*, 250–256. [CrossRef]
37. Ma, J.; Chen, F.; Liu, Y.; Xu, Y.; Chen, K.; Liu, M. The analysis of global ecological regionalization factors based on geographical detector models. *Sci. Rep.* **2020**, *10*, 16443.
38. Song, Y.Z.; Wang, J.F.; Ge, Y.; Xu, C.D. An optimal parameters-based geographical detector model enhances geographic characteristics of explanatory variables for spatial heterogeneity analysis: Cases with different types of spatial data. GISci. *Remote Sens.* **2020**, *57*, 593–610. [CrossRef]
39. Meng, X.; Gao, X.; Lei, J.; Li, S. Development of a multiscale discretization method for the geographical detector model. *Int. J. Geogr. Inf. Sci.* **2021**, *35*, 1650–1675. [CrossRef]
40. Zheng, S.; Tang, Y.; Chan, F.K.S.; Cao, L.Y.; Song, R.X. The demographic implication for promoting sponge city initiatives in the Chinese megacities: A case of Wuhan. *Water* **2022**, *14*, 883. [CrossRef]
41. Xu, L.; Zhang, Z.; Tan, G.M.; Zhou, J.; Wang, Y. Analysis on the Evolution and Resilience of Ecological Network Structure in Wuhan Metropolitan Area. *Sustainability* **2022**, *14*, 8580. [CrossRef]
42. Ehrlich, D.; Freire, S.; Melchiorri, M.; Kemper, T. Open and Consistent Geospatial Data on Population Density, Built-Up and Settlements to Analyse Human Presence, Societal Impact and Sustainability: A Review of GHSL Applications. *Sustainability* **2021**, *13*, 7851. [CrossRef]
43. Yang, J.; El-Kassaby, Y.; Guan, W. The Effect of Slope Aspect on Vegetation Attributes in a Mountainous Dry Valley, Southwest China. *Sci. Rep.* **2020**, *10*, 16465. [CrossRef]
44. Eisfelder, C.; Asam, S.; Hirner, A.; Reiners, P.; Holzwarth, S.; Bachmann, M.F.; Gessner, U.; Dietz, A.; Huth, J.; Bachofer, F.; et al. Seasonal Vegetation Trends for Europe over 30 Years from a Novel Normalised Difference Vegetation Index (NDVI) Time-Series—The TIMELINE NDVI Product. *Remote Sens.* **2023**, *15*, 3616. [CrossRef]
45. Stoyanov, A. Application of Tasseled Cap Transformation of Sentinel-2—MSI Data for Forest Monitoring and Change Detection on Territory of Natural Park "BLUE STONES". *Environ. Sci. Proc.* **2022**, *22*, 42. [CrossRef]
46. Han, W.; Duan, S.-B.; Tian, H.; Lian, Y. Estimation of land surface temperature from AMSR2 microwave brightness temperature using machine learning methods. *Int. J. Remote Sens.* **2023**, 1–22. [CrossRef]
47. Azad, R.; Balzter, H.; Rasul, G.R.F.I.; Hameed, H.M.; Wheeler, J.; Adamu, B.; Ibrahim, S.; Najmaddin, P.M. Applying Built-Up and Bare-Soil Indices from Landsat 8 to Cities in Dry Climates. *Land* **2018**, *7*, 81. [CrossRef]
48. Li, Q.; Yu, F.F.; Mu, X. Evaluation of the Ecological Environment Quality of the Kuye River Source Basin Using the Remote Sensing Ecological Index. *Int. J. Environ. Res. Public Health* **2022**, *19*, 12500. [CrossRef] [PubMed]
49. Jiang, X.; Guo, X.; Wu, Y.; Xu, D.; Liu, Y.; Yang, Y.; Lan, G. Ecological vulnerability assessment based on remote sensing ecological index (RSEI): A case of Zhongxian County, Chongqing. *Front. Environ. Sci.* **2023**, *10*, 1074376.
50. Gong, C.; Wang, S.; Lu, H.; Liu, J. Research Progress on Spatial Differentiation and Influencing Factors of Soil Heavy Metals Based on Geographical Detector. *Huan Jing Ke Xue = Huanjing Kexue* **2023**, *44*, 2799–2816. [PubMed]
51. Xu, T.Q.; Chen, Y.P. Eco-Efficiency Assessment of Wuhan Based on Data Envelopment Analysis Approach. In Proceedings of the 27th Chinese Control and Decision Conference (2015 CCDC), Qingdao, China, 23–25 May 2015.
52. Xiong, H.; Hu, H.; Han, P.; Wang, M. Integrating Landscape Ecological Risks and Ecosystem Service Values into the Ecological Security Pattern Identification of Wuhan Urban Agglomeration. *Int. J. Environ. Res. Public Health* **2023**, *20*, 2792. [CrossRef] [PubMed]
53. Xie, F.; Liu, S.; Gao, Y.; Zhu, Y.; Wu, K.; Qi, M.; Duan, S.; Tahir, A.M. Derivation of Supraglacial Debris Cover by Machine Learning Algorithms on the GEE Platform: A Case Study of Glaciers in the Hunza Valley. *ISPRS Ann. Photogramm. Remote Sens. Spat. Inf. Sci.* **2020**, *V-3*, 417–420. [CrossRef]
54. Papaiordanidis, S.; Gitas, I.Z.; Katagis, T. Soil Erosion Prediction Using the Revised Universal Soil Loss Equation (RUSLE) in Google Earth Engine (GEE) Cloud-Based Platform. *Dokuchaev Soil Bull.* **2020**, *100*, 36–52. [CrossRef]
55. Wang, H.; Hua, L. Dynamic Monitoring of Ecological Environment Quality in Xiamen Based on the GEE Platform. In Proceedings of the Fourth International Conference on Geoscience and Remote Sensing Mapping (GRSM 2022), Changchun, China, 4–6 November 2022.
56. Wu, S.; Cao, L.; Xu, D.; Zhao, C. Historical Eco-Environmental Quality Mapping in China with Multi-Source Data Fusion. *Appl. Sci.* **2023**, *13*, 8051. [CrossRef]
57. An, M.; Li, W.; Wu, H.; An, H.; Huang, J. The Local Coupling and Telecoupling of Urbanization and Ecological Environment Quality Based on Multisource Remote Sensing Data. *J. Environ. Manag.* **2022**, *327*, 116921. [CrossRef]
58. Xu, H.; Wang, M.; Shi, T.; Guan, H.; Fang, C.; Lin, Z. Prediction of Ecological Effects of Potential Population and Impervious Surface Increases Using a Remote Sensing Based Ecological Index (RSEI). *Ecol. Indic.* **2018**, *93*, 730–740. [CrossRef]

59. Cai, W.; Xu, F. The Impact of the New Environmental Protection Law on Eco-Innovation: Evidence from Green Patent Data of Chinese Listed Companies. *Environ. Sci. Pollut. Res.* **2021**, *29*, 10047–10062. [CrossRef] [PubMed]
60. Hu, S.; Deng, J.; Li, X. Practice of Water Ecological Restoration of Large Urban Eutrophication Lake—A case of study of Donghu Lake, Wuhan. *E3S Web Conf.* **2023**, *394*, 01011. [CrossRef]
61. Wang, K.; Zhou, W.Q.; Li, W. Impacts of population spatio-temporal dynamics on ecosystem quality during fast urbanization in Beijing, China. *J. Appl. Ecol.* **2016**, *27*, 2137–2144.
62. Zhang, Y.Z.; Jiang, Z.Y.; Li, Y.Y.; Yang, Z.G.; Wang, X.H.; Li, X.B. Construction and Optimization of an Urban Ecological Security Pattern Based on Habitat Quality Assessment and the Minimum Cumulative Resistance Model in Shenzhen City, China. *Forests* **2021**, *12*, 847. [CrossRef]
63. Li, J.; Gong, J.; Guldmann, J.-M.; Yang, J. Assessment of Urban Ecological Quality and Spatial Heterogeneity Based on Remote Sensing: A Case Study of the Rapid Urbanization of Wuhan City. *Remote Sens.* **2021**, *13*, 4440. [CrossRef]
64. Huang, M.H.; Chen, J.J. Spatial and Temporal Change Analysis of Urban Heat Island Effect in Wuhan City. *ISPRS—Int. Arch. Photogramm. Remote Sens. Spat. Inf. Sci.* **2020**, *XLII-3/W10*, 705–712. [CrossRef]
65. Halbac-Cotoara-Zamfir, R.; Marucci, A.; Salvia, R.; Quaranta, G.; Sateriano, A.; Cecchini, M.; Bianchini, L. Caring of the Fringe? Mediterranean Desertification between Peri-Urban Ecology and Socioeconomics. *Sustainability* **2022**, *14*, 1426. [CrossRef]
66. Zipperer, W.C.; Northrop, R.J.; Andreu, M.G. Urban development and environmental degradation. In *Oxford Research Encyclopedia of Environmental Science*; Oxford University Press: Oxford, UK, 2020; ISBN 978-0-19-938941-4.
67. Xie, Q.J.; Han, Y.D.; Zhang, L.M.; Han, Z. Dynamic Evolution of Land Use/Land Cover and Its Socioeconomic Driving Forces in Wuhan, China. *Int. J. Environ. Res. Public Health* **2023**, *20*, 3316. [CrossRef]
68. Zhang, J.L.; Hou, Y.; Dong, Y.F.; Wang, C.; Chen, W.P. Land Use Change Simulation in Rapid Urbanizing Regions: A Case Study of Wuhan Urban Areas. *Int. J. Environ. Res. Public Health* **2022**, *19*, 8785. [CrossRef]
69. Wang, Z.; Zeng, J.; Chen, W. Impact of Urban Expansion on Carbon Storage Under Multi-Scenario Simulations in Wuhan, China. *Environ. Sci. Pollut. Res.* **2022**, *29*, 45507–45526. [CrossRef] [PubMed]
70. Pan, Y.; Gong, J.; Li, J. Assessment of Remote Sensing Ecological Quality by Introducing Water and Air Quality Indicators: A Case Study of Wuhan, China. *Land* **2022**, *11*, 2272. [CrossRef]
71. Zhang, M.; Kafy, A.; Ren, B.; Zhang, Y.; Tan, S.; Li, J. Application of the Optimal Parameter Geographic Detector Model in the Identification of Influencing Factors of Ecological Quality in Guangzhou, China. *Land* **2022**, *11*, 1303. [CrossRef]
72. Yin, H.; Chen, C.N.; Dong, Q.; Zhang, P.; Chen, Q.L.; Zhu, L. Analysis of Spatial Heterogeneity and Influencing Factors of Ecological Environment Quality in China's North-South Transitional Zone. *Int. J. Environ. Res. Public Health* **2022**, *19*, 2236. [CrossRef]
73. Chen, H.; Liu, Y.; Hu, L.; Zhang, Z.; Chen, Y.; Tan, Y. Constructing a Flood-Adaptive Ecological Security Pattern from the Perspective of Ecological Resilience: A Case Study of the Main Urban Area in Wuhan. *Int. J. Environ. Res. Public Health* **2022**, *20*, 385. [CrossRef]

sustainability

MDPI

Article

Research on the Spatio-Temporal Changes of Vegetation and Its Driving Forces in Shaanxi Province in the Past 20 Years

Ming Shi [1,2,3], Fei Lin [1,2,3], Xia Jing [3], Bingyu Li [3], Jingsha Qin [4], Manqi Wang [4], Yang Shi [1,2,5] and Yimin Hu [1,2,5,*]

1 Institute of Intelligent Machines, Hefei Institutes of Physical Science, Chinese Academy of Sciences, Hefei 230031, China; mshi@iim.ac.cn (M.S.); feilin@iim.ac.cn (F.L.); shiyang@iim.ac.cn (Y.S.)
2 Intelligent Agriculture Engineering Laboratory of Anhui Province, Hefei 230031, China
3 College of Geomatics, Xi'an University of Science and Technology, Xi'an 710054, China; jingxiaxust@163.com (X.J.); 22110010001@stu.xust.edu.cn (B.L.)
4 School of Resources and Environmental Engineering, Anhui University, Hefei 200601, China; 19334088548@163.com (J.Q.); wangmanqi777@163.com (M.W.)
5 Hefei Institutes of Collaborative Research and Innovation for Intelligent Agriculture, Hefei 231131, China
* Correspondence: ymhu@iim.ac.cn

Abstract: (1) Background: Vegetation is an important component of ecosystems. Investigating the spatio-temporal dynamic changes in vegetation in various Shaanxi Province regions is crucial for the preservation of the local ecological environment and sustainable development. (2) Methods: In this study, the KNDVI vegetation index over the 20-year period from 2003 to 2022 was calculated using MODIS satellite image data that was received from Google Earth Engine (GEE). Sen and MK trend analysis as well as partial correlation analysis were then utilized to examine the patterns in vegetation change in various Shaanxi Province regions. This paper selected meteorological factors, such as potential evapotranspiration (PET), precipitation (PRE), and temperature (TMP); human activity factors, such as land-use type and population density; and terrain factors, such as surface elevation, slope direction, and slope gradient, as the influencing factors for vegetation changes in the research area in order to analyze the driving forces of vegetation spatio-temporal changes. These factors were analyzed using a geo-detector. (3) Results: The vegetation in the research area presented a growth trend from 2003 to 2022, and the area of vegetation improvement was 189,756 km^2, accounting for 92.15% of the total area. Among them, the area of significantly improved regions was 174,262 km^2, accounting for 84.63% of the total area, and the area of slightly improved regions was 15,495 square kilometers, accounting for 7.52% of the total area. (4) Conclusions: The strengthening of bivariate factors and nonlinear enhancement were the main interaction types affecting vegetation changes. The combination of interaction factors affecting vegetation change in Shaanxi Province includes PRE ∩ PET as well as TMP ∩ PET. Therefore, climate conditions were the main driving force of KNDVI vegetation changes in Shaanxi Province. The data supported by this research are crucial for maintaining the region's natural ecosystem.

Keywords: KNDVI; trend analysis; MODIS; driver analysis

1. Introduction

A crucial component of ecosystems, vegetation, is essential to the global atmospheric and energy cycles, as well as to the flow of carbon and water [1]. It also plays an important role in global change monitoring, providing essential information for research on material cycles, biodiversity, land use, and climate change, as well as being a scientific basis for environmental protection and sustainable development. Thus, one of the hottest subjects in regional and global change study is tracking the dynamics of vegetation [2–4].

Currently employed as one of the vegetation indices, NDVI (Normalized Difference Vegetation Index) [5] is a useful predictor of vegetation growth state, biological activity, and geographical distribution [6]. It may accurately reflect the information about changes in land surface vegetation and has a strong link with measures such as aboveground biomass [7], leaf area index [8], chlorophyll fluorescence produced by sunlight [9], and GPP [10]. It has been applied by many scholars in monitoring vegetation dynamics. Using NDVI, Beck et al. [11] examined how the vegetation changed dynamically in high-latitude regions, while Pettorelli et al. [12] used NDVI to examine how plants react to changes in their environment. The Nenjiang River Basin's vegetation dynamic changes in response to multiscale drought stress were examined by Zhu Guanglei et al. [13]. The Brazilian Amazon region's main vegetation underwent dynamic changes, which were examined by Raquel Carvalho et al. [14]. Moreover, both natural and man-made causes have an impact on vegetation alterations. The growth and spread of plants are strongly correlated with several environmental elements, including terrain, climate, and others [15–18]. Among climate factors, temperature and precipitation have a significant impact on vegetation. For instance, Zhou et al. [19] discovered that in some high-latitude regions of the northern hemisphere, precipitation is the primary factor influencing variations in plant cover. According to Suzuki et al. [20], rising temperatures have been shown to lengthen the growth season and increase vegetation production. Regarding human influences, they frequently affect vegetation dynamics in both good and bad ways. For instance, Ma Haiyun et al. [21] discovered that changes in southwest China's plant cover are positively impacted by human activity. According to Wang et al. [22], ecological initiatives such as converting farms back into forests and grasslands may greatly expand the amount of vegetation in an area. According to Maeda et al. [23] and Nunes et al. [24], local vegetation cover will be significantly reduced as a result of land development, urbanization, excessive forest logging, and other human activities.

The current research mainly focuses on the impact of single factors (such as climate, topography, human activities, etc.) on vegetation. The effect of human and environmental causes on vegetation is not as well studied. When monitoring vegetation dynamics, commonly used vegetation indices such as NDVI and NIRv are often used as monitoring indicators. Nevertheless, photosynthesis itself is not reflected in the nonlinear, saturated connection between NDVI and aboveground biomass. Interactions between human and natural elements frequently affect vegetation [25,26]. A skewed interpretation of vegetation changes and an overestimation of the significance of the elements under research may result from focusing solely on the response of vegetation to a particular factor and ignoring the causes that induce vegetation changes. Therefore, in addition to considering traditional climate factors, the driving forces of vegetation changes must also comprehensively consider the influence of natural elements such as topography and human activities. In the past, techniques including trend analysis, partial correlation analysis, and residual analysis were mostly utilized in the investigation of the mechanisms behind changes in vegetation. Nevertheless, complex nonlinear interactions may also be a part of the process of driving variables for vegetation changes, in addition to a straightforward linear connection [27]. The nonlinear linkages between many influencing elements, particularly the one between human causes and climate change, cannot be explained by the aforementioned approaches. Wang Jinfeng et al. [28] proposed a statistical method called geographic detector, which can quantitatively identify the driving forces of single factors, the interaction between two factors, and risk zone detection. This method does not assume linearity and can better explain the interaction between factors and analyze variables. Currently, this model has been widely used in the study of vegetation NDVI driving mechanisms [29–33]. For example, Yao Bo et al. [34] examined the spatial patterns and underlying causes of vegetation dynamics in the Chongqing region of the Yangtze River Basin using geographic detector analysis. The results indicate that the locations experiencing trends in vegetation growth are largely found in the Chong-qing urban areas of the Wuling Mountain region and the Three Gorges Reservoir region. The three main factors influencing vegetation changes were human activity, climate, and geography. The factors that had the most influence were elevation, the average annual temperature, and the amount of light present at night.

Pei Hongze et al. [35] used geographic detector to study the net ecosystem productivity (NEP) of the Loess Plateau region between 2000 and 2020, with a particular emphasis on the factors that drive it and its spatio-temporal structure. The results showed that the main reasons of NEP in the west, center, and east sub-regions of the research area had distinct geographical differentiation features. Precipitation, relative humidity, and other moisture conditions were the main climatic factors affecting the central and western regions. Combinations of geography, climate, and human activity most impacted the eastern area, with land use serving as the most prominent human component.

GEE is a cloud platform for planetary-scale geospatial analysis in terms of data gathering [36]. It significantly cuts down on the time needed for the collecting and processing of remote sensing data by offering rich open-source data and robust computer resources for regional and global change studies. For these reasons, the study constructed a time series of KNDVI spanning from 2003 to 2022 for the province of Shaanxi and used the Google Earth Engine platform to gather monthly NDVI datasets for the study region. The vegetation dynamics and changes in the research region over a 20-year period were then examined using the Sen and MK trend analysis techniques. In order to offer a theoretical foundation and methodological reference for the vegetation dynamics, evaluation, and ecosystem preservation in Shaanxi Province, partial correlation analysis and geographic detector were then employed to examine the driving forces behind the temporal variations in KNDVI data.

2. Study Area

Shaanxi Province is located in central China in the center of the Yellow River. It borders the higher levels of the Jialing River in the Qinba Mountain region as well as the southern portion of the Han River Basin, which is a Yangtze River tributary. Sichuan Province and Chongqing Municipality to the south, Hubei Province and Henan Province to the southeast, Ningxia Hui Autonomous Region and Gansu Province to the west, Shanxi Province to the east across the Yellow River, and Inner Mongolia Autonomous Region to the north are its borders. There are 206,000 square kilometers in all. The province is mostly made up of several types of topography, with a tendency toward higher elevations in the north and south and lower elevations in the center. These terrains include plains, mountains, plateaus, and basins. The climate of Shaanxi Province's north and south varies significantly, as do the kinds and amounts of flora in each region. Northern Shaanxi, Guanzhong, and southern Shaanxi are the province's three naturally occurring geographical areas, separated by variations in terrain, landforms, and flora types. Shaanxi spans three climatic zones, with the northern part of northern Shaanxi and along the Great Wall belonging to the temperate zone, southern Shaanxi belonging to the northern subtropical zone, and Guanzhong and most of northern Shaanxi belonging to the warm temperate zone. The province's yearly mean temperature ranges from 0 to 16 °C, progressively dropping from east to west and from south to north. The province experiences between 340 and 1240 mm of precipitation on average every year, with the south receiving more precipitation than the north. The regions of Guanzhong, northern Shaanxi, and southern Shaanxi are semi-arid, semi-humid, and humid, respectively. There are large disparities in the distribution of the province's complex and varied flora types. The region north of the Great Wall in Shaanxi's northern region is near the desert, where desert plants predominate and there is little vegetation. The southern part of Yulin and the northern part of Yan'an, south of the Great Wall, are typical loess plateau regions where soil erosion is severe and vegetation coverage is low, mainly consisting of shrubs. The Beishan Mountains are distributed with deciduous broad-leaved forests, with higher vegetation coverage. The Guanzhong region is characterized by a large number of agricultural fields, and urban development has led to lower vegetation coverage. The Qinling Mountains and the northern part of southern Shaanxi are dominated by warm temperate deciduous broad-leaved forests, while the Bashan region has evergreen broad-leaved forests and deciduous broad-leaved forests, with good vegetation coverage. Figure 1 depicts the study area's location, land-use types, and elevation distribution, where

Figure 1A indicates the location, Figure 1B illustrates the distribution of land-use types in the study area in 2022, and Figure 1C depicts the distribution of surface elevation in the study area.

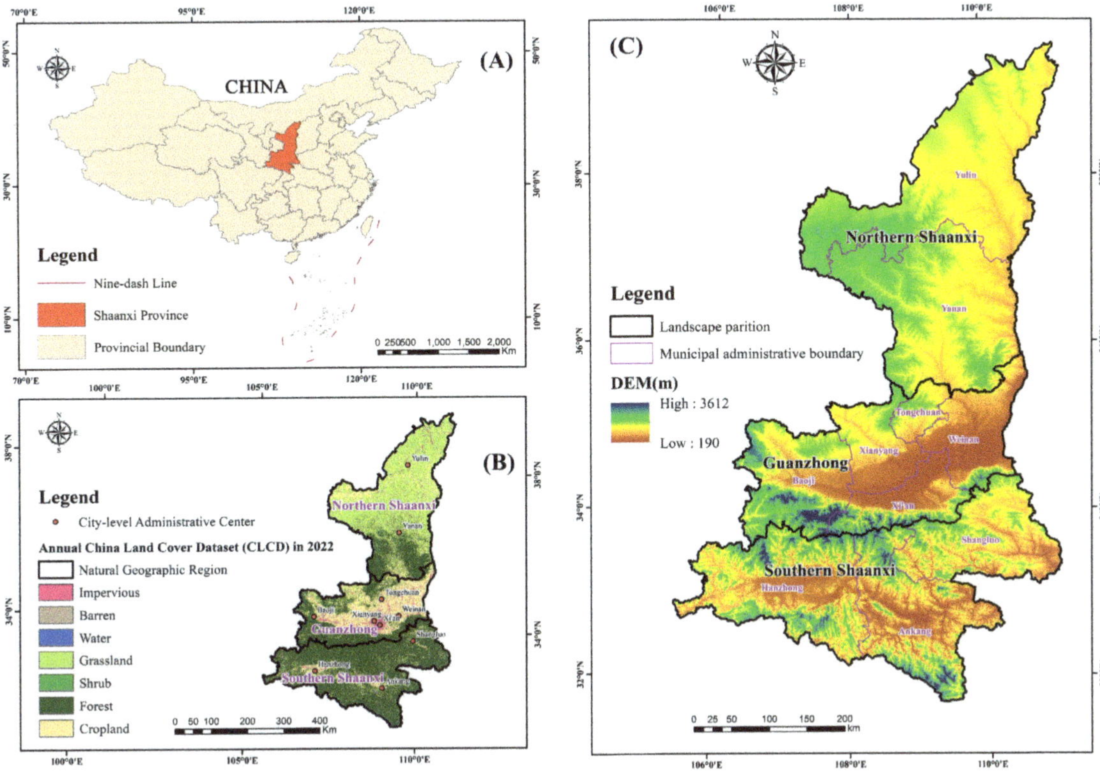

Figure 1. (**A–C**) Map of the study area location and land use.

3. Materials and Methods

3.1. Data Preprocessing and Acquisition

This article selects monthly MODIS data from 2003 to 2022 as the data source for calculating the KNDVI vegetation index. The Loess Plateau branch of the NESS Data Center (htxxp://loess.geodata.cn, accessed on 31 December 2022) provided the monthly average temperature (TMP) and monthly average precipitation (PRE), specifically the 1 km resolution average temperature and monthly precipitation datasets for China from 1901 to 2022. East View Cartographic contributed the population density data (PD), which come from the LandScan global population dataset created by ORNL (Oak Ridge National Laboratory, Oak Ridge, TN, USA). LandScan is the most accurate and reliable global population dynamic statistical analysis database based on geographic location, using innovative methods such as remote sensing and GIS, and it has the best resolution and distribution models (htxxps://landscan.ornl.gov/, accessed on 31 December 2022). The 1 km monthly potential evapotranspiration dataset for China from 1901 to 2022 is sourced from the National Tibetan Plateau Data Center (data.tpdc.ac.cn, accessed on 31 December 2022). The yearly China Land Cover Dataset (CLCD), which is a 30 m yearly land cover dataset and its dynamics in China from 1985 to 2022 (zenodo.org, accessed on 31 December 2022), was created by Huang Xin et al. from Wuhan University using 335,709 scenes of Landsat data on Google Earth Engine as the basis for the land-use dataset [37]. The DEM data is derived from the Geographic Spatial Data Cloud's 90 m resolution SRTM data. Table 1 displays the particular parameters of each dataset.

Table 1. Data sources and description.

Satellite Data	Parameter	Unit	Spatial Resolution/m
MOD13Q1	Vegetation Indices	/	250
CLCD	Land Cover	/	30
DEM	Digital Elevation Model	m	90
Landscan/PD	Population Density	Population density/km^2	1000
PET	Potential Evapotranspiration	mm	1000
PRE	Precipitation	mm	1000
TMP	Temperature	°C	1000

The GEE platform database is the source of the MODIS data that was previously discussed. Through the use of an internet database, we were able to obtain the MODIS data and resample it to a spatial resolution of 1000 m. Each month's KNDVI is computed, and the yearly KNDVI data are then obtained by performing the maximum value synthesis. In order to match the spatial resolution of other data, the DEM data's spatial resolution is resampled to 1000 m and utilized to compute the research area's slope and aspect information. The ArcGIS closest neighbor approach is used to resample the CLCD land-use type data to a geographic resolution of 1000 m. A uniform projection transformation is applied to all data in order to guarantee coordinate systems consistency.

Figure 2 shows the mean distribution of temperature, precipitation, and potential evapotranspiration (PET, PRE, and TMP, respectively) in the study area over a period of 20 years. Panels (a)–(c) depict the 20-year average distribution of potential evapotranspiration (PET), precipitation (PRE), and temperature (TMP), respectively. From Figure 2, it is evident that the spatial distribution of the three meteorological factors exhibits significant heterogeneity. In Panel (a), the 20-year mean of potential evapotranspiration ranges from 45.79 mm to 105.98 mm. The central region (Guanzhong) has higher values of evapotranspiration, while values are smaller in northern and southern Shaanxi. Panel (b) illustrates that the 20-year mean precipitation ranges from 26.93 mm to 97.88 mm. The southern Shaanxi region has the highest precipitation, followed by the Guanzhong region, and the lowest is in northern Shaanxi, especially in the northwest region, which, being close to the desert, has low vegetation coverage and scarce precipitation. Panel (c) reveals that the 20-year mean temperature ranges from −0.98 °C to 16.83 °C. The northern Shaanxi region has the lowest average temperature, while the Guanzhong and southern Shaanxi regions have relatively higher average temperatures. The urbanized Guanzhong region, characterized by a high proportion of impervious surfaces, exhibits elevated temperatures, while the southern Shaanxi region, boasting high elevation and abundant sunlight, also experiences higher temperatures.

3.2. KNDVI Calculation

The most used indicator for tracking vegetation changes is the NDVI; however, it has two main drawbacks. First, there is a nonlinear and saturating relationship between NDVI and green biomass [38]. The enhanced vegetation index (EVI) and other indices have attempted to use additional band information to construct vegetation indices to compensate for this issue, but the saturation phenomenon still exists. Second, when constructing vegetation indices, they respond to the presence of green leaves but do not directly reflect the process of photosynthesis in green vegetation. This means that GPP can decrease without leaf loss (i.e., reduced LAI) or a decrease in leaf chlorophyll [39]. In 2021, scholars from multiple countries proposed a non-linear vegetation index, KNDVI, in SCIENCE ADVANCES [40]. This index maximizes the utilization of spectral information and employs a machine-learning perspective, using kernel analysis to linearize NDVI and effectively prevent its saturation and sluggish response to photosynthesis. It addresses the long-standing problem of satellite observation of the terrestrial biosphere and can more accurately reflect the dynamic changes between land carbon sources and sinks. Compared to traditional NDVI, NIRv, and other vegetation indices, this method demonstrates greater stability and robustness. The method is shown in Equations (1)–(4).

$$KNDVI = \frac{k(n,n) - k(n,r)}{k(n,n) + k(n,r)} \tag{1}$$

The reflectance of the red band is denoted by r in the equation, the reflectance of the near-infrared band by n, and the correlation between the bands is represented by $k(n,n)$ and $k(n,r)$.

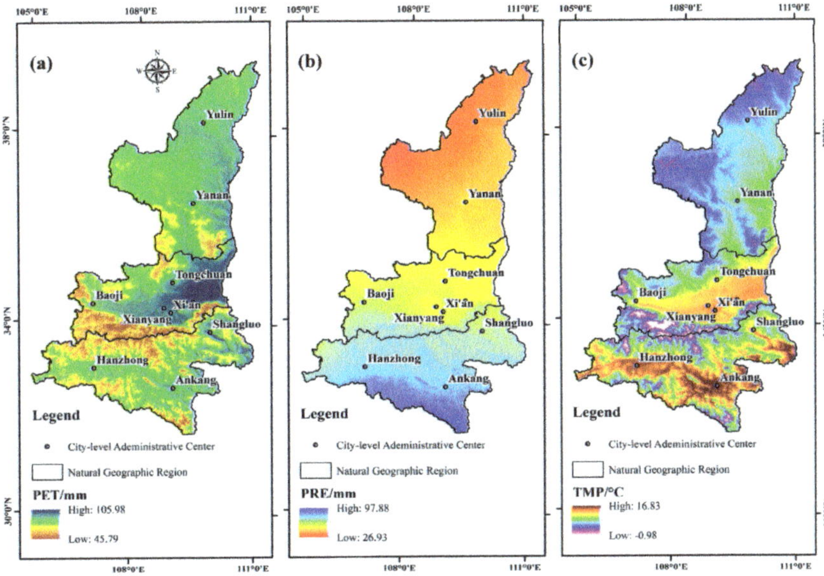

Figure 2. Mean distribution maps of three meteorological factors in the study area from 2013 to 2022. Specifically, panels (**a**–**c**) illustrate the mean values of potential evapotranspiration (PET), precipitation (PRE), and temperature (TMP), respectively.

Furthermore, a radial basis function (RBF) is used to describe the correlation between the bands.

$$k(n,r) = \frac{\exp(-(n-r)^2)}{2\sigma^2} \tag{2}$$

The near-infrared and red bands' separation from one another is determined by the equation's parameter σ.

$$KNDVI = \frac{1 - k(n,r)}{1 + k(n,r)} = \tanh((\frac{n-r}{2\sigma})^2) \tag{3}$$

The average distance between the red and near-infrared bands, or $\sigma = 0.5(n+r)$, is fixed as the length scale parameter σ in order to further simplify the index. The index functions well in practice thanks to this simplification, which enables it to be adaptable for every pixel. Equation (4) displays the outcome of the final computation.

$$KNDVI = \tanh(NDVI^2) \tag{4}$$

3.3. Methods

Theil-Sen Median and Mann-Kendall trend analysis techniques have been used in the quantitative study of vegetation change trends in Shaanxi Province over the previous 20 years using temporal KNDVI data. This study used elevation, slope, and aspect as environmental parameters and land-use type and population density as anthropogenic elements in accordance with previous research [41–45]. The meteorological parameters that were selected included the yearly average temperature, the annual average precipitation, and the annual average potential evapotranspiration. The association between the KNDVI data and each component was examined and evaluated using the partial correlation analysis approach. The reactions and underlying causes of interannual vegetation changes to each condition were also examined using the geo-detector. The study's flowchart is displayed in Figure 3.

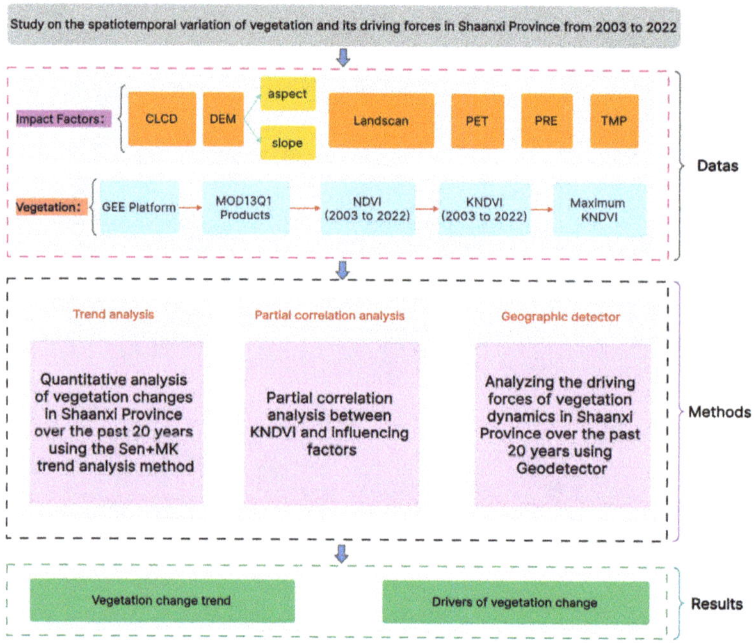

Figure 3. Experimental flowchart.

3.3.1. Trend Analysis

Sen's slope estimation, sometimes referred to as the Theil-Sen median method, is a reliable non-parametric statistical approach for determining trends. The technique is less susceptible to outliers and measurement mistakes and has a very high computing efficiency. It is frequently used to examine trends in data from lengthy time periods [46–48].

$$S_{KNDVI} = mean(\frac{x_j - x_i}{j - i}), (\forall j > i) \quad (5)$$

In the equation, S_{KNDVI} represents the slope of vegetation change and x_i and x_j represent long time-series KNDVI data. $S_{\text{KNDVI}} > 0$ and $S_{\text{KNDVI}} < 0$ indicate vegetation improvement and degradation trends, respectively. Mann–Kendall is a commonly used method for non-parametric statistical testing. Its advantages are that it does not require the measured values to follow a normal distribution, does not assume a linear trend, and is not affected by missing values and outliers. It has been widely used in the trend significance testing of long time-series data [49–52]. For a time series $X_{i,j} = 1, 2, \ldots, i, \ldots, j, \ldots, n$, the standardized test statistic, Z, is defined as

$$Z = \begin{cases} \frac{S}{\sqrt{Var(S)}} & ,(S > 0) \\ 0 & ,(S = 0) \\ \frac{S+1}{\sqrt{Var(S)}} & ,(S < 0) \end{cases} \quad (6)$$

$$S = \sum_{i=1}^{n-1} \sum_{j=i+1}^{n} sign(x_j - x_i) \quad (7)$$

$$sign(KNDVI_i - KNDVI_j) = \begin{cases} -1 & ,if(KNDVI_i - KNDVI_j) < 0 \\ 0 & ,if(KNDVI_i - KNDVI_j) = 0 \\ 1 & ,if(KNDVI_i - KNDVI_j) > 0 \end{cases} \quad (8)$$

In the formula, n represents the number of data points, while x_i and x_j stand for long time-series KNDVI data. In this work, we examined 20 years' worth of Shaanxi Province vegetation KNDVI data, where n is greater than or equal to 8. With mean and variance, the test statistic S has an approximation normally distributed distribution:

$$E(S) = 0 \quad (9)$$

$$Var(S) = \frac{n(n-1)(2n+5)}{18} \quad (10)$$

At the significance level α, if $|Z| > Z_{1-\alpha/2}$, it indicates a significant change trend in the time-series data. $Z_{1-\alpha/2}$ represents the value corresponding to the standard normal distribution function at a confidence level of α. Based on the significance testing method and referring to relevant literature [53–55], $|Z_s| = 1.96$ is chosen as the criterion for significance division. When $|Z_s| \leq 1.96$, it indicates that the vegetation change is not significant, and when $|Z_s| > 1.96$ it indicates that the vegetation change is significant.

3.3.2. Partial Correlation Analysis

In order to assess the correlations between land use, population density, annual average temperature, yearly average precipitation, annual average potential evapotranspiration (which are regarded as five parameters), and KNDVI, this study used the partial correlation analysis approach. The link between each component and KNDVI was examined independently by adjusting for other factors. The relationship between land use, population density, annual average temperature, yearly average precipitation, annual average potential evapotranspiration, and KNDVI is shown by the positive or negative value of the partial correlation coefficient [56–58].

$$r_{xy} = \frac{\sum_{i=1}^{n} [(x_i - \overline{x})(y_i - \overline{y})]}{\sqrt{\sum_{i=1}^{n} (x_i - \overline{x})^2 \sum_{i=1}^{n} (y_i - \overline{y})^2}} \quad (11)$$

reflects the correlation between variables x and y in the equation, where sample number is denoted by i. The vegetation's KNDVI value for the i-th year is represented by the symbol x_i, and one of the contributing elements, such as the annual average temperature or the annual average precipitation for the corresponding time, is represented by the symbol y_i. $\overline{x}$ represents the average value of KNDVI for the study area from 2003 to 2022 and $\overline{y}$ represents the value of the influencing factor for the corresponding time period.

3.3.3. Analysis by Geographic Detector

Wang Jinfeng et al. created the Geodetector statistical technique, which can be used to analyze geographical differentiation and identify its causes [59]. By using the viewpoint of spatial stratified heterogeneity, it ascertains how comparable the spatial distributions of two variables are [60–63]. Four components make up the Geodetector framework: factor detection, interaction detection, ecological detection, and danger detection. We used Geodetector's factor and interaction detection features in this investigation.

The spatial differentiation of the dependent variable (Y), which in this study is the KNDVI, and the explanatory power (q) of the driving factors (X), which in this study are the KNDVI, potential evapotranspiration, annual temperature, precipitation, and CLCD, on the spatial differentiation of the dependent variable, are investigated using factor detection. Its goal is to investigate how driving factors affect the KNDVI's spatial variation and differentiation. Equations (12) and (13) present the computation formulas:

$$q = 1 - \frac{\sum_{h=1}^{L} N_h \sigma_h^2}{N\sigma^2} = 1 - \frac{SSW}{SST} \quad (12)$$

$$SSW = \sum_{h=1}^{L} N_h \sigma_h^2,\ SST = N\sigma^2 \quad (13)$$

Higher q values in the equation signify a factor's stronger explanatory power; the q value ranges from [0, 1]; L represents the strata of the dependent variable Y or the factor X; N_h and N represent the number of units in stratum h and the entire region, respectively; σ_h^2 and σ^2 represent the variance of Y values in stratum h and the entire region, respectively; and SSW and SST represent the sum of within-stratum variances and the total variance of the entire region.

The purpose of interaction detection is to determine whether the various influencing factors, X_s, work in concert to affect the dependent variable Y. It assesses whether there is a difference in the explanatory power of the dependent variable Y when different factors interact compared to when they act individually. This is done by separately calculating the $q(X_1)$ and $q(X_2)$ for different factors such as X_1 and X_2 on the dependent variable Y and then calculating their interaction term $q(X_1 \cap X_2)$. Finally, $q(X_1)$, $q(X_2)$, and $q(X_1 \cap X_2)$ are compared. Various types of interactions are shown in Table 2.

Table 2. Information on interaction types.

Description	Interaction
Weaken, nonlinear	$q(X_1 \cap X_2) < \min[q(X_1), q(X_2)]$
Weaken, uni-	$\min[q(X_1), q(X_2)] < q(X_1 \cap X_2) < \max[q(X_1), q(X_2)]$
Enhance, bi-	$q(X_1 \cap X_2) > \max[q(X_1), q(X_2)]$
Independent	$q(X_1 \cap X_2) = q(X_1) + q(X_2)$
Enhance, nonlinear	$q(X_1 \cap X_2) > q(X_1) + q(X_2)$

4. Results

4.1. Temporal Analysis of Mean Value of KNDVI

The average KNDVI of Shaanxi Province and its geographical sub-regions between 2003 and 2022 was subjected to statistical analysis; the findings are displayed in Figure 4. The figure shows that Shaanxi Province's average KNDVI varied between 0.42 and 0.52 over the given period. The spatial distribution of KNDVI was categorized into three groups based on earlier research [64–66]: medium-low (0.2~0.4), medium (0.4~0.6) and medium-high (0.6~0.8). Medium-low and medium vegetation cover categories were the most common in Shaanxi Province. The average KNDVI in the southern Shaanxi region was greater than in other parts of the province, followed by the Guanzhong region and the northern Shaanxi region, based on the geographic sub-regions. The average KNDVI ranged from 0.58 to 0.61 in southern Shaanxi from 2003 to 2022, from 0.46 to 0.53 in the Guanzhong area, and from 0.24 to 0.42 in northern Shaanxi. The amount of vegetation varied clearly by region, with the cover falling toward the north. Shaanxi Province and every geographic sub-region had positive slopes in the linear regression analysis on the annual KNDVI values, suggesting an overall trend toward increased plant cover. The regression function's slope was 0.0046 throughout Shaanxi, 0.0073 in northern Shaanxi, 0.0027 in the Guanzhong region, and 0.003 in southern Shaanxi. This suggests that the northern Shaanxi region had the highest rate of vegetation cover expansion, followed by the Guanzhong region and the southern Shaanxi region.

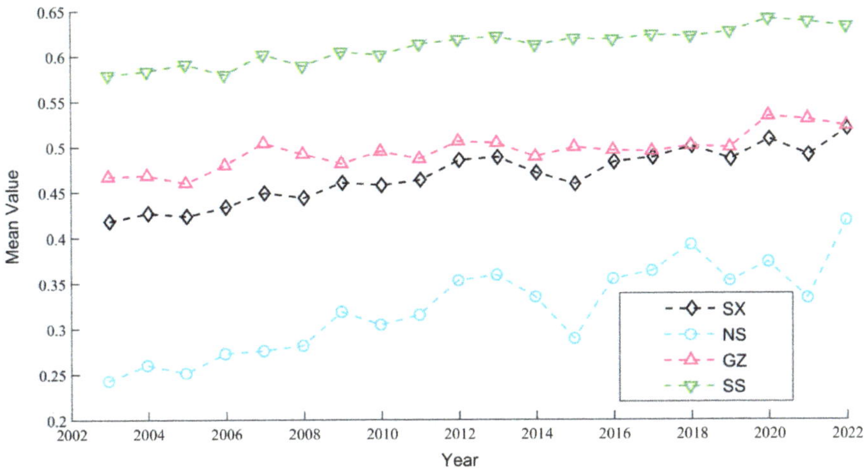

Figure 4. The research area's yearly average KNDVI values from 2003 to 2022. Shaanxi Province is represented by SX, whereas the areas of Guanzhong, northern Shaanxi, and southern Shaanxi are represented by GZ, NS, and SS, respectively.

4.2. Trend Analysis of KNDVI

The temporal KNDVI data slopes were estimated using the Theil–Sen median method, as shown in Figure 5. The analysis reveals an overall improvement in vegetation cover in Shaanxi Province, with localized areas exhibiting a declining trend. Urban regions such as Xi'an, Baoji et al. in the Guanzhong urban cluster, significant cities in southern Shaanxi such as Hanzhong and Ankang, and urban areas in northern Shaanxi such as Yulin and Yan'an are the main locations of vegetation degradation. The area with improved vegetation growth covers 189,756 km^2, accounting for 92.15% of the total area; the area with stable vegetation growth covers 3977 km^2, accounting for 1.93% of the total area; and the area with deteriorating vegetation growth covers 12,184 km^2, accounting for 5.92% of the total area. In terms of geographic regions, the area with improved vegetation growth has the highest proportion in the northern Shaanxi region, accounting for 98.92% of the northern Shaanxi area, followed by the southern Shaanxi region, accounting for 95.71% of the southern Shaanxi area, and finally the Guanzhong region, accounting for 78.08% of the Guanzhong area. The area with stable vegetation growth has the highest proportion in the Guanzhong region, accounting for 4.58% of the Guanzhong area, followed by the southern Shaanxi region, accounting for 1.32% of the southern Shaanxi area, and finally the northern Shaanxi region, accounting for 0.61% of the northern Shaanxi area. In the Guanzhong region, the area with declining vegetation growth makes up the largest proportion (17.34%), followed by the southern Shaanxi region (2.98%) and the northern Shaanxi region (0.47%) of the Guanzhong region.

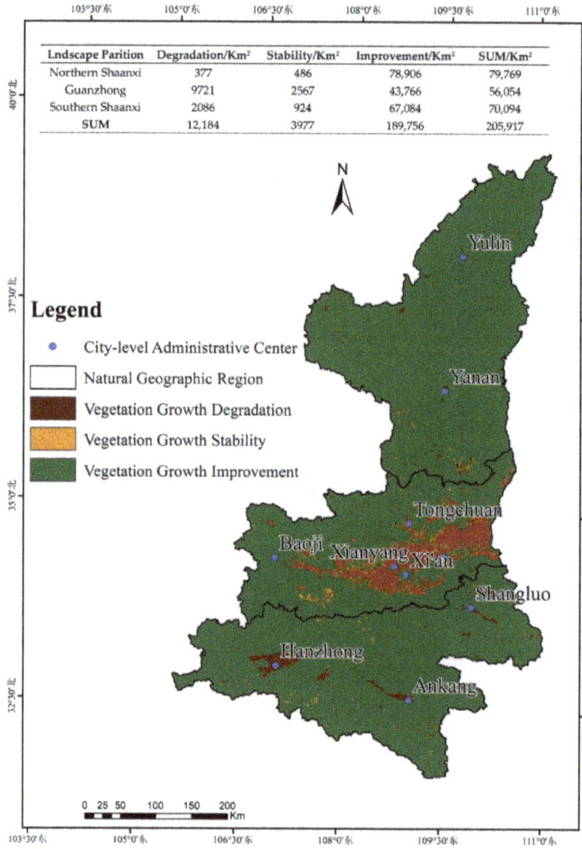

Figure 5. Temporal trend analysis of KNDVI based on the Theil–Sen method.

Figure 6 illustrates the results of the MK significance test. It is evident from the figure that the vegetation growth throughout Shaanxi Province shows significant spatial heterogeneity, with degradation primarily concentrated in urban areas, particularly in large cities such as Xi'an and Xianyang. Consistent with earlier study findings, the improvement in vegetation growth is greatest in the northern and southern parts of Shaanxi and least in the center region. By statistically analyzing the area of different trends in vegetation change, the area showing an improvement trend is 189,757 km^2, of which the area of significant improvement is 174,262 km^2, accounting for 84.63% of the total area. The area of slight improvement is 15,495 km^2, accounting for 7.52% of the total area. The area of stable and unchanged vegetation is 3977 km^2, accounting for 1.93% of the total area. The area showing a degradation trend is 12,184 km^2, of which the area of slight degradation is 5767 km^2, accounting for 2.8% of the total area, and the area of significant degradation is 6417 km^2, accounting for 3.12% of the total area. In terms of different geographical regions, for the northern region of Shaanxi, the area of improvement accounts for 98.92% of the total area, of which the area of significant improvement accounts for the highest proportion, 93.06% of the total area, followed by the area of improvement, accounting for 5.86% of the total area. In the central region, the area of deterioration makes up 17.34% of the whole area, with the area of considerable degradation being 5126 km^2, and the area of improvement is 78.08% of the total area, with the area of improvement being 36,321 km^2. In Shaanxi's southern region, the area of improvement makes up 95.71% of the total area, while the area of deterioration makes up 2.98%.

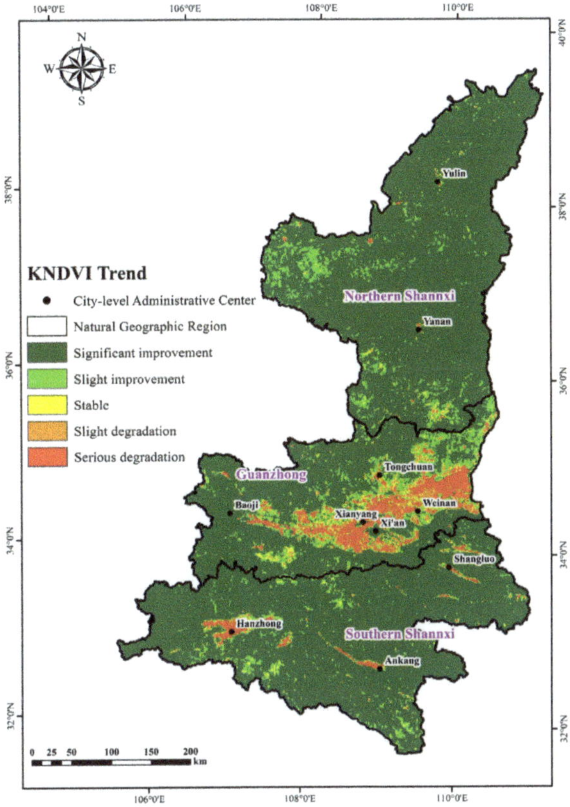

Figure 6. Significance analysis of temporal KNDVI change trend.

4.3. Partial Correlation Analysis of Influencing Factors

Due to different hydrothermal conditions in different regions, human activities have varying effects and degrees of impact on nature, resulting in spatial variations in vegetation growth. In this study, the corresponding KNDVI data were used as the dependent variable for a partial correlation analysis, and the land-use types (a), population density (b), annual average potential evapotranspiration (c), annual average precipitation (d), and annual average temperature (e) data from 2003 to 2022 were used as independent variables. Figure 7 presents the findings. The KNDVI ranges from −0.97 to 0.98 for population density data, from −0.82 to 0.91 for yearly average potential evapotranspiration, from −0.95 to 0.88 for yearly average precipitation, and from −0.92 to 0.83 for yearly average temperature. The partial correlation coefficients between land-use types and KNDVI range from −0.99 to 0.95.

The significant pixel area at the significance level for all land-use categories in the province is 73,944 km^2. Of this total area, 48.35% is made up of positively correlated pixels, while 51.65% is made up of negatively correlated pixels. According to population density, the relevant pixel area at the significance level is 73,944 km^2, of which 37.26% and 62.74% are positively correlated and negatively correlated, respectively, of the entire area of this kind of pixel. At the significance level, the yearly average potential evapotranspiration has a significant pixel area of 73,908 km^2. Of this type of pixel, the positively correlated pixel area makes up 50.35% of the entire area, while the negatively correlated pixel area makes up 49.65%. The relevant pixel area for yearly average precipitation at the significance level is 73,908 km^2, of which the positively correlated pixel area makes up 27.95% and the negatively correlated pixel area accounts for 72.05% of the total area of this type of pixel. At the significance level, the yearly average temperature has a significant pixel size of 73,908 km^2. Of this type of pixel, the positively correlated pixel area makes up 24.7% of the overall area, while the negatively correlated pixel area makes up 75.3%.

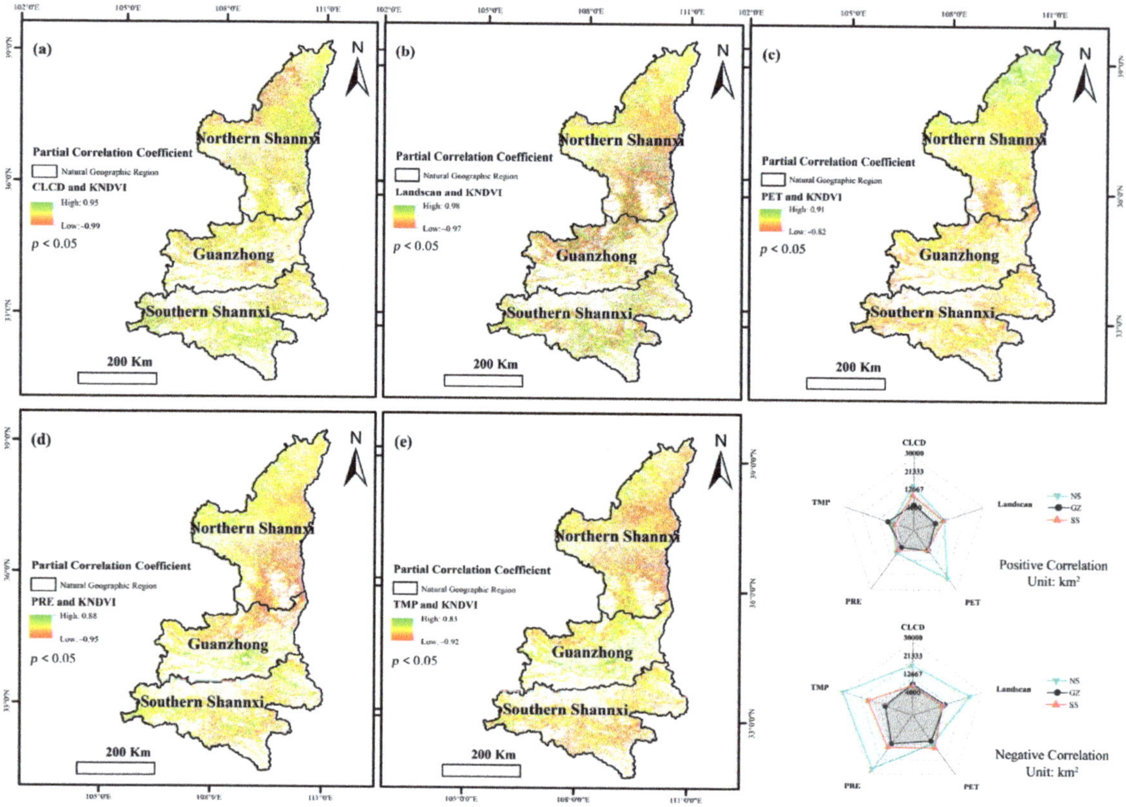

Figure 7. Partial correlation analysis of influencing factors and KNDVI, where NS, GZ, and SS represent northern Shaanxi, Guanzhong, and southern Shaanxi, respectively. (**a**–**e**) respectively represent CLCD, Landscan, PET, PRE, and TMP.

For different geographical regions, the proportion of the area where KNDVI is negatively correlated with land-use types in the northern Shaanxi and Guanzhong regions (23.68%, 17.81%) is greater than the proportion of the area where it is positively correlated (20.7%, 13.01%). In contrast, in southern Shaanxi, the proportion of the area where KNDVI is positively correlated with land-use types (17.04%) is greater than the proportion of the area where it is negatively correlated (13.28%). KNDVI is negatively correlated with population density in the northern Shaanxi, Guanzhong, and southern Shaanxi regions, with the proportions being 30.69%, 20.11%, and 15.17%, respectively. KNDVI is positively correlated with annual average potential evapotranspiration in northern Shaanxi (29.33%), while it is negatively correlated in the Guanzhong and southern Shaanxi regions (18.43%, 20.52%). There is a negative correlation between each region and annual average precipitation, with the proportions being northern Shaanxi (33.82%), Guanzhong (21.59%), and southern Shaanxi (20.2%). Additionally, there is a negative association between the yearly average temperature and each region; the proportions are as follows: Guanzhong (16.01%), southern Shaanxi (24.21%), and northern Shaanxi (37.23%).

Figure 8 computes and displays the annual count of interactions between different variables from 2003 to 2022. Bivariate enhancement and nonlinear enhancement are two examples of the interactions between influencing elements that are depicted in the graph. In graph (a), the interaction effects among influencing factors leading to changes in vegetation KNDVI in Shaanxi Province were relatively balanced between the two types from 2003 to 2011. However, from 2012 to 2022, the number of bivariate enhancement interaction types exceeds that of nonlinear enhancement. From graph (b), it can be seen that, in the northern Shaanxi region, the number of nonlinear enhancement interaction types between influencing factors is greater than that of bivariate enhancement. From graph (c), it can be seen that, in the Guanzhong region, the bivariate enhancement is the dominant type of interaction between influencing factors. From graph (d), it can be seen that, for the southern Shaanxi region, the interaction types between influencing factors are similar to those in the Guanzhong region, with bivariate enhancement being the main type.

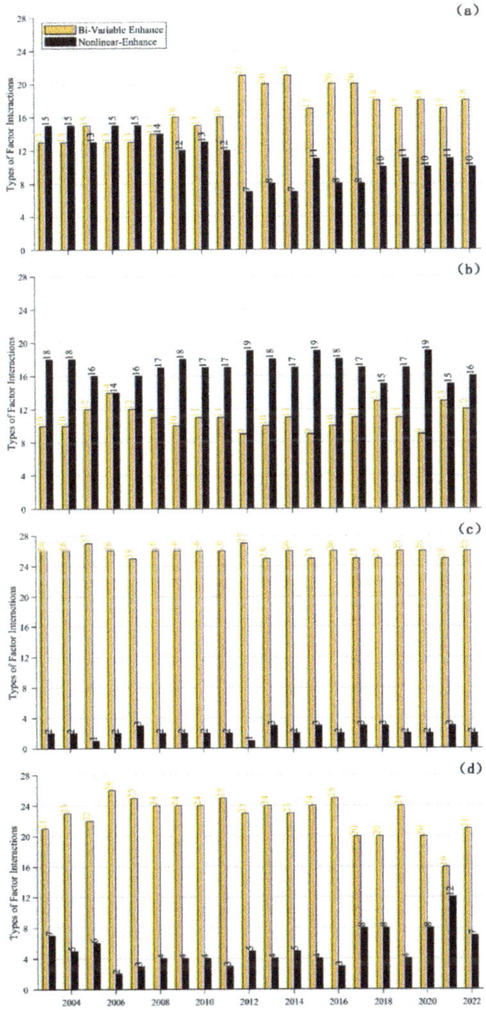

Figure 8. Statistics of interaction types between influencing factors in Shaanxi Province and its geographical regions from 2003 to 2022. Graphs (**a–d**) represent the Shaanxi Province, northern Shaanxi, Guanzhong, and southern Shaanxi regions, respectively.

5. Discussion

5.1. Response of KNDVI to Influencing Factors

In this study, the relationship between the eight influencing factors—land-use type (CLCD), elevation (DEM), slope (Slope), aspect (Aspect), population density (Landscan/PD), annual average potential evapotranspiration (PET), annual average precipitation (PRE), and annual average temperature (TMP) and the changes in the KNDVI in Shaanxi Province from 2003 to 2022 was examined using the Geographic Detector. Figure 9 displays the findings of the Geographic Detector's single-factor study. Among them, (a), (b), (c), and (d) represent the single-factor detection results for Shaanxi Province, northern Shaanxi, Guanzhong, and southern Shaanxi, respectively. From Figure 9a, it can be seen that there are significant differences in the contribution values (q values) of each factor to the KNDVI of vegetation in the entire province of Shaanxi. By calculating the average q values for each factor over the years and sorting them, the ranking is as follows: CLCD (0.655) > PRE (0.584) > PET (0.423) > Slope (0.382) > TMP (0.133) > DEM (0.093) > Landscan (0.023) > Aspect (0.007).

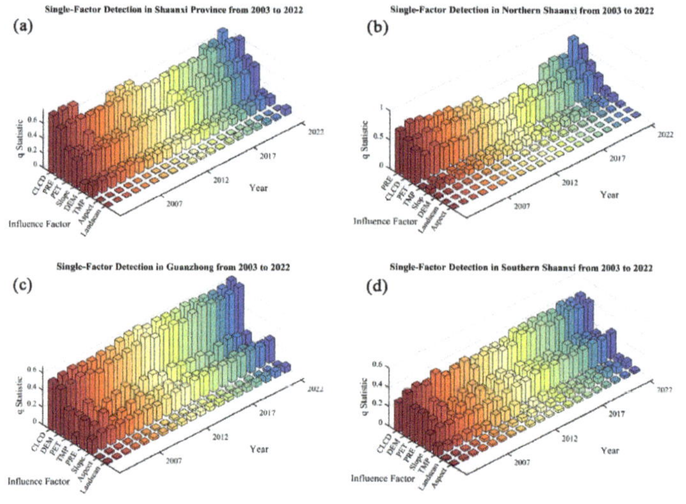

Figure 9. (**a**–**d**) Results of single-factor detection. Graphs (**a**), (**b**), (**c**), and (**d**) respectively represent the single-factor detection results of Shaanxi Province, Northern Shaanxi, Guanzhong, and Southern Shaanxi.

The changes in land-use type have caused significant variations in vegetation KNDVI, which may be related to anthropogenic factors such as urban expansion, afforestation, and reforestation. Precipitation is the most important climate factor affecting vegetation KNDVI changes, as adequate rainfall can promote vegetation growth. When analyzing the effects of different factors on vegetation KNDVI changes, we found that the year 2015 (q = 0.678) had the greatest impact of land-use type changes on vegetation k-NDVI values, 2003 (q = 0.117) for surface elevation, 2017 (q = 0.406) for slope, 2022 (q = 0.01) for aspect, 2022 (q = 0.085) for population, 2005 (q = 0.53) for annual potential evapotranspiration, 2004 (q = 0.789) for annual average precipitation, and 2007 (q = 0.176) for annual average temperature.

Different climatic conditions, natural environments, and vegetation types in different geographical regions result in varying effects of different influencing factors on vegetation KNDVI changes. For the northern Shaanxi region, the relationship of the effects of different influencing factors on regional vegetation KNDVI changes is as follows: CLCD (0.538) > PRE (0.524) > PET (0.386) > Slope (0.1) > TMP (0.098) > DEM (0.03) > Aspect (0.006) > Landscan (0.001). The year 2003 (q = 0.669) was found to have the greatest influence of land-use type changes on vegetation k-NDVI values, 2016 (q = 0.04) for surface elevation, 2012 (q = 0.118) for slope, 2022 (q = 0.008) for aspect, 2022 (q = 0.002) for population, 2003 (q = 0.587) for annual potential evapotranspiration, 2021 (q = 0.829) for annual average precipitation, and 2012 (q = 0.182) for annual average temperature. Thus, it can be seen that land-use type and precipitation are the most important influencing factors causing vegetation k-NDVI changes in the northern Shaanxi region.

For the Guanzhong region, the relationship of different influencing factors on regional vegetation KNDVI changes is as follows: CLCD (0.621) > DEM (0.587) > PET (0.496) > TMP (0.429) > Slope (0.342) > PRE (0.277) > Aspect (0.041) > Landscan (0.01). In the year 2022, land-use type, surface elevation, and slope were found to have the greatest influence on vegetation KNDVI changes, with the respective q values of 0.676, 0.658, and 0.408. The year 2019 (q = 0.051) had the greatest influence of aspect on vegetation KNDVI changes, 2007 (q = 0.016) for population, 2003 (q = 0.557) for annual potential evapotranspiration, 2005 (q = 0.351) for annual average precipitation, and 2012 (q = 0.495) for annual average temperature. Thus, it can be seen that land-use type, elevation, and evapotranspiration are the most important influencing factors causing vegetation KNDVI changes in the Guanzhong region.

For the southern Shaanxi region, the relationship of different influencing factors on regional vegetation KNDVI changes is as follows: DEM (0.43) > CLCD (0.378) > PET, PRE (0.217) > Slope (0.215) > TMP (0.068) > Landscan (0.041) > Aspect (0.021). The year 2020 (q = 0.452) was found to have the greatest influence of land-use type changes on vegetation KNDVI values, 2006 (q = 0.502) for surface elevation, 2022 (q = 0.262) for slope, 2014 (q = 0.025) for aspect, 2020 (q = 0.071) for population, 2003 (q = 0.338) for annual potential evapotranspiration, and 2011 (q = 0.123) for annual average temperature. Thus, it can be seen that elevation, land-use type, evapotranspiration, and precipitation are important influencing factors causing vegetation KNDVI changes in the southern Shaanxi region.

5.2. *Examination of the Factors Influencing Vegetation KNDVI*

The dynamic and multifaceted process of vegetation change is impacted by a wide range of variables. China has been implementing ecological measures since 2000, such as preserving natural forests and converting farms back to forests and grasslands. These actions have increased the amount of plant cover and promoted beneficial ecological growth [67–69]. According to this study, there is a general tendency toward improvement as the KNDVI values of the vegetation in Shaanxi Province steadily rise from north to south. Shaanxi's northern region, which makes up 98.92% of the territory's total land, has seen the greatest increase in vegetation. The southern region, which makes up 95.71% of the region's total area, is next in line. Lastly, 78.08% of its land is made up of the Guanzhong region. The northern Shaanxi region has shown the greatest improvement in vegetation, which is in line with earlier research [70–72]. The primary factors impacting the development of vegetation are slope, evapotranspiration, precipitation, and land use. Shaanxi Province's environment has become warmer and more humid over time, which might be good for the growth and recovery of vegetation [73–76].

From 2003 to 2022, the areas in Shaanxi Province with higher KNDVI values are mainly located in high-altitude regions such as the Qinling Mountains. These areas have suitable temperatures, sufficient rainfall, low human activity intensity, predominantly forest vegetation types, strong resistance to natural disasters such as soil erosion, and good vegetation stability, exhibiting low fluctuation. Therefore, these areas exhibit high KNDVI values. The areas with lower vegetation KNDVI values are primarily located in urban areas with intensive human activities, such as the Guanzhong urban agglomeration, including cities such as Xi'an, Xianyang, and Baoji, or in environmentally harsh desertification areas, such as regions near the Mu Us Desert in northern Shaanxi.

This work maps the findings of the interactions among numerous factors in Shaanxi Province, as shown in Figure S1 (see Supplementary Materials), in order to analyze the interactions between different factors in different years. With less noticeable interactions with elevation, slope, aspect, and population density, the figure shows a substantial association between land-use categories, yearly average precipitation, annual average evaporation, and annual average temperature. To elaborate, the association with other parameters like DEM,

Slope, Aspect, and PD is less prominent, even if the correlation with CLCD, PRE, PRE, and TMP is strong. The strongest interacting factors for different years are summarized in Table 3.

Table 3 shows that TMP ∩ PET and PRE ∩ PET are the main interaction variables influencing the vegetation KNDVI variations in Shaanxi Province. Over the 20-year period, there were 6 years of interaction between annual precipitation and annual potential evapotranspiration, and 9 years of interaction between annual average temperature and annual potential evapotranspiration. The statement makes it abundantly evident that the research area's plant KNDVI fluctuations are mostly caused by the local climate.

The climate conditions vary in different regions of Shaanxi Province, with significant differences in vegetation types. This results in noticeable variations in the influencing factors among different natural geographical zones. Therefore, an analysis of the influencing factors and driving forces of vegetation KNDVI changes in different geographical zones is conducted. Figure S2 (see Supplementary Materials) illustrates the interactions of influencing factors in different geographical zones from 2003 to 2022.

Table 3. Maximum Interaction of Influencing Factors in Shaanxi Province from 2003 to 2022.

Year	Max Value	Type	Year	Max Value	Type
2003	0.892	PRE ∩ PET	2013	0.732	TMP ∩ CLCD
2004	0.89	PRE ∩ PET	2014	0.83	TMP ∩ PET
2005	0.899	PRE ∩ PET	2015	0.824	TMP ∩ PET
2006	0.791	TMP ∩ PET	2016	0.782	TMP ∩ PET
2007	0.805	TMP ∩ PET	2017	0.791	TMP ∩ PET
2008	0.789	TMP ∩ PET	2018	0.744	TMP ∩ PET
2009	0.811	PRE ∩ PET	2019	0.795	PRE ∩ CLCD
2010	0.83	PRE ∩ CLCD	2020	0.809	PRE ∩ PET
2011	0.834	PRE ∩ PET	2021	0.853	PRE ∩ CLCD
2012	0.786	TMP ∩ PET	2022	0.761	PRE ∩ CLCD

The land-use type (CLCD), elevation (DEM), annual average precipitation (PRE), annual average evapotranspiration (PRE), and annual average temperature (TMP) show rather substantial interactions with other affecting elements, as shown in Figure S2 (see Supplementary Materials). Conversely, the relationships between population density (PD), aspect (Aspect), and slope (Slope) are less strong. Table 4 provides a summary of the most influential elements for each year.

For the northern part of Shaanxi, the primary interacting factors affecting vegetation KNDVI changes are TMP ∩ PET and PRE ∩ PET. Over the 20-year period, there were 6 years of interaction between annual precipitation and annual potential evapotranspiration and 6 years of interaction between annual average temperature and annual potential evapotranspiration. This suggests that climatic conditions are the most significant driving force for vegetation KNDVI changes in the northern region.

The primary interacting elements affecting vegetation KNDVI changes in the Guanzhong area are DEM ∩ CLCD. There was 18 years of interaction between land-use type and elevation throughout the 20-year timeframe. This suggests that the main factors influencing plant KNDVI variations in the Guanzhong region are elevation and human activity.

The combination of elevation and land-use type was the largest interaction element during the 20-year period, with DEM ∩ CLCD being the key variables impacting vegetation KNDVI changes for the southern half of Shaanxi. This statement implies that elevation and human activity are the main drivers of vegetation KNDVI changes in the southern area.

Table 4. Strongest interacting factors and interaction types in different geographical zones from 2003 to 2022.

Northern Shaanxi					
Year	**Max Value**	**Type**	**Year**	**Max Value**	**Type**
2003	0.856	PRE ∩ PET	2013	0.638	TMP ∩ CLCD
2004	0.863	PRE ∩ CLCD	2014	0.746	TMP ∩ PET
2005	0.869	PRE ∩ PET	2015	0.64	TMP ∩ PET
2006	0.673	TMP ∩ CLCD	2016	0.654	TMP ∩ PET
2007	0.708	TMP ∩ CLCD	2017	0.676	TMP ∩ PET
2008	0.696	TMP ∩ CLCD	2018	0.621	TMP ∩ PET
2009	0.677	PRE ∩ PET	2019	0.651	PRE ∩ PET
2010	0.721	PRE ∩ CLCD	2020	0.68	PRE ∩ PET
2011	0.669	PRE ∩ CLCD	2021	0.797	PRE ∩ PET
2012	0.655	TMP ∩ PET	2022	0.61	PRE ∩ CLCD
Guanzhong					
Year	**Max Value**	**Type**	**Year**	**Max Value**	**Type**
2003	0.667	DEM ∩ CLCD	2013	0.696	DEM ∩ CLCD
2004	0.681	DEM ∩ CLCD	2014	0.73	DEM ∩ CLCD
2005	0.701	DEM ∩ CLCD	2015	0.727	DEM ∩ CLCD
2006	0.676	DEM ∩ CLCD	2016	0.707	DEM ∩ CLCD
2007	0.65	DEM ∩ CLCD	2017	0.739	DEM ∩ CLCD
2008	0.681	DEM ∩ CLCD	2018	0.755	DEM ∩ CLCD
2009	0.68	DEM ∩ CLCD	2019	0.739	DEM ∩ CLCD
2010	0.707	PET ∩ CLCD	2020	0.732	DEM ∩ CLCD
2011	0.731	PET ∩ CLCD	2021	0.74	DEM ∩ CLCD
2012	0.749	DEM ∩ CLCD	2022	0.774	DEM ∩ CLCD
Southern Shaanxi					
Year	**Max Value**	**Type**	**Year**	**Max Value**	**Type**
2003	0.539	DEM ∩ CLCD	2013	0.557	DEM ∩ CLCD
2004	0.585	DEM ∩ CLCD	2014	0.553	DEM ∩ CLCD
2005	0.543	DEM ∩ CLCD	2015	0.629	DEM ∩ CLCD
2006	0.612	DEM ∩ CLCD	2016	0.608	DEM ∩ CLCD
2007	0.533	DEM ∩ CLCD	2017	0.629	DEM ∩ CLCD
2008	0.565	DEM ∩ CLCD	2018	0.593	DEM ∩ CLCD
2009	0.596	DEM ∩ CLCD	2019	0.599	DEM ∩ CLCD
2010	0.478	DEM ∩ CLCD	2020	0.661	DEM ∩ CLCD
2011	0.569	DEM ∩ CLCD	2021	0.628	DEM ∩ CLCD
2012	0.578	DEM ∩ CLCD	2022	0.643	DEM ∩ CLCD

5.3. Changes in Land Use and Their Effects on Vegetation KNDVI

In order to investigate the effects of land-use changes in Shaanxi Province during the previous 20 years on vegetation KNDVI changes, this research used land-use data from the years 2003, 2013, and 2022 as the data source. The land cover scenario for the years 2019, 2020, and 2021 is shown in Figure 10.

Table 5 illustrates that, between 2003 and 2013, the greatest area of land in Shaanxi Province that was converted to other land uses was agricultural land, accounting for 24,881 km^2. Of them, 11,371 km^2 was the greatest area converted to grassland, making up 45.7% of the total area transformed in that category. At 90,339 km^2, or 43.9% of the entire area, the province's forest area was the greatest in 2013. The area covered by forests grew by 16,409 km^2 in comparison to 2003. Table 6 presents an examination of land-use categories in Shaanxi Province from 2013 to 2022.

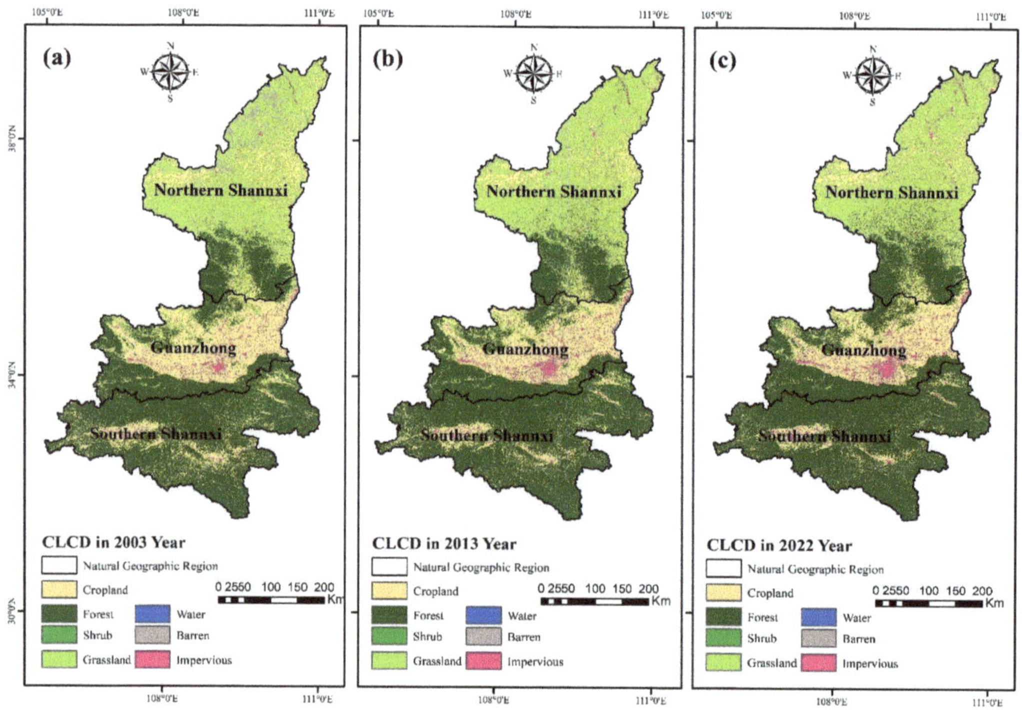

Figure 10. The land cover distribution in Shaanxi Province for the years 2003, 2013, and 2022, where (**a–c**) represent the years 2003, 2013, and 2022, respectively.

Table 5. Land-use transition matrix for Shaanxi Province from 2003 to 2013.

	2013CLCD							
2003CLCD	**Barren**	**Cropland**	**Forest**	**Grassland**	**Impervious**	**Shrub**	**Water**	**SUM/km^2**
Barren	90	85	0	1538	13	0	12	1738
Cropland	15	34,704	10,309	11,371	2817	46	323	59,585
Forest	0	7081	73,930	2924	104	314	51	84,404
Grassland	123	9827	5146	40,025	409	42	162	55,734
Impervious	1	1515	129	231	1029	0	55	2960
Shrub	0	96	746	99	0	17	0	958
Water	7	196	79	59	61	0	137	539
SUM/km^2	236	53,504	90,339	56,247	4433	419	740	205,918

Table 6. Land-use transition matrix for Shaanxi Province from 2013 to 2022.

	2022CLCD							
2013CLCD	**Barren**	**Cropland**	**Forest**	**Grassland**	**Impervious**	**Shrub**	**Water**	**SUM/km^2**
Barren	72	30	0	120	8	0	6	236
Cropland	1	45,897	3325	3304	913	1	63	53,504
Forest	0	1269	89,026	19	1	24	0	90,339
Grassland	115	5220	2506	48,223	159	16	8	56,247
Impervious	0	4	0	0	4383	0	46	4433
Shrub	0	20	287	36	0	76	0	419
Water	10	95	1	6	55	0	573	740
SUM/km^2	198	52,535	95,145	51,708	5519	117	696	205,918

Table 6 shows that 8024 km^2 was the largest amount of grassland in Shaanxi Province that was changed to other land-use categories between 2013 and 2022. Of all of them, 65.1% of the total conversion area fell into the group where the largest area was turned into cultivated land. With 95,145 km^2 of forest covering 46.2% of the province's total area in 2022, it was the largest in the province. The amount of forest land grew by 6119 km^2 in comparison to 2013. According to the aforementioned conclusions, the province's total forest covering has increased, which has raised vegetation KNDVI values in line with earlier research.

This article calculates the land-use transition matrix for the northern Shaanxi, Guanzhong, and southern Shaanxi areas from 2003 to 2022 in order to assess changes in land use in various geographical locations, as indicated in Table 7.

Table 7. Land-use transition matrix for different geographical regions in Shaanxi Province from 2003 to 2022.

NS	2022CLCD							
2003CLCD	**Barren**	**Cropland**	**Forest**	**Grassland**	**Impervious**	**Shrub**	**Water**	**SUM/km^2**
Barren	52	182	0	1463	26	0	13	1736
Cropland	12	5583	1135	9192	180	1	45	16,148
Forest	0	799	10,040	995	18	22	4	11,878
Grassland	121	8326	3341	36,817	460	6	135	49,206
Impervious	3	78	15	168	84	0	7	355
Shrub	0	18	197	42	1	0	0	258
Water	7	44	6	51	19	0	62	189
SUM/km^2	195	15,030	14,734	48,728	788	29	266	79,770

GZ	2022CLCD							
2003CLCD	**Barren**	**Cropland**	**Forest**	**Grassland**	**Impervious**	**Shrub**	**Water**	**SUM/km^2**
Barren	0	0	0	0	0	0	2	2
Cropland	1	21,946	2514	1093	2930	5	156	28,645
Forest	0	1518	17,907	549	26	21	6	20,027
Grassland	1	2223	1696	839	70	1	12	4842
Impervious	1	1187	26	10	950	0	37	2211
Shrub	0	32	108	6	0	0	0	146
Water	0	81	15	5	33	0	47	181
SUM/km^2	3	26,987	22,266	2502	4009	27	260	56,054

SS	2022CLCD						
2003CLCD	**Cropland**	**Forest**	**Grassland**	**Impervious**	**Shrub**	**Water**	**SUM/km^2**
Cropland	5834	8298	91	468	4	97	14,792
Forest	4108	47,932	276	101	43	39	52,499
Grassland	306	1262	92	13	9	4	1686
Impervious	154	108	1	124	0	7	394
Shrub	53	479	17	0	5	0	554
Water	63	66	1	16	0	23	169
SUM/km^2	10,518	58,145	478	722	61	170	70,094

From 2003 to 2022, the northern Shaanxi region witnessed substantial land-use changes, notably the conversion of 8326 km^2 of grassland to cultivated land, representing the largest transformation. By 2022, this region encompassed 48,728 km^2 of grassland, constituting 61.1% of its total area.

Between 2003 and 2022, Guanzhong saw the most land-use type conversion area, which was the conversion of cultivated land to other land-use types. Of these, 2930 km^2 accounted for 43.7% of the total conversion area in that category. This is connected to the Guanzhong region's rising rate of urbanization and urban growth. With 48.1% of Guanzhong's total area under cultivation in 2022, the Guanzhong region possessed the most amount of land.

From 2003 to 2022, the land-use type in southern Shaanxi that had the greatest area conversion was the conversion of cultivated land to other land uses, with 8298 km^2 of that land changed to forest land, or 92.6% of the total area converted in that category. In 2022, the southern Shaanxi region possessed the most area of forest land, with 58,145 km^2, or 83% of the region's entire area.

In 2022, the total amount of vegetation, which included grassland, shrubland, and forests, was calculated for various geographic locations; 79.6%, 44.2%, and 83.7% of the total area in each region were represented by the areas for the northern Shaanxi, Guanzhong, and southern Shaanxi regions, which were 63,491 km2, 24,795 km^2, and 58,684 km^2, respectively. It is evident that different geographic locations have varying levels of plant covering, which results in various KNDVI values. In line with the earlier findings, the Guanzhong area has the lowest vegetation coverage, while the southern and northern Shaanxi regions have the greatest coverage.

The Guanzhong region's densely populated urban cluster is seeing a notable growth rate in its urban regions due to the accelerating urbanization process. This expansion is accompanied by a notable rise in building land and a decrease in the area of existing agriculture. Both urban and rural regions see an increase in water demand when there is a concentration of people. The Guanzhong region's vegetation covering is growing slowly, which is explained by the significant influence of human activity on this cover.

The Guanzhong area has seen tremendous expansion in both agricultural and industrial development throughout the last 20 years. The region's vegetation sustainability has been significantly impacted by this development. Consequently, the explanation of the Guanzhong region's vegetation sustainability will be the main goal of this research.

Firstly, the Guanzhong region's vegetation cover has been trending downward as a result of increased industrial and agricultural activity. There is a decline in the amount of woodland and grassland regions as a result of the extensive land usage for farming and factory construction. As a result, the environment is under strain, endangering the preservation and protection of biodiversity. Numerous plant species have been harmed, upsetting the ecological equilibrium.

Second, both the survival and growth of plants have been adversely impacted by water contamination resulting from industrial and agricultural operations. Groundwater and surface water have been contaminated by the release of wastewater from factories and the use of chemical pesticides and fertilizers on agricultural land. This has tainted plant water supplies, limiting the development of the plants. Certain delicate plant species might not be able to withstand this environmental stress, which would cause their populations to decline or perhaps go extinct.

In addition, climate change has had an impact on the Guanzhong region's capacity to sustain its flora. Changes in temperature and precipitation patterns brought forth by global warming might affect plant lifecycles and growth seasons. There might be a decline in the population of some plant species if they are unable to adjust to these changes.

Future population growth, economic expansion, and the resulting increased demand for land and water resources might present the Guanzhong area with ever-greater issues. These elements may make water pollution and deforestation worse. The viability of the vegetation may also be further jeopardized by worsening climate change, which might expose the area to increasingly frequent and severe extreme weather events such as floods and droughts. The area's capacity to preserve its vegetation has been weakened throughout the last 20 years of rapid industrial and agricultural growth. Reduced vegetation covering and the adverse impacts of water pollution and climate change on plant development and survival are possible outcomes. Implementing sustainable agriculture techniques, strengthening land conservation initiatives, and raising environmental awareness are all necessary to address these problems.

6. Conclusions

While certain localized locations in Shaanxi Province are showing a deteriorating trend, overall, the province's vegetation covering is improving: 92.15% of the entire area, or 189,756 km^2, is covered by the enhanced vegetation growth area; 3977 km^2, or 1.93% of the total area, are covered by areas with steady vegetation growth; while 12,184 km^2, or 5.92% of the total area, are covered by areas with falling vegetation growth. This research shows that although plant growth has improved over a large region of the province, there has been a dramatic decrease in vegetation cover in a smaller but important area.

The types of interaction factors include two categories: bivariate enhancement and nonlinear enhancement. The main interactive factors affecting the variation of vegetation k-NDVI in Shaanxi Province are TMP∩PET and PRE∩PET. Climatic conditions serve as the primary driving force for the variation of vegetation k-NDVI in Shaanxi Province.

Supplementary Materials: The following supporting information can be downloaded at: https://www.mdpi.com/article/10.3390/su152316468/s1.

Author Contributions: Conceptualization, M.S.; methodology, M.S. and X.J.; software, F.L.; validation, M.S.; formal analysis, M.S. and F.L.; investigation, F.L.; resources, Y.H.; data curation, X.J.; writing—original draft preparation, M.S.; writing—review and editing, F.L., Y.S. and X.J. visualization, B.L., J.Q. and M.W.; supervision, Y.H.; project administration, Y.H.; funding acquisition, Y.H. All authors have read and agreed to the published version of the manuscript.

Funding: This research was funded by National Key Research and Development Program (grant number 2021YFD2000200), Agricultural non-point source pollution control and supervision guidance project in Feidong County (grant number GH20230401), and National Natural Science Foundation of China (grant number 42171394).

Institutional Review Board Statement: Not applicable.

Informed Consent Statement: Not applicable.

Data Availability Statement: No new data were created or analyzed in this study. Data sharing is not applicable to this article.

Acknowledgments: We would like to extend our sincere thanks to Yitong Qin of Sun Yat-sen University. The authors thanks him for guidance and assistance in code writing.

Conflicts of Interest: The authors declare no conflict of interest.

References

1. Verrall, B.; Pickering, C.M. Alpine vegetation in the context of climate change: A global review of past research and future directions. *Sci. Total Environ.* **2020**, *748*, 141344. [CrossRef]
2. Li, Z.; Chen, Y.; Li, W.; Deng, H.; Fang, G. Potential impacts of climate change on vegetation dynamics in Central Asia. *J. Geophys. Res. Atmos.* **2015**, *120*, 12345–12356. [CrossRef]
3. Jiang, L.; Jiapaer, G.; Bao, A.; Guo, H.; Ndayisaba, F. Vegetation dynamics and responses to climate change and human activities in Central Asia. *Sci. Total Environ.* **2017**, *599*, 967–980. [CrossRef]
4. Kidane, Y.; Stahlmann, R.; Beierkuhnlein, C. Vegetation dynamics, and land use and land cover change in the Bale Mountains, Ethiopia. *Environ. Monit. Assess.* **2012**, *184*, 7473–7489. [CrossRef]
5. Walsh, S.J.; Moody, A.; Allen, T.R.; Brown, D.G. Scale dependence of NDVI and its relationship to mountainous terrain. In *Scale in Remote Sensing and GIS*; Routledge: London, UK, 2023; pp. 27–55.
6. Hendricks, A.S.; Bhatt, U.S.; Frost, G.V.; Walker, D.A.; Bieniek, P.A.; Raynolds, M.K.; Lader, R.T.; Epstein, H.E.; Pinzon, J.E.; Tucker, C.J.; et al. Decadal variability in spring sea-ice concentration linked to summer temperature and NDVI on the Yukon-Kuskokwim Delta. *Earth Interact.* **2023**, *27*, e230002. [CrossRef]
7. Anuar, N.I.; Khalid, N.; Tahar, K.N.; Othman, A.N. Analyze the Relationship between Aboveground Biomass and NDVI Values Derived from UAV Multispectral Imagery. In Proceedings of the IOP Conference Series: Earth and Environmental Science, Al Diwaniyah, Iraq, 17–18 May 2023; IOP Publishing: Bristol, UK, 2023; Volume 1240, p. 012015.
8. Furlanetto, J.; Dal Ferro, N.; Longo, M.; Sartori, L.; Polese, R.; Caceffo, D.; Nicoli, L.; Morari, F. LAI estimation through remotely sensed NDVI following hail defoliation in maize (Zea mays L.) using Sentinel-2 and UAV imagery. *Precision Agric.* **2023**, *24*, 1355–1379. [CrossRef]

9. Zhang, R.; Zhou, Y.; Hu, T.; Sun, W.; Zhang, S.; Wu, J.; Wang, H. Detecting the Spatiotemporal Variation of Vegetation Phenology in Northeastern China Based on MODIS NDVI and Solar-Induced Chlorophyll Fluorescence Dataset. *Sustainability* **2023**, *15*, 6012. [CrossRef]
10. Wang, Q.; Moreno-Martínez, Á.; Muñoz-Marí, J.; Campos-Taberner, M.; Camps-Valls, G. Estimation of vegetation traits with kernel NDVI. *ISPRS J. Photogramm. Remote Sens.* **2023**, *195*, 408–417. [CrossRef]
11. Beck, P.S.A.; Atzberger, C.; Høgda, K.A.; Johansen, B.; Skidmore, A.K. Improved monitoring of vegetation dynamics at very high latitudes: A new method using MODIS NDVI. *Remote Sens. Environ.* **2006**, *100*, 321–334. [CrossRef]
12. Pettorelli, N.; Vik, J.O.; Mysterud, A.; Gaillard, J.M.; Tucker, C.J.; Stenseth, N.C. Using the satellite-derived NDVI to assess ecological responses to environmental change. *Trends Ecol. Evol.* **2005**, *20*, 503–510. [CrossRef] [PubMed]
13. Zhu, G.; Zhao, C.; Tong, S.; Zhu, W. Response of vegetation dynamic change to multi-scale drought stress in the high-latitude Nenjiang River basin in China. *Front. Ecol. Evol.* **2022**, *10*, 1074199. [CrossRef]
14. Carvalho, R.; de Aguiar, A.P.D.; Amaral, S. Diversity of cattle raising systems and its effects over forest regrowth in a core region of cattle production in the Brazilian Amazon. *Reg. Environ. Change* **2020**, *20*, 1–15. [CrossRef]
15. Peng, W.; Kuang, T.; Tao, S. Quantifying influences of natural factors on vegetation NDVI changes based on geographical detector in Sichuan, western China. *J. Clean. Prod.* **2019**, *233*, 353–367. [CrossRef]
16. Zhao, H.; Gu, B.; Lindley, S.; Zhu, T. Regulation factors driving vegetation changes in China during the past 20 years. *J. Geogr. Sci.* **2023**, *33*, 508–528. [CrossRef]
17. Guo, B.; Liu, Y.; Fan, J.; Lu, M.; Zang, W.; Liu, C.; Wang, B.; Huang, X.; Lai, J.; Wu, H. The salinization process and its response to the combined processes of climate change–human activity in the Yellow River Delta between 1984 and 2022. *Catena* **2023**, *231*, 107301. [CrossRef]
18. Wang, F.; Liu, J.; Fu, T.; Gao, H.; Qi, F. Spatial-Temporal Variations in of Soil Conservation Service and Its Influencing Factors under the Background of Ecological Engineering in the Taihang Mountain Area, China. *Int. J. Environ. Res. Public Health* **2023**, *20*, 3427. [CrossRef]
19. Zhou, L.; Tucker, C.J.; Kaufmann, R.K.; Slayback, D.; Shabanov, N.V.; Myneni, R.B. Variations in northern vegetation activity inferred from satellite data of vegetation index during 1981 to 1999. *J. Geophys. Res. Atmos.* **2001**, *106*, 20069–20083. [CrossRef]
20. Suzuki, R.; Masuda, K.; Dye, D.G. Interannual covariability between actual evapotranspiration and PAL and GIMMS NDVIs of northern Asia. *Remote Sens. Environ.* **2007**, *106*, 387–398. [CrossRef]
21. Ma, H.; Zhang, L.; Wei, X.; Shi, T.; Chen, T. Spatio-temporal changes in land use and vegetation cover in the Southwest region of China from 2000 to 2015. *Chin. J. Appl. Ecol.* **2021**, *32*, 618–628.
22. Wang, J.; Wang, K.; Zhang, M.; Zhang, C. Impacts of climate change and human activities on vegetation cover in hilly southern China. *Ecol. Eng.* **2015**, *81*, 451–461. [CrossRef]
23. Maeda, E.E.; Pellikka, P.K.E.; Siljander, M.; Clark, B.J.F. Potential impacts of agricultural expansion and climate change on soil erosion in the Eastern Arc Mountains of Kenya. *Geomorphology* **2010**, *123*, 279–289. [CrossRef]
24. Nunes, A.N.; De Almeida, A.C.; Coelho, C.O.A. Impacts of land use and cover type on runoff and soil erosion in a marginal area of Portugal. *Appl. Geogr.* **2011**, *31*, 687–699. [CrossRef]
25. Fang, L.; Wang, L.; Chen, W.; Jia, S.; Cao, Q.; Wang, S.; Wang, L. Identifying the impacts of natural and human factors on ecosystem service in the Yangtze and Yellow River Basins. *J. Clean. Prod.* **2021**, *314*, 127995. [CrossRef]
26. Zhu, L.; Meng, J.; Zhu, L. Applying Geodetector to disentangle the contributions of natural and anthropogenic factors to NDVI variations in the middle reaches of the Heihe River Basin. *Ecol. Indic.* **2020**, *117*, 106545. [CrossRef]
27. Quillet, A.; Peng, C.; Garneau, M. Toward dynamic global vegetation models for simulating vegetation–climate interactions and feedbacks: Recent developments, limitations, and future challenges. *Environ. Rev.* **2010**, *18*, 333–353. [CrossRef]
28. Wang, J.; Xu, C. Geographical Detectors: Principles and Prospects. *Acta Geogr. Sin.* **2017**, *72*, 116–134.
29. Solly, B.; Jarju, A.M.; Sonko, E.; Yaffa, S.; Sawaneh, M.; Jarju, A. Detection of recent changes in Gambia vegetation cover using time series MODIS NDVI. *BELGEO* **2021**, *2021*, 1–14. [CrossRef]
30. Dong, Y.; Yin, D.; Li, X.; Huang, J.; Su, W.; Li, X.; Wang, H. Spatial–temporal evolution of vegetation NDVI in association with climatic, environmental and anthropogenic factors in the loess plateau, China during 2000–2015: Quantitative analysis based on geographical detector model. *Remote Sens.* **2021**, *13*, 4380. [CrossRef]
31. Yan, L.; He, R.; Kašanin-Grubin, M.; Luo, G.; Peng, H.; Qiu, J. The dynamic change of vegetation cover and associated driving forces in Nanxiong Basin, China. *Sustainability* **2017**, *9*, 443. [CrossRef]
32. Meng, N.; Wang, N.; Cheng, H.; Liu, X.; Niu, Z. Impacts of climate change and anthropogenic activities on the normalized difference vegetation index of desertified areas in northern China. *J. Geogr. Sci.* **2023**, *33*, 483–507. [CrossRef]
33. Cao, Z.; Li, Y.; Liu, Y.; Chen, Y.; Wang, Y. When and where did the Loess Plateau turn "green"? Analysis of the tendency and breakpoints of the normalized difference vegetation index. *Land Degrad. Dev.* **2018**, *29*, 162–175. [CrossRef]
34. Yao, B.; Ma, L.; Si, H.; Li, S.; Gong, X.; Wang, X. Spatial Pattern of Changing Vegetation Dynamics and Its Driving Factors across the Yangtze River Basin in Chongqing: A Geodetector-Based Study. *Land* **2023**, *12*, 269. [CrossRef]
35. Pei, H.; Zhao, Y.; Zhang, T. Spatial and temporal patterns and driving forces of Net Ecosystem Productivity (NEP) on the Loess Plateau from 2000 to 2020. *Arid. Zone Res.* **2023**, 1–14. Available online: https://link.cnki.net/urlid/65.1095.X.20231016.1736.004 (accessed on 1 November 2023).

36. Zhao, Q.; Yu, L.; Li, X.; Peng, D.; Zhang, Y.; Gong, P. Progress and trends in the application of Google Earth and Google Earth Engine. *Remote Sens.* **2021**, *13*, 3778. [CrossRef]
37. Yang, J.; Huang, X. The 30 m annual land cover dataset and its dynamics in China from 1990 to 2019. *Earth Syst. Sci. Data* **2021**, *13*, 3907–3925. [CrossRef]
38. Gao, S.; Zhong, R.; Yan, K.; Ma, X.; Chen, X.; Pu, J.; Gao, S.; Qi, J.; Myneni, R.B. Evaluating the saturation effect of vegetation indices in forests using 3D radiative transfer simulations and satellite observations. *Remote Sens. Environ.* **2023**, *295*, 113665. [CrossRef]
39. Wang, R.; Gamon, J.A.; Emmerton, C.A.; Springer, K.R.; Yu, R.; Hmimina, G. Detecting intra- and inter-annual variability in gross primary productivity of a North American grassland using MODIS MAIAC data. *Agric. For. Meteorol.* **2020**, *281*, 107859. [CrossRef]
40. Camps-Valls, G.; Campos-Taberner, M.; Moreno-Martínez, Á.; Walther, S.; Duveiller, G.; Cescatti, A.; Mahecha, M.D.; MarÍ, J.M.; GarcÍa-haro, F.J.; Running, S.W.; et al. A unified vegetation index for quantifying the terrestrial biosphere. *Sci. Adv.* **2021**, *7*, eabc7447. [CrossRef]
41. Zhang, S.; Zhou, Y.; Yu, Y.; Li, F.; Zhang, R.; Li, W. Using the Geodetector Method to Characterize the Spatiotemporal Dynamics of Vegetation and Its Interaction with Environmental Factors in the Qinba Mountains, China. *Remote Sens.* **2022**, *14*, 5794. [CrossRef]
42. Shi, S.; Wang, X.; Hu, Z.; Zhao, X.; Zhang, S.; Hou, M.; Zhang, N. Geographic detector-based quantitative assessment enhances attribution analysis of climate and topography factors to vegetation variation for spatial heterogeneity and coupling. *Glob. Ecol. Conserv.* **2023**, *42*, e02398. [CrossRef]
43. Shen, F.; Yang, L.; Zhang, L.; Guo, M.; Huang, H.; Zhou, C. Quantifying the direct effects of long-term dynamic land use intensity on vegetation change and its interacted effects with economic development and climate change in jiangsu, China. *J. Environ. Manag.* **2023**, *325*, 116562. [CrossRef] [PubMed]
44. Assede, E.S.P.; Orou, H.; Biaou, S.S.H.; Geldenhuys, C.J.; Ahononga, F.C.; Chirwa, P.W. Understanding drivers of land use and land cover change in Africa: A review. *Curr. Landsc. Ecol. Rep.* **2023**, 1–11. [CrossRef]
45. Penny, J.; Ordens, C.M.; Barnett, S.; Djordjević, S.; Chen, A.S. Small-scale land use change modelling using transient groundwater levels and salinities as driving factors–An example from a sub-catchment of Australia's Murray-Darling Basin. *Agric. Water Manag.* **2023**, *278*, 108174. [CrossRef]
46. Zaman, T.; Alakuş, K. Integrating Jackknife into the Theil-Sen Estimator in Multiple Linear Regression Model. *REVSTAT-Stat. J.* **2023**, *21*, 97–114.
47. Zhou, J.; Deitch, M.J.; Grunwald, S.; Screaton, E. Do the Mann-Kendall test and Theil-Sen slope fail to inform trend significance and magnitude in hydrology? *Hydrol. Sci. J.* **2023**. *just-accepted*. [CrossRef]
48. Malisch, R.; Fürst, P.; Šebková, K.; Sapunova, D.; Kalina, J. Time Trends in Human Milk Derived from WHO-and UNEP-Coordinated Exposure Studies, Chapter 3: Perfluoroalkyl Substances (PFAS). In *Persistent Organic Pollutants in Human Milk*; Springer International Publishing: Cham, Switzerland, 2023; pp. 543–600.
49. Hamed, K.H. Trend detection in hydrologic data: The Mann–Kendall trend test under the scaling hypothesis. *J. Hydrol.* **2008**, *349*, 350–363. [CrossRef]
50. Xue, Y.; Song, D.; Chen, J.; Li, Z.; He, X.; Wang, H.; Zhou, C.; Sobolev, A. Integrated rockburst hazard estimation methodology based on spatially smoothed seismicity model and Mann-Kendall trend test. *Int. J. Rock Mech. Min. Sci.* **2023**, *163*, 105329. [CrossRef]
51. De Beurs, K.M.; Henebry, G.M. Trend analysis of the Pathfinder AVHRR Land (PAL) NDVI data for the deserts of Central Asia. *IEEE Geosci. Remote Sens. Lett.* **2004**, *1*, 282–286. [CrossRef]
52. Ngwira, G.M.; Bolaane, B.; Parida, B.P. Investigating the trend of road traffic fatalities in Malawi using Mann-Kendall statistic. *Heliyon* **2023**, *9*, e13700. [CrossRef]
53. Chukwuka, A.V.; Ogbeide, O.; Otomo, P.V. Trend relationship between mountain normalized difference vegetation index (NDVI) and aerosol optical depth (AOD) across two decades: Implication for water quality within the Lesotho Highlands, Drakensberg, South Africa. *Environ. Monit. Assess.* **2023**, *195*, 584. [CrossRef]
54. Adan, M.; Tonnang, H.E.Z.; Greve, K.; Borgemeister, C.; Goergen, G. Use of time series normalized difference vegetation index (NDVI) to monitor fall armyworm (Spodoptera frugiperda) damage on maize production systems in Africa. *Geocarto Int.* **2023**, *38*, 2186492. [CrossRef]
55. Wang, J.; Rich, P.M.; Price, K.P. Temporal responses of NDVI to precipitation and temperature in the central Great Plains, USA. *Int. J. Remote Sens.* **2003**, *24*, 2345–2364. [CrossRef]
56. Gong, C.; Wang, S.X.; Lu, H.C.; Chen, Y.; Liu, J. Research Progress on Spatial Differentiation and Influencing Factors of Soil Heavy Metals Based on Geographical Detector. *Huan Jing Ke Xue* **2023**, *44*, 2799–2816. [PubMed]
57. Wang, H.; Xu, Y.; Wei, X. Rural Resilience Evaluation and Influencing Factor Analysis Based on Geographical Detector Method and Multiscale Geographically Weighted Regression. *Land* **2023**, *12*, 1270. [CrossRef]
58. Li, C.; Wang, J.; Wang, Q.; Wang, D.; Song, X.; Wang, Y.; Huang, W. Estimating Wheat Grain Protein Content Using Multi-Temporal Remote Sensing Data Based on Partial Least Squares Regression. *J. Integr. Agric.* **2012**, *11*, 1445–1452. [CrossRef]
59. Ma, Y.; Ren, X.; Hu, H.; Liu, M.; Meng, Q. Vegetation dynamics and its driving force in Otindag Sandy Land based on Geodetector. *J. Desert Res.* **2021**, *41*, 195–204.

60. Xie, W. A Novel Hybrid Method for Landslide Susceptibility Mapping-Based GeoDetector and Machine Learning Cluster: A Case of Xiaojin County, China. *Int. J. Geo-Inf.* **2021**, *10*, 93. [CrossRef]
61. Kang, Y.; Guo, E.; Wang, Y.; Bao, Y.; Bao, Y.; Mandula, N. Monitoring vegetation change and its potential drivers in Inner Mongolia from 2000 to 2019. *Remote Sens.* **2021**, *13*, 3357. [CrossRef]
62. Liu, X.; Zhu, X.; Pan, Y.; Zhao, A. Spatiotemporal changes in vegetation coverage in China during 1982—2012. *Acta Ecol. Sin.* **2015**, *35*, 5331–5342.
63. Li, S.; Wang, J.; Zhang, M.; Tang, Q. Characterizing and attributing the vegetation coverage changes in North Shanxi coal base of China from 1987 to 2020. *Resour. Policy* **2021**, *74*, 102331. [CrossRef]
64. Duo, A.; Zhao, W.; Qu, X.; Jing, R.; Xiong, K. Spatio-temporal variation of vegetation coverage and its response to climate change in North China plain in the last 33 years. *Int. J. Appl. Earth Obs. Geoinf.* **2016**, *53*, 103–117.
65. Nie, T.; Dong, G.; Jiang, X.; Lei, Y. Spatio-temporal changes and driving forces of vegetation coverage on the loess plateau of Northern Shaanxi. *Remote Sens.* **2021**, *13*, 613. [CrossRef]
66. Li, S.; Yan, J.; Liu, X.; Wan, J. Response of vegetation restoration to climate change and human activities in Shaanxi-Gansu-Ningxia Region. *J. Geogr. Sci.* **2013**, *23*, 98–112. [CrossRef]
67. LLi, S.; Yang, S.; Liu, X.; Liu, Y.; Shi, M. NDVI-based analysis on the influence of climate change and human activities on vegetation restoration in the Shaanxi-Gansu-Ningxia Region, Central China. *Remote Sens.* **2015**, *7*, 11163–11182. [CrossRef]
68. Wei, H.; Fan, W.; Ding, Z.; Weng, B.; Xing, K.; Wang, X.; Lu, N.; Ulgiati, S.; Dong, X. Ecosystem services and ecological restoration in the Northern Shaanxi Loess Plateau, China, in relation to climate fluctuation and investments in natural capital. *Sustainability* **2017**, *9*, 199. [CrossRef]
69. Yuan, W.; Wu, S.; Hou, S.; Xu, Z.; Lu, H. Normalized Difference Vegetation Index-based assessment of climate change impact on vegetation growth in the humid-arid transition zone in northern China during 1982–2013. *Int. J. Climatol.* **2019**, *39*, 5583–5598. [CrossRef]
70. Hao, L.; Sun, G.; Liu, Y.; Gao, Z.; He, J.; Shi, T.; Wu, B. Effects of precipitation on grassland ecosystem restoration under grazing exclusion in Inner Mongolia, China. *Landsc. Ecol.* **2014**, *29*, 1657–1673. [CrossRef]
71. El Kateb, H.; Zhang, H.; Zhang, P.; Mosandl, R. Soil erosion and surface runoff on different vegetation covers and slope gradients: A field experiment in Southern Shaanxi Province, China. *Catena* **2013**, *105*, 1–10. [CrossRef]
72. Bai, J.; Bai, J.; Wang, L. Spatio-temporal Change of Vegetation NDVI and Its Relations with Regional Climate in Northern Shaanxi Province in 2000–2010. *Sci. Geogr. Sin.* **2014**, *34*, 882–888.
73. Vernet, W.J.L. Vegetation, sedimentary deposits and climates during the Late Pleistocene and Holocene in eastern Morocco. *Palaeogeogr. Palaeoclimatol. Palaeoecol.* **1992**, *94*, 141–167. [CrossRef]
74. Trivedi, A.; Tang, Y.-N.; Qin, F.; Farooqui, A.; Wortley, A.H.; Wang, Y.-F.; Blackmore, S.; Li, C.-S.; Yao, Y.-F. Holocene vegetation dynamics and climatic fluctuations from Shuanghaizi Lake in the Hengduan Mountains, southwestern China. *Palaeogeogr. Palaeoclimatol. Palaeoecol.* **2020**, *560*, 110035. [CrossRef]
75. Kato, K.; Tanizoe, C.; Beiles, A.; Nevo, E. Geographical Variation in Heading Traits in Wild Emmer Wheat, Triticum Dicoccoides. II. Variation in Heading date and Adaptation to Diverse Eco-Geographical Conditions. *Hereditas* **2004**, *128*, 1601–5223. [CrossRef]
76. Sun, W.; Lu, H.; Wang, Y.; Han, F.; Wang, H.; Li, Y.; Wu, G.; Li, S.; Jiang, P. Deposits and Palaeoenvironmental Record of the Eocene Honghe Formation in Lantian, Weihe Basin, Central China. *Geol. J. China Univ.* **2017**, *23*, 533–544.

MDPI AG
Grosspeteranlage 5
4052 Basel
Switzerland
Tel.: +41 61 683 77 34

Sustainability Editorial Office
E-mail: sustainability@mdpi.com
www.mdpi.com/journal/sustainability

www.ingramcontent.com/pod-product-compliance
Lightning Source LLC
Chambersburg PA
CBHW061520210926
51850CB00008B/1985
* 9 7 8 3 7 2 5 8 6 1 2 1 7 *